Contents

Acknowledgements

The Publishers are grateful to the following for permission to reproduce photographs in the text:

Crown Copyright courtesy of Tropical Development and Research Institute for Fig. 6.34; Neville Grant for Fig. 6.16d; Frants Hartmann for Figs. 3.8b, 6.4, 6.16b and 6.31; The Hutchison Library for Fig. 8.1; David Keith Jones for Figs. 5.8, 5.12, 5.13, 5.20a, 6.8, 6.11 and 6.13; Dr Walter Leuthold for Fig. 6.10; Natural History Photo Agency/Stephen Dalton for Fig. 6.21; D.C.H. Plowes for Figs. 4.9b and 4.9c.

All remaining photographs have been provided by the authors.

Cover photograph kindly supplied by Natural History Photo Agency.

The Publishers are also grateful to the following for permission to reproduce figures in the text:

Academic Press (per J. F. Monk): Fig. 6.35 from *The Palaearctic-African Bird Migration System* (1972) by R. E. Moreau; Blandford Press Ltd: Fig. 6.23 from *Elephants* (1982) by S. K. Eltringham; Blackwell Scientific Publications Ltd: Fig. 4.3 from *Journal of Applied Ecology*, 18, 571–587 (1981) by D. Whitehead et al; Fig. 7.10 from *Journal of Applied Ecology*, 14, 465–476 (1977) by D. E. Pomeroy; Fig. 7.25 from *Journal of Animal Ecology*, 47, 219–247 (1978) by Anderson and May; Fig. 4.10 from *Journal of Ecology*, 70, 791–806 (1982) by D. Lieberman. The Company of Biologists Ltd: Fig. 5.7 from the *Journal of Experimental Biology*, 49, 1–13 (1968) by J. Loveridge; Blackie, Glasgow and London: Figure for Box 5.1 from *Mammal Ecology* (1982) by M. J. Delaney; Cambridge University Press: Fig. 2.12 from *The Tropical Rain Forest – An Ecological Study* (1952) by P. W. Richards; Harper & Row, Publishers, Inc.: Fig. 7.24a from *Ecology: The Experimental Analysis of Distribution and Abundance*, 2nd Edition by Charles J. Krebs. Copyright © 1978 by Charles J. Krebs, reprinted by permission; Her Britannic Majesty's Stationery Office: Fig. 7.8d from *Span*, 82, 98–100 (1979), reproduced with the permission of the Controller of Her Britannic Majesty's Stationery Office © British Crown copyright 1979; Journal of the East Africa Natural History Society: Fig. 4.11 from *Journal of the East African History Society* and *National Museum* no. 175, 1–12 (1982) by M. Fenner; Macmillan Journals Limited: Fig. 7.15 reprinted by permission from Nature, No. 182, 330–31, Copyright © 1958; National Geographic: Fig. 5.2 is redrawn with permission of the National Geographic Cartographic Division; Oxford University Press: Fig. 2.14 from *Plant Life in West Africa* (1966) by G. W. Lawson; Plenum Publishing Corp.: Figs. 3.5 and 3.6 from *Human Ecology*, 1, 303–332 (1973) by H. Leith; Poyser Ltd: Figure for Box 5.1 by permission from Newton *Population Ecology of Raptors* (1979); Royal Entomological Society of London: Fig. 5.14 from *Transactions of the Royal Entomological Society of London* 122, 101–43; Publishing House of the Swedish Research Councils: Fig. 2.18 from *Svensk Botanisk Tidskrift*, 45, 140–202 (1951) by O. Hedberg; UNESCO: Fig. 4.12 from *Tropical Forest Ecosystems* – © UNESCO/UNEP (1978); Fig. 2.6 from UNESCO publication (1978); Weidenfeld and Nicolson Ltd: Figure for Box 5.6 from *Mimicry in plants and animals* (1968) by W. Wickler; University of Chicago: Fig. 7.22 from *Serengeti: Dynamics of an Ecosystem* by A. R. E. Sinclair and M. Noton-Griffiths © 1979 by the University of Chicago, all rights reserved. John Wiley and Sons Inc.: Fig. 7.24(b) from *Introduction to Ecology* by Colinvaux, Copyright © 1973 John Wiley and Sons Inc., reprinted by permission of John Wiley and Sons Inc.

Tropical ecology

D. E. Pomeroy, *Department of Zoology, Makerere University, Uganda; formerly Kenyatta University, Kenya*

M. W. Service, *Department of Medical Entomology, Liverpool School of Tropical Medicine, England; formerly Malaria Service, Nigeria*

Contributions from:

J. K. E. Egunjobi, *Department of Botany, University of Ibadan, Nigeria*

N. Dickinson, *Department of Botany, Kenyatta University, Kenya*

Longman Scientific & Technical
Longman Group UK Limited
Longman House, Burnt Mill, Harlow
Essex CM20 2JE, England
and Associated Companies throughout the world.

Published in the United States of America
by Longman Inc.
First published in 1986

British Library Cataloguing in Publication Data
Pomeroy, D.E.
Tropical Ecology.
1. Ecology—Africa 2s Ecology—Tropics
I. Title II. Service, M.W.
574.5 2623'0967 QH194

Library of Congress Cataloging-in-Publication Data
Pomeroy, D.E.
Tropical Ecology.
Includes index.
1. Ecology—Tropics. 1. Service, Mike W.
II. Title
QH 541.5.T7P65 1986 574.5'2623 86-10632
ISBN 0-58-264353-8

ISBN 0 582 64353 8

Set in Linotron 202 9/11 pt Plantin

Produced by Longman Group (FE) Ltd
Printed in Hong Kong

Preface

Our aim in writing this book has been to outline current ecological knowledge and thinking, and to illustrate this with examples from tropical Africa. To a certain extent we have integrated the ecology of plants with that of animals; some would declare total integration to have been better. Whilst agreeing with this in principle, we find that in most University courses they are taught separately rather than as a single subject. But we have included frequent cross-references where there are parallels to be drawn between plants and animals.

Ecology as a subject is expanding rapidly and we have drawn attention to some topics of particular interest in the 1980s. One of these is undoubtedly behavioural ecology, and we make no excuse for including a chapter (5) describing some of the more interesting interactions between ecology and behaviour. Chapter 8 gives an outline of the ecology of our own species, a subject of ever-increasing concern.

Responsibility for preparing all the chapters has been shared between us. We are indebted to our colleagues, Drs J. K. E. Egunjobi of the University of Ibadan, Nigeria and N. Dickinson of Kenyatta University, Kenya, for their contributions on plant ecology.

Many friends, colleagues and students have been kind enough to provide us with information and to comment upon the manuscript at various stages. There were times when they disagreed with our point of view, but their perceptive criticisms were much appreciated nonetheless. We would particularly like to thank Michael Fenner for reading the entire manuscript, and the following for reading parts of it: H. G. Andrewartha, R. Davis, A. G. Hildrew, P. Jewell, J. M. Lock, G. H. G. Martin, S. Nsibirwa, J. Phillipson, B. Tengecho, D. J. Thompson, K. G. van Orsdol and K-K. Waciiro. Nevertheless, any remaining faults are the authors' responsibility and they would appreciate any information on errors or misleading statements.

D. E. Pomeroy, Makerere University, Uganda
M. W. Service, Liverpool School of Tropical Medicine, England

1 Introduction

Why should Africa need its own ecological textbook? Well, we wrote this book because our students – in various parts of Africa – had difficulty in finding a textbook that was relevant to the ecology of their own continent and countries. There are many excellent books on ecology but most of them concentrate on ecological examples from North America or from other temperate regions of the world, with little space devoted to African situations. Such books that discuss deer, robins and oak trees seem remote when outside your window, weavers are noisily nesting in a fig tree and at night you are disturbed by the annoying buzz of a persistent hungry mosquito. This book discusses ecological principles and theories in the context of African ecology.

What is ecology?

'Ecology is a pleasant science' – so opens Colinvaux's book, one of those we recommend as a general text (see below). Almost everybody is familiar with the word 'ecology', but the layman usually has only a vague notion of its meaning, and is unlikely to be very clear as to what ecologists *do*. Ecology is 'the scientific study of living things in relation to each other and to their environment'; so says the Oxford Paperback Dictionary. 'Environment' is a word we often see these days – the dictionaries describe it as our 'surroundings' and that will do for a start. In a later chapter (5) we shall see that some ecologists at least use the word in a more restricted sense. Why only 'some ecologists'? Because ecology, being a young and vigorous science, is still at the stage when the meanings and significance of many of the words it uses have yet to acquire generally accepted definitions. Until they do, we have to face the fact that words such as 'community', 'regulation' and 'competition', and many others taken from everyday English, are used differently by different people, including ecologists.

The great names of the early days of ecology – such as Charles Elton, Eugene Odum and A. G. Tansley – were people who began as naturalists, enjoying nature for itself, but later trying to apply scientific methodology to what they observed. Ecology is really scientific natural history.

The word ecology has even entered politics! Several highly industrialized countries, such as the USA, England, Germany and Japan, have political candidates standing for an 'Ecology Party' or 'Green Party'. Such groups have arisen out of people's concern over the destruction of unspoilt habitats for factories, the increasing environmental pollution, and so on (see chapter 8). But, of course, such candidates are not necessarily real ecologists, and unfortunately their arguments on environmental problems are often unsound and may serve to discredit ecology amongst influential government officials. Nevertheless, it is becoming increasingly recognized by thinking people, and governments, that all is not well with 'the environment' and that ecologists do have something to offer, and so their advice is being sought on a variety of ecologically related problems of everyday life.

Ecology is first and foremost an outdoors science, concerned with how plants and animals interact and survive 'in the field'. Observations in the field sometimes lead ecologists to carry out experiments under controlled laboratory conditions (and we shall describe a few of those). But laboratory experiments will be judged useful only if they increase our understanding of what happens in nature. Nowadays, experiments in the field are gaining in importance. Using the field as an outdoor laboratory is not easy, because it is not as controllable as the normal laboratory environment, but then nature is not tidy. However, well-planned field experiments can often allow ecologists to manipulate some of the variables that affect the plants and animals they are studying, and so allow them to make important observations on them in their natural surroundings.

The scales of approach to ecology

Although in this book our main concern is with Africa, it is useful to begin with a broader view, and here we introduce a few topics that will help to set the scene for later chapters.

The diameter of the world is nearly 13 000 km, but the biosphere, that part of the world where living things are found, is, by comparison, a very thin layer over its surface. Few organisms live more than 5 or 10 m below the earth's surface, or at greater depths than 50 to 100 m in lakes and seas. Similarly, there is little life 100 m above the earth's surface, although a few creatures, mainly birds, occasionally reach heights of up to a few kilometres. But the great majority of living organisms – almost all of the biomass – is contained in a layer much less than one kilometre thick.

Biologists sometimes divide the land surface of the world into **biomes**, major areas covered by characteristic plant-types and containing characteristic animals. Examples are tropical rainforests, savannas, and sand-deserts. Each of these occurs in Africa (see chapter 2) but others, such as tundras, and boreal forests, are found only at high latitudes. Biomes are so large that in practice we find it easier to deal with smaller units, usually called **ecosystems** (chapter 3). Ecosystems are ecological systems of defined limits that can be considered, for practical purposes, to be self-contained. Thus an *Acacia* woodland, or a lake, may be conveniently sized units for an ecological study. On a smaller scale, plant and animal communities or the populations of particular species, are the subjects of study.

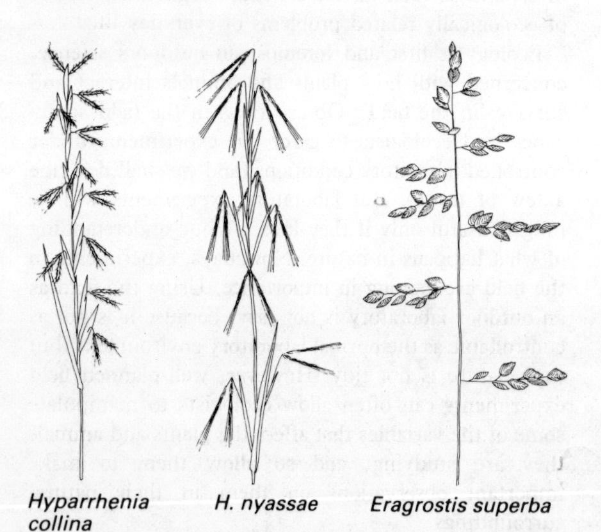

Figure 1.1 Closely related species resemble each other more than distantly related ones. The two *Hyparrhenia* species have a basically similar inflorescence of paired spikes emerging from sheaths, but differ in length of awns, and the angle at which spikes are held, as well as in microscopic features such as hairiness of spikelets. The *Eragrostis* species has a rather different inflorescence, a loose panicle.
(Courtesy of M. Fenner)

We shall be constantly referring to **species** of plants and animals, and it is very important for ecologists to have a clear idea of what a species is. One definition, based on E. Mayr, 1963, *Animal Species and Evolution*, Harvard University Press, Cambridge, USA is that 'a species consists of all the individuals which can successfully reproduce with each other, and which do so in nature'. That is, any two species are reproductively isolated from each other. In practice, individuals of a species usually look more or less the same as each other, which is how we normally distinguish them from other species. Members of closely related species differ only slightly in appearance (Figure 1.1). With fossil species, and with those plants and animals that do not reproduce sexually, we cannot apply the test of reproductive isolation (Figure 1.2). They are mainly distinguished on morphological characteristics.

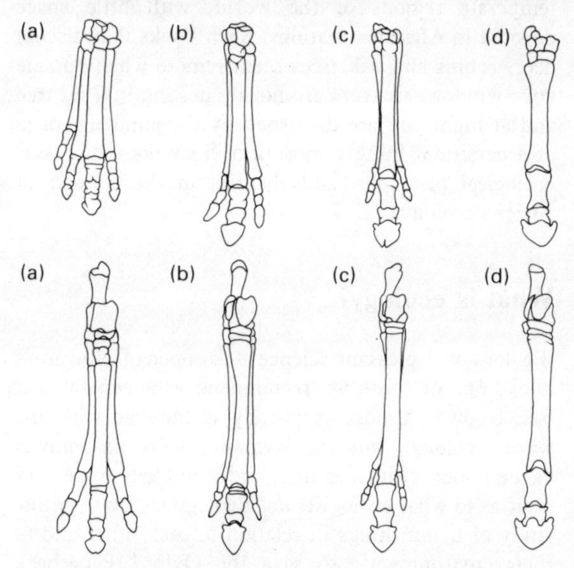

Figure 1.2 Palaeontologists have the difficult task of putting together fossil remains to decide what constituted a genus or a species. They are, however, able to make reasonably good judgement as to what these were, and they use them in tracing evolutionary trends. Of course they cannot distinguish closely related species that did not differ in the hard parts, which are all that are usually preserved as fossils. Consequently we will never be able to recognize all past species.

In this figure the evolution of the feet in horses can be traced by studying fossils of the front feet (top line) and hind feet (bottom line). (a) *Eohippus*, a primitive Eocene form having four toes in front and three behind; (b) *Miohippus*, a three-toed Miocene form; (c) the next stage in development, *Merychippus*, a late Miocene horse with reduced lateral toes; finally (d) feet of *Equus*, which originated in the Pleistocene and is the present-day horse. These examples all show large enough differences to be considered as distinct genera.
(Adapted from Romer, A. S. 1954. '*Man and the Vertebrates*', vol 1, Pelican Books, Middlesex, England.)

But we must remember that speciation has not ceased – new species are still evolving, although relatively slowly. It is not surprising, therefore, that during the process of one species evolving into two, the two new ones at first resemble each other very closely. We may know that they are incapable of interbreeding, or if they do so their offspring are infertile, yet they may look exactly the same. Sometimes, however, they can be separated by examining their chromosomes (cytogenetics) or by small differences in their enzymes (biochemical taxonomy: taxonomy is the science of classifying living things).

There are probably some five to ten million species of plants and animals living in the world today, although human activity is resulting in some of them becoming extinct, and this is happening at an ever-increasing rate (chapter 8).

All species of plants and animals are given a scientific name when they are first described, these being the generally familiar binomials like *Homo sapiens* (man) or *Zea mays* (maize). But in tropical Africa, many species which are known to exist have been poorly described, incorrectly identified and named, or not yet described. This applies especially to non-flowering plants and invertebrate animals. The process of naming and describing new species – taxonomy – is complex, and there are too few taxonomists. An unfortunate consequence is that it is sometimes difficult, or even impossible, for ecologists to know what species they are studying!

Tropical environments, because of their high temperatures, are potentially very favourable for life; but lack of water limits the distributions of many species. In places which are both warm and moist, like tropical rainforests, life is particularly abundant and varied. The tropical rainforest biome is found in South America, South-east Asia and north-east Australia, as well as in Africa. But the species of plants and animals in the tropical rainforest biomes of the four continents are different. To a large extent this applies to all biomes. During the nineteenth century, when biologists began collecting plants and animals on a large scale from all over the world, they became aware of these differences, noticing that each of the continents had its characteristic flora and fauna.

Various attempts were made to divide the world into biogeographical regions (Figure 1.3). The most commonly accepted name for Africa, south of the Sahara, was until recently the Ethiopian Region but because of confusion with the country now called Ethiopia, the name Afrotropical Region has come to be generally used and this is the area which forms the subject of this book. The island of Madagascar, which

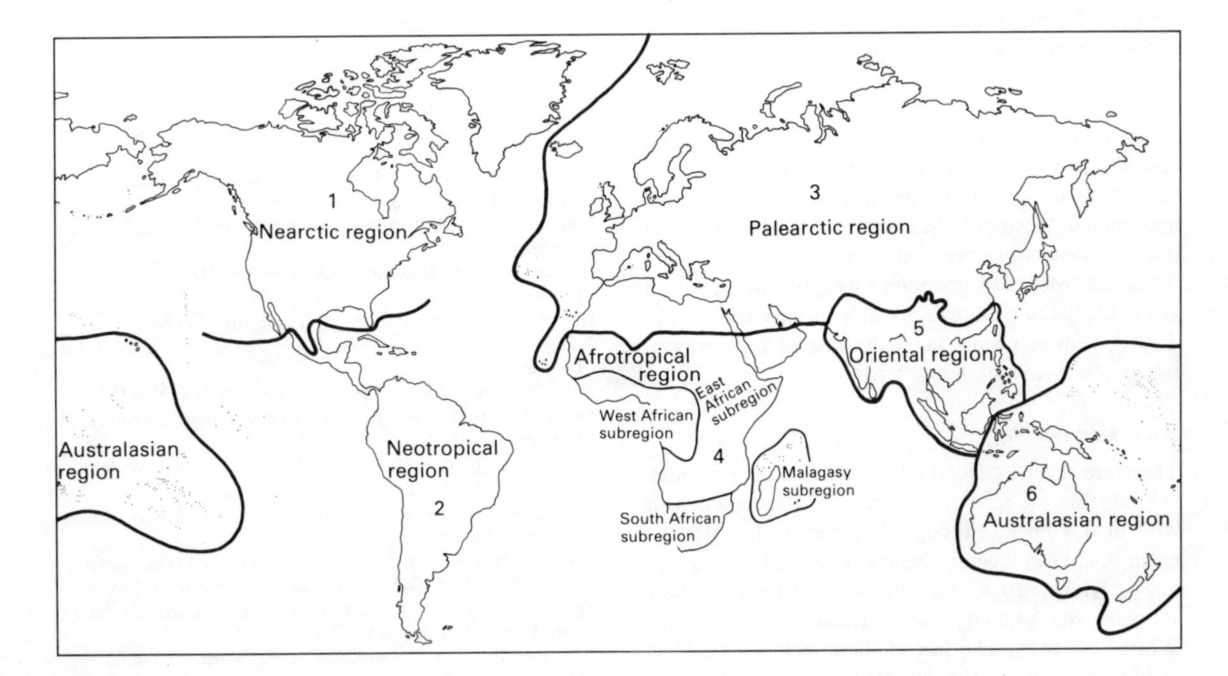

Figure 1.3 The six major zoogeographical regions of the world. Each region can be divided into subregions, but in this map such divisions are shown only for the Afrotropical region. Within each region are found distinctive assemblages of plants and animals, differing substantially from the other regions. Some species, however, are found in more than one region, although only a very few have world-wide distributions.

we shall mention occasionally, is often considered as a separate subregion. It has been separated from mainland Africa for at least sixty million years, and its flora and fauna are quite distinctive, although with important resemblances to those of mainland Africa.

Guide to this book

Each chapter of the book is preceded by a short **synopsis**, highlighting the main topics and putting them into perspective. Within the chapter, the text is illustrated with figures and tables, and it is important to study these. We have used **boxes** to give a detailed example illustrating a particular topic, to give emphasis to a particular point of view, or to summarize factual material useful for reference. However, the text can be read without the boxes. Original sources of the observations and information we present are not usually mentioned in the text, so as to avoid interrupting the narrative. This should not hinder the reader keen to explore a topic in greater depth, for whom we provide a section headed **Suggested reading** at the end of each chapter. The references in that section are mostly to books, which are usually more readily available in libraries than periodicals; but periodicals are very important, because it is in them that most results of research are first published.

Each chapter concludes with a few **Essays and problems**, intended to extend the reader's thinking. Often, effective answers to these will need to draw upon information from more than one section of the book, and sometimes also from library sources.

Because animals depend upon plants, it is appropriate that the structure of some chapters (2 and 3), reflects this relationship. But there is no denying that animals differ from plants in such fundamental respects as nutrition and locomotion. For this reason, as well as for practical convenience, we treat plants and animals separately in chapters 4 to 7, and consequently some concepts – such as competition – are discussed twice. In some ways this apparent repetition is useful, because animals are able to respond in ways denied to most plants, as in the case of competition, by aggression, or by going away.

Suggested reading

There are several general textbooks on ecology, which are valuable for reference to most of the topics dealt with in this book, although they contain few African examples. They include Colinvaux, 1973, Collier *et al.*, 1973, Krebs, 1978, Odum, 1971, Ricklefs, 1980, Putman & Wratten, 1984, Remmert, 1980, and Whittaker, 1975. The last of these includes a review of biomes of the world (in chapter 4).

Amongst the books that deal specifically with Africa are Ewer & Hall, 1972, 1978, Ewusie, 1980, and Owen, 1976. Delany & Happold, 1979, provide much information on the ecology of African mammals, as does Skaife, 1979, on insects.

The subject of biogeography is treated by Cox & Moore, 1980, and George, 1962, and that of taxonomy and nomenclature by Jeffrey, 1977.

Essays and problems

1 Consult various ecological books and see how they define 'biome' and 'ecosystem'. Following these, use an outline map of your country to show the distribution of the major biomes found within it. What are the characteristic features of these biomes?

2 What is meant by the terms species and subspecies? Why is correct identification important for ecologists? (Ewer & Hall, 1978, vol. 2, pp. 452–59, and Jeffrey, 1977, will be useful, especially for the first part of this question.)

3 How much ecology is taught in schools in your country? Comment on this in relation to your assessment of the importance of understanding our environment.

4 What major changes is man making to the environment in your home area? Summarize the likely consequences of these changes over the next decade or two, indicating beneficial effects, as well as those that could be considered harmful.

References to suggested reading

Colinvaux, P. A. 1973. *Introduction to Ecology*, Wiley: New York.
Collier, B. D., Cox, G. W., Johnson, A. W. & Miller, P. C. 1973. *Dynamic Ecology*, Prentice-Hall: New Jersey.
Cox, C. B. & Moore, P. D. 1980. *Biogeography: An Ecological and Evolutionary Approach*, (3rd edn). Blackwell: Oxford.
Delany, M. J. & Happold, D. C. D. 1979. *Ecology of African Mammals*, Longman: London.
Ewer, D. W. & Hall, J. B. (eds). 1: 1972, 2: 1978. *Ecological Biology*, Longman: London.
Ewusie, J. Y. 1980. *Elements of Tropical Ecology*, Heinemann: London.
George, W. 1962. *Animal Geography*, Heinemann: London.
Jeffrey, C. 1977. *Biological Nomenclature*, (2nd edn). Edward Arnold: London.
Krebs, C. J. 1978. *Ecology*, (2nd edn). Harper & Row: New York.
Owen, D. F. 1976. *Animal Ecology in Tropical Africa*, (2nd edn). Longman: London.
Odum, E. P. 1971. *Fundamentals of Ecology*, (3rd edn). W. B. Saunders: Philadelphia.
Putman, R. J. & Wratten, S. D. 1984. *Principles of Ecology*, Croom Helm: Beckenham.
Remmert, H. 1980. *Ecology: a Textbook*, Springer-Verlag: Berlin.
Ricklefs, R. E. 1980. *Ecology*, (2nd edn). Thomas Nelson: London.
Skaife, S. H. (revised by J. Ledger) 1979. *African Insect Life*, (2nd edn). Country Life: London.
Whittaker, R. H. 1975. *Communities and Ecosystems*, (2nd edn). Macmillan: New York.

2 The natural ecosystem

Within the area we know as tropical Africa is a great range of environments – from deserts to jungles, and snow-capped mountains to coral reefs. In this chapter we introduce the factors responsible for this diversity, which are primarily climate and soil, and consider the main types of vegetation resulting from their interaction. This approach gives us a large-scale description of our part of the biosphere, within the framework of which we shall be concerned later with the ecology of plants and animals and their interactions.

Africa is huge, with an area of 30 300 000 km², which is nearly 20 per cent of the land surface of the earth. It lies across the equator, extending to 37°N in Tunisia and to 34°S at Cape Agulhas, South Africa. This book is mainly concerned with the area lying between the Tropic of Cancer (23° 28′ N) and the Tropic of Capricorn (23° 28′S), but areas outside them, both in Africa and elsewhere, are mentioned where relevant.

The natural environment of any part of Africa primarily determines the type of vegetation in the region. Climate and soils are the most important determinants, and a background understanding of these factors and their variations in Africa is essential. This introductory chapter outlines the essential features of the soils and vegetation.

Seasonal climatic changes are evident almost everywhere, and we begin by looking in some detail at rainfall and temperature and several closely related features of climate. This is followed by an examination of the physical factors that determine the immediate environment of individual plants and animals, and we consider the importance of radiation balance and microclimates.

The range of soil types found in tropical Africa and their formation are broadly described and the small-scale soil factors that directly influence individual plant or animal species are examined. They are the physical, chemical and biological processes involved in soil formation and the maintenance of fertility.

Essentially it is the climate and soil that determine the potential of the vegetation of a region. The remainder of the chapter is concerned with the types of natural vegetation found throughout tropical Africa, classified according to whether climate or soil are the primary influences. The importance of man's activities as probably the main determinant of the present-day vegetation is discussed.

Climate

Seasons

The climate of almost every part of Africa varies during the year to give seasonal patterns of rainfall and temperature. This variation can be related to changes in the noon position of the sun. At any time during the year, the sun is exactly overhead at noon along a particular line of latitude. For example, at the **equinoxes** (21 March and 23 September) the sun is overhead along the equator. If the axis of rotation of the earth were exactly at right angles to the line joining the earth and the sun, then the sun would be overhead at noon at the equator throughout the year; consequently, there would be no seasons. Because the earth's axis of rotation is not at right angles to the sun–earth line, the noon position of the sun moves progressively during the year (Figure 2.1). During the **summer solstice** (21 June), it is overhead along the Tropic of Cancer at 23° 28′ N. It then appears to move south, crossing the equator again on 23 September to reach its southernmost position along the Tropic of Capricorn on 22 December. The pattern is then reversed, and the sun appears to move north again to reach the equator on 21 March.

When the sun is overhead at noon, its rays pass through the smallest possible thickness of atmosphere before reaching the ground. They therefore lose the minimum amount of energy by absorption, reflection and scattering, and heat the ground and the lower layers of the atmosphere more strongly than in regions to the north and south, where the incoming rays have to pass through greater thicknesses of the atmosphere. Moreover, when the sun is overhead, its radiation reaches the earth's surface at right angles, but elsewhere the angle is lower, and the heating effect per unit area is proportionally less.

Warm air expands and rises, and is replaced by air that moves in from the north and south. Because of the rotation of the earth, these northerly and southerly winds are deflected and blow from the north-east and south-east. They are known as the Trade Winds. Along the zone of greatest heating, where the air is rising, rain tends to fall, because as the air rises it expands and cools, and the amount of water that it can

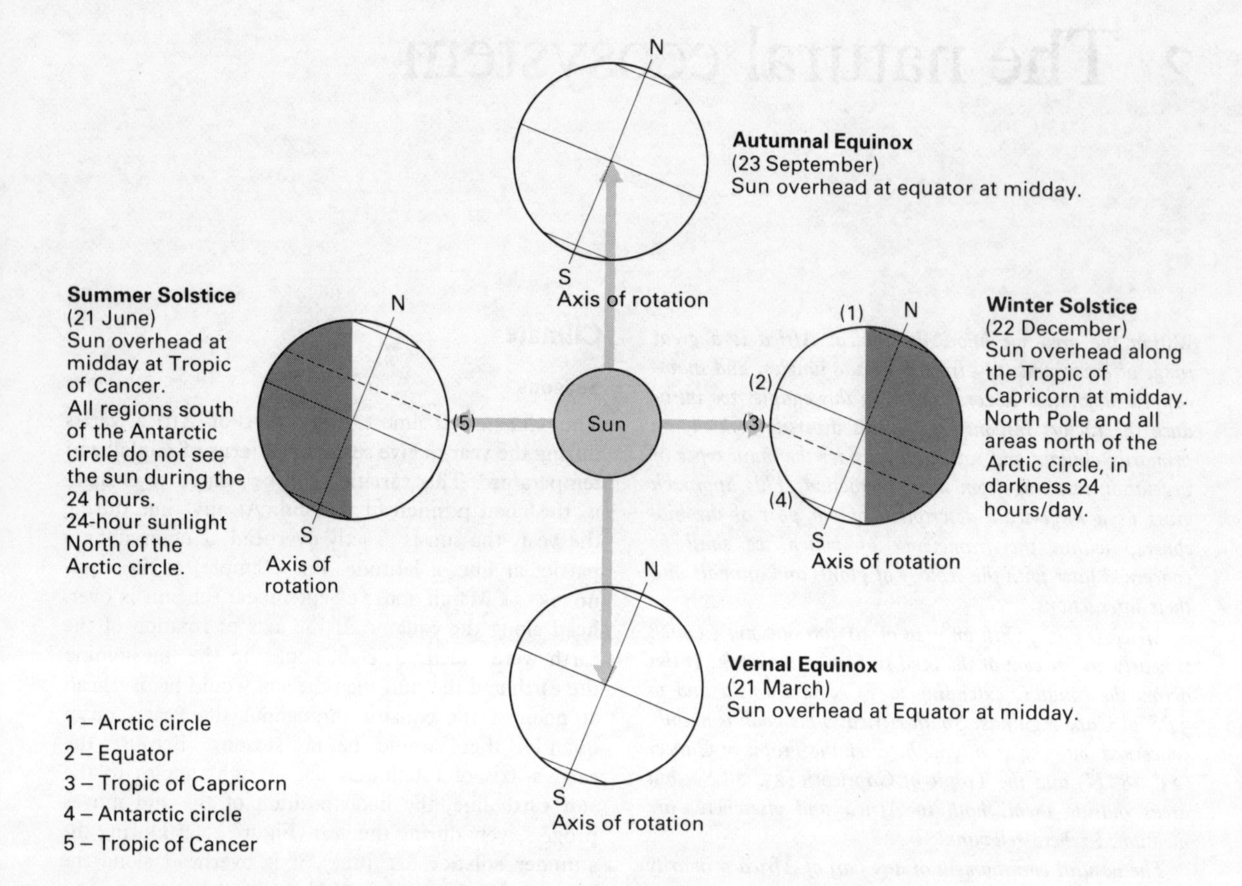

Figure 2.1 How the seasons result from the relationship between the earth's axes of rotation and its orbit around the sun. The axis of rotation is tilted at 23°28′ from the vertical, hence the Tropics of Cancer and Capricorn lie at this latitude north and south of the equator. Between them the sun is overhead twice during the year; at them, once. North of the Arctic circle, and south of the Antarctic circle, there are periods of the year with either daylight or night for 24 hours each day.

hold decreases. Formerly this belt where the Trade Winds converge to produce rain was believed to be a narrow front (the Inter-Tropical Front), separating two distinct air masses, but it seems more accurate to regard it as a broad zone of mixing – the **Inter-Tropical Convergence Zone** (ITCZ). Under normal circumstances it runs more or less east-west, although East Africa's weather is complicated by moist air masses which move eastwards from the Gulf of Guinea and the Zaire River basin.

Until recently, meteorological stations have been very sparsely scattered in Africa, and consequently our understanding of day-to-day **weather** – as opposed to annual variations in **climate** – has been poor. Recently, however, it has been recognized that West Africa, at least, is the 'breeding ground' of weather systems which affect the whole of the Tropical Atlantic Ocean. Since some of these weather systems can develop into destructive hurricanes, much effort has recently been devoted to studies of African weather. Meteorological satellites can show the whole continent in one picture, and this has made Africa's weather easier to appreciate, if not to understand.

Climatic factors

Rain – how much and when?

We have seen how the places on the earth's surface where the sun is overhead at noon are affected by converging winds and rising air, and thus tend to have rain. The rains, in fact, follow the sun on its apparent travels north and south, with a lag of about a month between the time when the sun is overhead at noon, and the time of maximum rainfall. Since the sun is overhead at the equator twice each year, it follows that here there are two rainfall peaks each year. Further north and south, the sun is also overhead twice each year, but the further from the equator one goes, the less is the interval between these two occasions. This means that the two rainfall peaks tend to join to form one major peak in regions more than 10° north or south of the equator. Figure 2.2 shows the distribution of the main rainfall seasons in Africa, and also shows clearly that very few places, even the wettest, lack some seasonal variation in rainfall.

Figure 2.3 shows the distribution of total annual precipitation in Africa; considered together with the seasonal distribution, this defines the rainfall climate of any region. Several features should be noted. On the western side of the continent, rainfall increases in

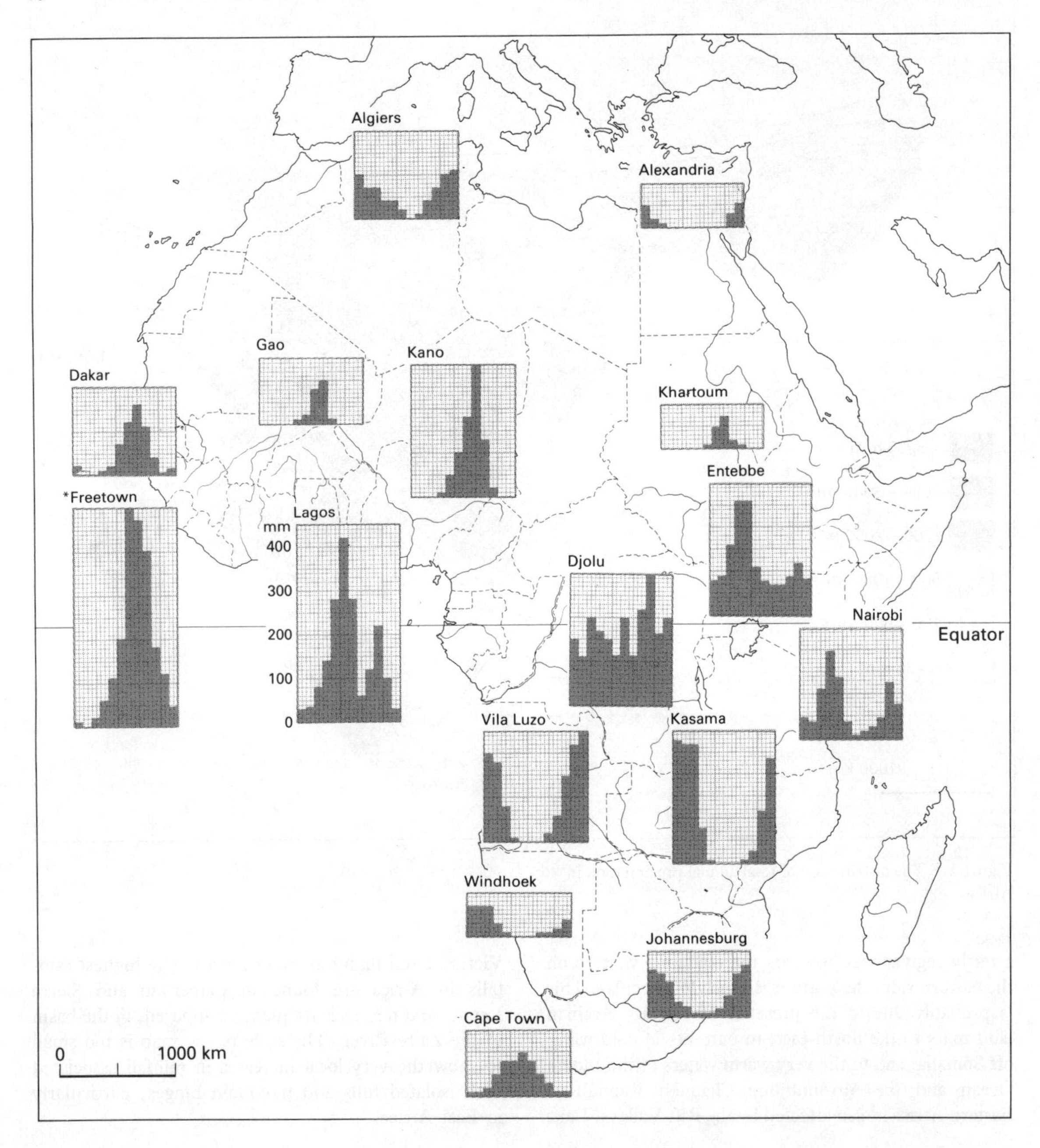

Figure 2.2 Monthly distribution of rainfall in Africa. Note the tendency towards a double-peaked pattern close to the equator, and towards a single-peaked pattern further north and south. Also note the winter rainfall of North Africa and Cape Province. *The peak for Freetown should be at 950 mm; nearly twice as high as drawn here.

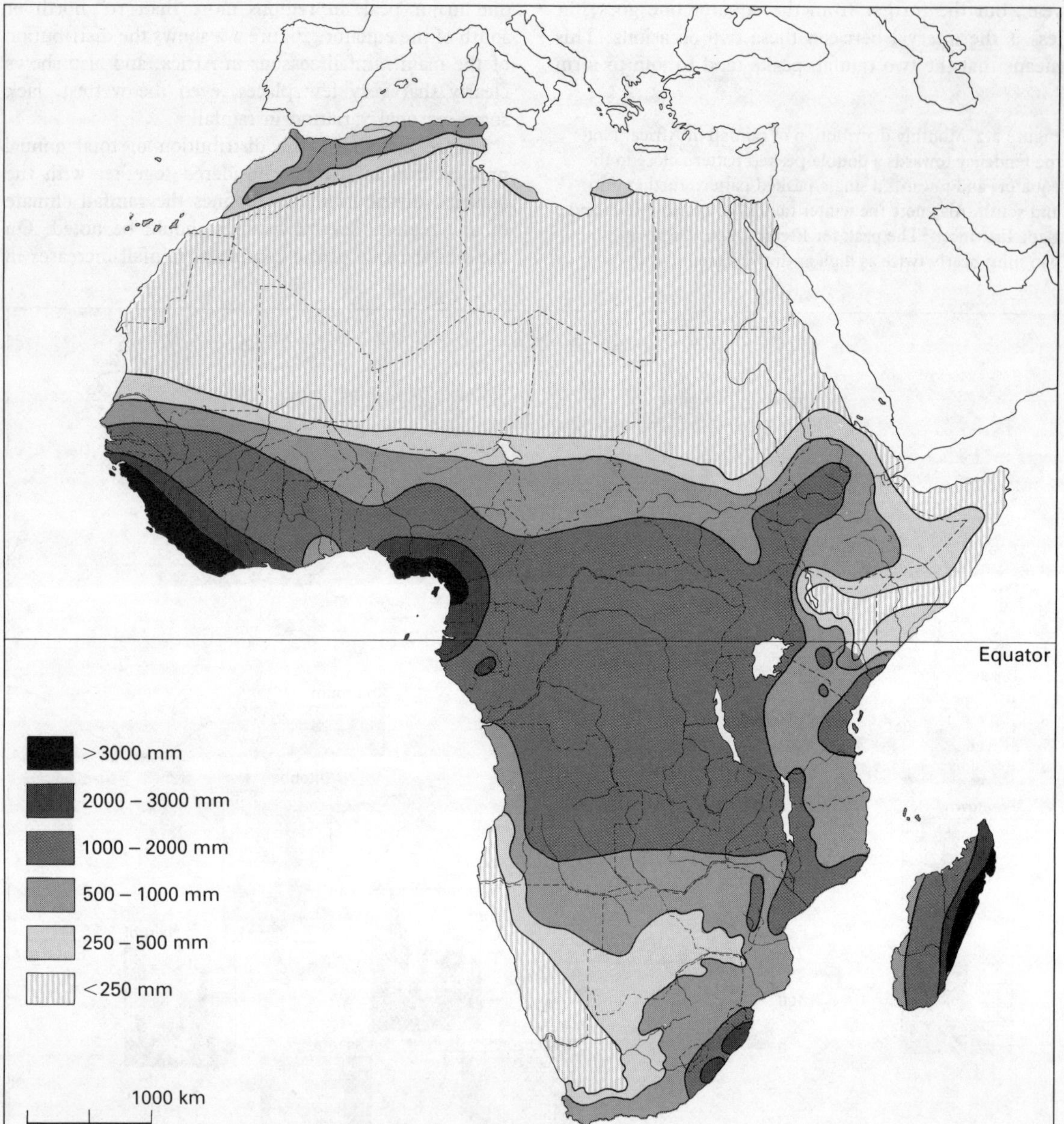

Figure 2.3 The distribution of total annual precipitation in Africa.

a fairly regular way towards the equator, whereas on the eastern side, the pattern is much less regular. This is probably due to the presence of the dry Arabian land mass to the north-east, to currents of cold water off Somalia, and to the very warm waters of the Indian Ocean and the Mozambique Channel. Rainfall in Eastern Africa is also affected by the Rift Valleys, Lake Victoria, and high mountain ranges. The highest rainfalls in Africa are found in Cameroun and Sierra Leone, and not, as is frequently supposed, in the basin of the Zaire River. The scale of the map is too small to show the very local increases in rainfall associated with isolated hills and mountain ranges, particularly in East Africa.

Temperature

Other climatic factors are less important than rainfall because they vary much less through the year. This points to an important contrast between the tropics, where seasons are defined mainly by rainfall, and the temperate zones, where seasons are defined by temperature. Near the equator, there is a relatively small fluctuation in maximum or minimum temperature during the year; the difference between the maximum and minimum temperatures during a day is usually much greater than the difference between the hottest and coldest months (Figure 2.4). Seasonal variations increase with distance from the equator because of the increasing annual changes in the amount of incoming solar radiation. Seasonal variation also increases away from the coast, because the sea heats up and cools down relatively slowly and exercises a buffering effect on the temperature of the air near to it. Temperature declines with increasing altitude; the rate of decline (the lapse rate) is about 1 °C for every 100 m of altitude.

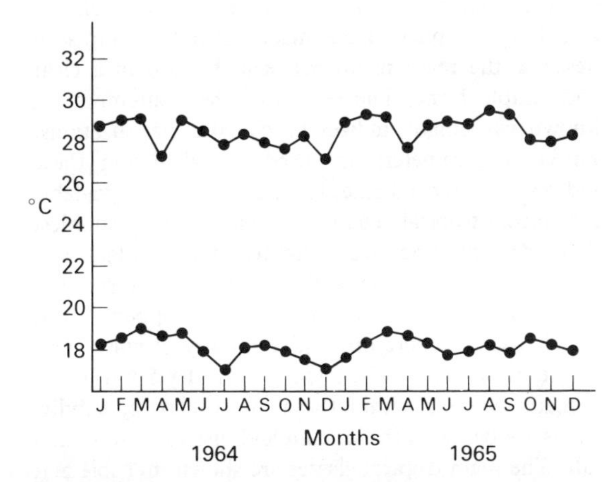

Figure 2.4 Monthly means of daily maximum and minimum temperatures at Mweya in Uganda during 1964 and 1965. The daily temperature range is usually about 10 °C which is considerably greater than the annual range.

Humidity, saturation deficit, and evaporation

The ability of the air to hold water vapour increases with increasing temperature. The water content of the air is its humidity, and this is normally expressed as the actual water content of the air, as a percentage of the water content of saturated air at the same temperature – this percentage figure is the relative humidity (RH). Ecologists often measure and discuss relative humidity with great enthusiasm and give it greater importance than it merits. Relative humidity is unsatisfactory as a measure of the environment of an animal or plant because it tells us little about the desiccating power of the air. This is because relative humidity depends upon the temperature of the air. At 5 °C, for instance, only a small addition of water vapour is needed to raise the RH of air from 50 to 100 per cent. At 30 °C, the same change would require the addition of very much more water. Thus the tendency for water to be lost from an organism would be much greater at 30 °C than at 5 °C, even at the same RH of 50 per cent. This difficulty can be overcome by using another measure, saturation deficit, which is a measure of the amount of water vapour that must be added to air at a particular temperature in order to saturate it.

Loss of water from the ground, evaporation, depends upon temperature, humidity of the air, and wind speed. Actual evaporation is difficult to measure, and instead an estimate, known as potential evapotranspiration, is often calculated. It is defined as the amount of water that could be lost from the ground due to evaporation and from plants due to transpiration if an unlimited supply of water was available. Potential evapotranspiration cannot be measured directly but can be calculated from measurements of temperature, precipitation, humidity and wind speed. Potential evapotranspiration tends to be highest in arid areas because of the low humidity of the air, and lowest in the wettest regions because these are areas with high cloud cover and high humidity. As a broad generalisation, when annual potential evapotranspiration is less than rainfall, then the natural vegetation is likely to be a forest, whereas if it exceeds rainfall then the vegetation will probably be characteristic of savanna or desert areas.

The presentation of climatic information

Klimadiagrams

It is not always easy to appreciate the basic characteristics of the climate of a region by looking at a list of averages. A simple method of presenting climatic data has been developed by Walter & Leith, 1967, in which temperature and rainfall are plotted on one diagram, while figures for absolute maximum and minimum temperatures, annual mean temperature, and other numerical data are presented at the side of the diagram. In preparing the diagram, a scale is used in which 10 °C is equivalent to 20 mm of rainfall. Using this convention, months in which the curve for precipitation falls below the curve of temperature are regarded as arid months, while those in which the reverse is true are relatively well watered. Very humid months, in which the rainfall exceeds 100 mm, are plotted on a different scale, reduced by 10 : 1 in relation to the lower rainfall curve. The area under this reduced scale is conventionally shaded black to indicate a very wet period. The main purpose of the reduced scale is to keep the size of the complete diagram manageable; if a true scale was used throughout some diagrams would be extremely large. A typical diagram from a tropical site is shown in Figure 2.5, and the information presented on it is explained in the caption.

The basic information presented by a **klimadiagram**, as these diagrams are called, gives an instant picture of the climate of a particular station, and, by extrapolation, of the region in which that station lies. Clearly one must be careful to see that stations from which extrapolations are made are truly representative of the area around them; it is unfortunately true that many human settlements are in sites atypical of their region, often for very good reasons such as the proximity of water, or hills with a locally cool climate suitable for rest and recreation.

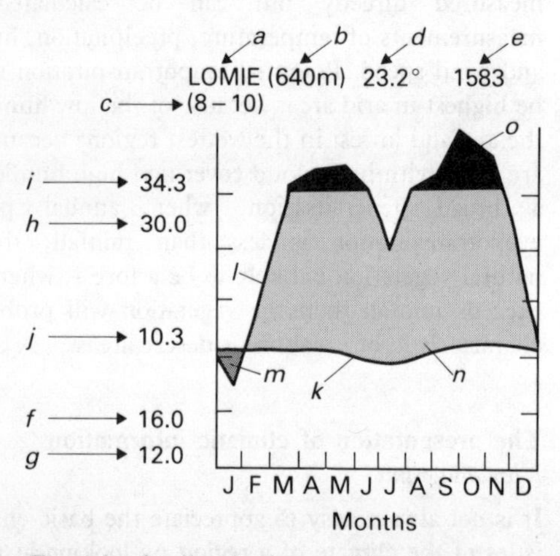

Figure 2.5 Example of a climatic diagram (klimadiagram, or climatogram) for Lomie in Cameroun. The symbols and figures on the diagram have the following meaning: *a* – station name; *b* – station height (m); *c* – number of years of observation (first figure – temperature, second – precipitation); *d* – mean annual temperature; *e* – mean annual precipitation (mm); *f* – mean daily minimum of coldest month; *g* – absolute minimum (lowest ever recorded); *h* – mean daily maximum of warmest month; *i* – absolute maximum; *j* – mean daily temperature range; *k* – curve of monthly mean temperatures (scale divisions are 10 °C); *l* – curve of monthly means of precipitation (scale divisions are 20 mm); *m* – drought period; *n* – humid period; *o* – monthly precipitation greater than 100 mm (scale 1/10) (from Walter H, 1971. *Ecology of Tropical and Subtropical Vegetation*, Oliver & Boyd: New York.)

A further convention which makes the comparison of klimadiagrams from north and south of the equator easier, is that diagrams from south of the equator are plotted using a horizontal axis which runs from July to June, instead of from January to December as is done in the northern hemisphere. This convention eliminates the effects of the seasonal movements of the sun which would otherwise make very similar climates appear dissimilar.

Klimadiagrams allow comparisons of climates throughout the world at a glance. They can be classified into a number of zonal climatic types, and Figure 2.6 shows the main types (zonobiomes) found in Africa, together with equivalents from other parts of the world. Comparisons like these allow assessments, for instance, of the likelihood of survival of a crop plant introduced into one part of the world from another. If the site of origin and the site of introduction have similar klimadiagrams, then there is a good chance that the plant will thrive in its new site.

Climatic indices

Climatic and vegetation patterns are closely related, and biogeographers have made several attempts to describe the relationship between the two in a clear and simple form. The results of these attempts are known as climatic indices or climatic classifications. Different parameters are used in calculating these indices; as a general rule, few parameters are available for most tropical countries and so the simplest methods, needing only figures for rainfall and temperature, are usually the most satisfactory. The most widely used index is that of Köppen, 1931, which uses temperature and humidity (Figure 2.6). There is a first division into hot, dry, and warm climatic classes; within these major classes, subdivisions are based on the amount and distribution of rainfall. The main tropical classes are shown in Table 2.1.

These climatic classes can be equated with vegetation types in many cases, but not in all. *Af* climates are those which naturally support tropical rainforest, but some of the drier forest types extend into *Am* climatic types. In West Africa, Guinea savanna occupies *Am* and *Aw* types; Sahel savanna lies within the *BSh* zone and desert in the *BWh* zone. In East Africa, there is a complication of differing altitude, and so it is not generally possible to produce very precise equivalents between vegetation and climatic type.

Relating climate and vegetation – critical factors

Climatic variations, as we have seen, are largely determined by latitude and altitude. They involve differences in many factors such as temperature, amount of

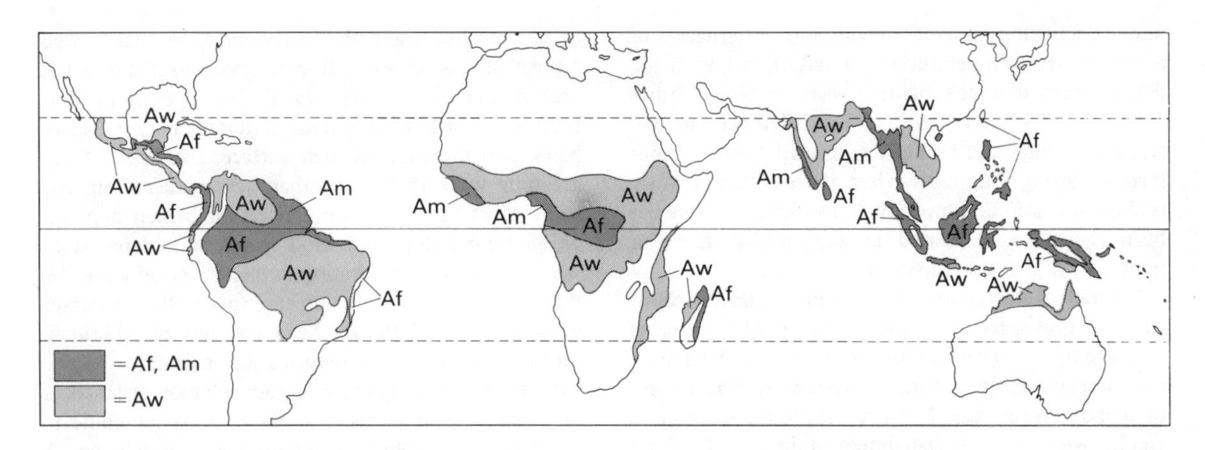

Figure 2.6 The world distribution of tropical climates according to classification of Köppen, W. 1931, *Grundrisse der Klimatunde*, Borntraager: Berlin. A_f = permanent wet rainforest: all months have sufficient precipitation. A_m = seasonally humid or subhumid evergreen rainforest with months with arid characteristics. A_w = dry period in the winter of the corresponding hemisphere: subhumid or exeromorphic forests, woodlands, shrublands or savannas (from UNESCO, 1978, reference in chapter 3.)

rainfall and distribution of rainfall through the year. One of these factors, or any combination of several, may be important in controlling the distribution of plants and vegetation types. Every species has optimum conditions for growth and upper and lower limits beyond which all growth stops.

One approach in plant ecology has been to attempt to classify the precise factors which control the distribution of individual species. Such descriptions exist for an increasing number of temperate species, but in the tropics knowledge is largely restricted to crop plants. Elucidation of the exact way in which climatic factors exert their influence on plant species usually requires detailed investigations. Climatic factors may act on any stage in the life-cycle of a species and may affect germination, establishment, growth, pollination, seed set, fruit ripening or competitive ability, and thus effectively delimit its distribution. Optimum temperature for germination varies between species. For example, maize seeds will not germinate below 8–10 °C or above 40–44 °C. Other seeds germinate best

Table 2.1 The main tropical climate classes, following W. Köppen (1931 *Grundrisse der Klimatkunde*, Borntraager, Berlin) which uses the mean temperature and humidity of each month for classification

	Temperature of coldest month	Temperature of warmest month	Mean annual temperature	Rainfall of driest month	Mean annual rainfall (r)
'A'	18	–	–	–	>20(t + 14)
Af	18	–	–	>60	>20(t + 14)
Am	18	–	–	>(100–r/25)	>20(t + 14)
Aw''	18	–	–	<60	>20(t + 14)
Aw	18	–	–	<60	>20(t + 14)
'B'	–	–	–	–	<20(t + 14)
BSh	–	–	–	–	<20(t + 14)
BWh	–	–	18	–	<10(t + 14)
'C'	−3	–	–	–	>20(t + 14)
Cfa	−3	>22	–	>30	>20(t + 14)
Cwa	−3	>22	–	<30	>20(t + 14)
Cwb	−3	<22	–	<30	>20(t + 14)

t – mean annual temperature, °C; *r* – mean annual rainfall, mm
Af – hot, moist; *Am* – hot, monsoonal; *Aw* – hot, dry winter;
Aw'' – has hot climate + two rainfall maxima
BSh – semi-arid climates; *BWh* – hot desert climates
Cfa – warm, moist; *Cwa* – warm, seasonal; *Cwb* – cooler, seasonal

under fluctuating temperate conditions. Alternation of day and night temperatures is a critical factor in the development of many plants. *Solanum* potato tuber formation is strongly affected by differences in day and night temperatures and day length: tubers are not formed during long days when the day temperature is above 30 °C. The flowering of a species of coffee has been shown to be caused by the sudden drops in temperature that accompany thunderstorms.

Rainfall characteristics have similar effects. Sisal, which is grown for hard fibre in low rainfall areas, is a markedly drought-resistant crop that can be grown commercially in areas with an average rainfall of less than 600 mm per year. However, it produces optimum yield, with a well-distributed rainfall of about 1250 mm, but in areas where the rainfall exceeds 1500 mm, it competes poorly with weeds. Many deciduous trees in the tropics have problems coping with two dry seasons per year. Seasonal patterns of leaf shedding, flowering and fruit development are particularly affected.

Clearly, climate is often the primary determinant of the type of vegetation that a region can support. There are, however, additional factors that must be taken into account. In East and Central Africa, land with more than 400 mm of rainfall can support a range of plant communities from short grassland to closed canopy woodland. Some of the other influential factors and reasons for patterns in plant communities are considered in chapter 4.

Radiation balance

All objects above a temperature of absolute zero (–273 °C) emit energy as electromagnetic radiation. A spectrum of wavelengths of radiation from the sun reaches the earth's outer atmosphere, consisting of 10 per cent ultraviolet (short-wave radiation), 45 per cent visible and 45 per cent infrared (long-wave radiation). Most of the harmful ultraviolet radiation is removed by the ozone layer in the upper atmosphere and never reaches the earth's surface.

Plants and animals receive direct and indirect radiation from the atmosphere and, at the same time, emit radiation themselves (Figure 2.7). During the day, they tend to receive more radiation than they emit and therefore increase in temperature: they are said to have a positive heat load. At night, the situation is reversed and they tend to cool down, with a negative heat load.

Exotherms (or poikilotherms – the so-called cold-blooded organisms) which include all plants, cannot regulate their body temperatures to the same extent as endotherms (or homoiotherms: see page 51). Animals can of course move into favourable positions of sun or shade, and the behaviour of reptiles is well known in this respect. Of significance to plants is the area of their surface exposed, the absorptivity of their surface (for example, colour differences or reflective hairs) and the angle of their surface. Essentially there are four ways that an exothermic organism can lose heat. Most important are convection, which depends on air temperature and wind speed, and evaporation, which is called transpiration in plants. Reradiation, by reflection, and conduction are generally of lesser importance. Endotherms have a variety of additional mechanisms including sweating and panting.

Due to the complexity of the radiation balance of an organism, air temperature is a very poor guide to the temperature that an organism is experiencing. A horizontal surface receiving solar radiation at 1000 W m^{-2}, on a typical sunny day in the tropics, and which only loses heat by long-wave radiation, has an equilibrium temperature of above 90 °C, which is well above the temperature limits of almost all organisms, therefore plants and animals have had to develop methods of reducing their temperatures. A variety of mechanisms are used to achieve this, such as sweating and panting, evaporation, reradiation, conduction and convection, as well as sheltering in shaded sites. Organisms in arid environments are faced with a dilemma of using water for evaporation as a heat sink or else conserving water and overheating.

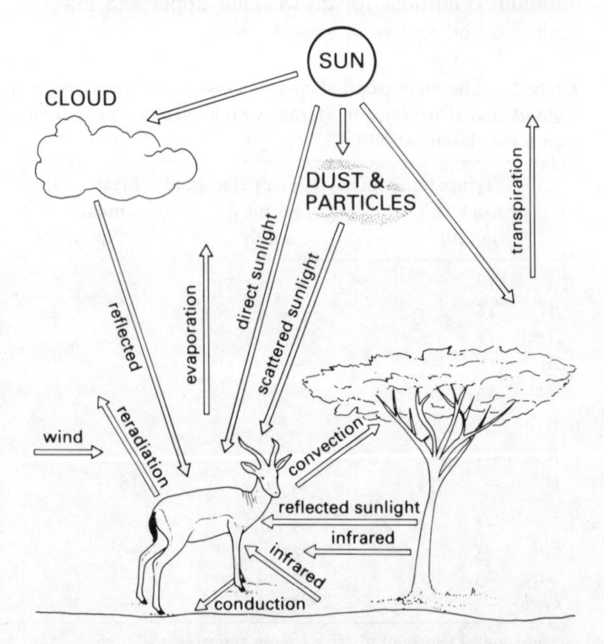

Figure 2.7 This animal receives radiation both directly and indirectly and emits radiation by convection, evaporation, reradiation and conduction.
(Information from Gates, D. M., 1980.)

Microclimates

The environment immediately surrounding an individual plant or animal differs considerably from the general climate of the region where it lives. The climate in the immediate vicinity of an organism is referred to as its microclimate. To a small seed or insect a raindrop may seem like a deluge, and a small rock or fallen leaf may determine the temperature, light and moisture conditions that the organism experiences. Plants often provide a microhabitat for other plants and animals, providing shelter, protection and an otherwise favourable microclimate. Fallen tree branches, boles of trees, cracks in rocks, leaf litter and heaps of elephant dung provide microhabitats for numerous organisms. The microclimate of plants is particularly important because, unlike animals, they cannot move to find more favourable conditions.

We have already seen (page 12) how the heat load that an organism experiences may cause its temperature to differ considerably from air temperature. Within an ecosystem this results in spatial differences in temperature, moisture, light availability, wind and CO_2 concentrations. These localized differences are primarily determined by the architecture of the vegetation stand. Vertical profiles of some of these factors are shown in Figure 2.8. Hot and cold extremes of the climate tend to be buffered by the vegetation stand. A seedling growing in exposed soil is subjected to much greater temperature extremes than one growing beneath a plant canopy. Wind speed is reduced considerably within a plant canopy with subsequently lesser effects on temperature and evaporation. Within the plant canopy, CO_2 concentrations change diurnally, being lower during the day as photosynthesis removes CO_2 and higher at night as CO_2 continues to be produced by respiration.

The many different shades of green within a forest are due to different combinations of photosynthetic pigments used to exploit differing intensities of light. Mottled leaves have been shown to be able to photosynthesize at higher leaf temperatures than can uniformly coloured leaves. Germination of many weed seeds is inhibited by the particular ratio of high far red to red wavelengths of light that is filtered through the leaves of a plant canopy, beneath which a new plant could not establish itself. Moisture conditions also vary within the canopy. The ground flora of a forest receives virtually no direct precipitation but obtains water from stem flow and leaf run off from trees. Water from these sources is rich in nutrients and is exploited by mosses, liverworts and ferns living as epiphytes or on the forest floor.

Differences in microclimate also occur on a very small scale. The microclimate close to a leaf surface differs from that surrounding the plant. The boundary layer, or phyllosphere, surrounding a leaf also has a different relative humidity and O_2 and CO_2 concentrations from those of the ambient environment. It has been demonstrated using small thermocouples that temperature can vary considerably on different parts of a single leaf. This microclimate is important for saprophytic microflora on the leaf surface and is of particular relevance to plant pathologists studying plant diseases caused by parasitic microorganisms.

The ecological significance of microclimates is important in many ways. When woody and herbaceous plants are removed from arid areas by overgrazing, burning, or cutting for fuel, the action of wind becomes more severe, soil temperature variations

Figure 2.8 Microclimatological profiles through a maize canopy at different times of the day, which are shown at the top of each profile. Intensity increases as we move to the right-hand side of each graph. For example, net radiation above the maize canopy (----) is most intense at 1300 hours and decreases towards evening. This difference is buffered closer to the soil surface. Wind speed differences do not show a strong correlation with time of day, but are buffered by the canopy in a similar way.
(Adapted from Uchijimi Z., 'Maize and Rice' in Monteith, 1976 volume 2.)

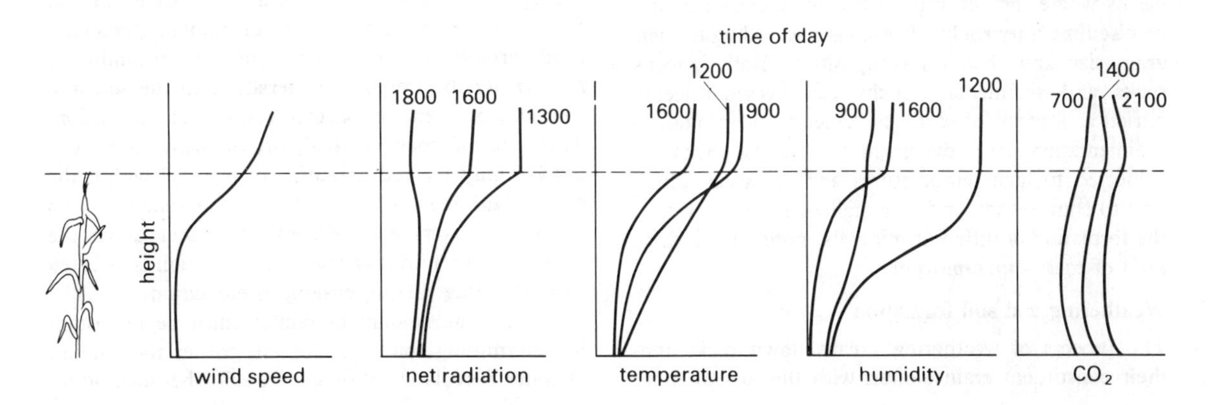

increase, water infiltration into the soil is affected and often only stress-tolerant xeromorphic plant species can survive. It has been suggested that the deterioration of microclimates is the main reason why many species of trees and shrubs regenerate very slowly under these conditions.

Soils

Rock types

Much of Africa consists of very ancient land surfaces which have been exposed to weathering for many millions of years. Weathering is the effect of climate on rocks. In very dry regions, mechanical weathering, such as the impact of wind-blown sand grains on rocks, is the most important means by which they are broken down. In contrast, in the humid or seasonally wet tropics, it is chemical weathering which is the predominant force in soil formation.

Some rocks originate in the molten interior of the earth, and reach the surface as a result of earth movements or volcanic activity. Rocks of this kind, which have crystallized from a molten state, are known as **igneous rocks**. They vary in two main ways, firstly in composition, and in particular in the amount of silica (quartz, SiO_2) which they contain, and secondly in their crystal size. Rocks with a large amount of silica are referred to as **acidic rocks**; those with relatively little as **basic rocks**. The crystal size of a rock depends on the speed at which it cooled from the molten state. Small surface lava flows cool quickly and produce very fine-grained rocks; large masses of molten rock or **magma** which have cooled slowly deep underground are usually made up of much larger crystals. The crystal size of a rock influences its rate of weathering, while its composition – basic or acidic – can determine the richness of the soil which results from the weathering.

Weathering of igneous rocks breaks them down into residual particles which are deposited in sedimentary basins where they eventually become compacted into new **sedimentary rocks**. These vary according to their grain size and chemical composition. Both igneous rocks and sedimentary rocks can become deeply buried as a result of earth movements and prolonged sedimentation. At these great depths, the rocks are subjected to high temperatures and pressures which lead to compression and recrystallization, and even to the formation of different minerals, producing a third kind of rock – **metamorphic**.

Weathering and soil formation

The process of weathering breaks down rocks into their constituent grains, often with the loss of more soluble fractions. Alternatively it may alter them *in situ* by differential solution and reaction to give a residue which often retains some of the features of the parent rock while having different properties and composition. The conversion of the weathering products of rock into a soil is caused by a number of factors working together; the process is known as **soil formation** (**pedogenesis**).

We must first examine the processes which occur during the weathering of rocks. Chemical weathering of rocks is a slow but continuous process. Rain-water is slightly acidic due to dissolved carbon dioxide, and water percolating through soil also contains organic acids from decomposed plant material. Some components of the rock are removed in solution by the percolating water; others are altered to other substances without being removed. The components of the rock which are most resistant to solution and alteration last longest, and it is these, the residual minerals, which form the basis of soils. In the temperate regions the usual residual mineral is quartz.

In the tropics, weathering proceeds somewhat differently. The solubility of silica in water rises rather steeply with increasing temperature, so that in tropical soils, silica tends to be dissolved out, albeit slowly, and carried away in water either to deeper soil layers or into rivers. The residual minerals which form the basis of many tropical soils are oxides and hydroxides of iron and aluminium. This is why so many of the world's deposits of iron and bauxite (the principal ore of aluminium) are in tropical regions – millions of years of weathering have concentrated them from the rocks in which they originally occurred. The large quantities of iron oxides in these soils give them the reddish colour which is so characteristic of many tropical soils.

How are the weathering products of rocks converted into soil? Some 'soils', such as those of deserts, are little more than finely divided rock, perhaps with some sorting of particles by the occasional rains. This example highlights the importance of water in soil formation. A certain amount of rainfall percolates right through the soil profile to join the groundwater and irreversibly removes materials from the soil into solution, or even, in extreme cases, into suspension. This removal may be total, or, in many cases, can involve only the redistribution of material within the profile, with metallic ions and clay being removed from the upper layers and redeposited further down the profile. The removal of material in solution is referred to as **leaching**: in suspension, as **eluviation**.

Clearly, the amount of rainfall must be important in determining soil type, for the greater the amount of water percolating through the soil, the more active will be the processes of leaching and eluviation. However, not all the rain that falls on a piece of ground penetrates the soil; some is redistributed by runoff. This is important in producing local differentiation of soils and will be considered later. As a general rule, high rainfall tends to produce a deep soil in which most of the positively charged metallic ions have been replaced by hydrogen, so that the soil is acid and poor in bases. The clay minerals tend to be ones such as kaolinite, with a low cation exchange capacity (see page 17). Under lower rainfall, the converse tends to be true; other things being equal, such soils are shallower, are less acid, and richer in bases than soils developed under a high rainfall. The clays tend to be ones such as illite and montmorillonite, which have a much higher cation exchange capacity than kaolinite. Often calcium is not completely leached out of the soil, but is deposited as a layer of secondary calcium carbonate at a depth in the soil which varies according to rainfall, being deeper the higher the rainfall.

All these examples assume a soil in which drainage is free, so that water flows straight through the soil. If water cannot penetrate the soil, or if the soil is constantly saturated from below by water originating elsewhere, then hydromorphic soils are formed, in which water is the dominant factor in soil formation. Iron compounds in the soil are reduced, giving grey, bluish or greenish colours, and oxidized iron may sometimes accumulate in patches, giving mottles of red–brown. Such soils are likely to be difficult to cultivate and may produce toxicity symptoms in unadapted plants. Other peculiar soils of agricultural importance occur in arid areas, where evaporation greatly exceeds precipitation. Here, water loss from the soil surface leads to accumulation of salt at the soil surface, sometimes giving a whitish crust. Such soils are often found naturally in deserts, but they can also be produced by the unwise use of water containing a small amount of dissolved salts for irrigation in arid areas. Soils of this type, once formed, are sterile as far as useful crops are concerned, and cannot easily be brought back into production.

In many parts of Africa there are vast areas of low-lying ground covered by black clay soils. These are often called 'cotton soils'. In the dry season they become extremely hard, and crack deeply. During the wet season, however, they quickly become very soft, and swell, closing the cracks. These properties are due to their high content of the clay mineral montmorillonite, which has a molecular structure which allows absorption of much water by separation of the molecular lattice layers. When dry, the upper soil layers, to depths of about 1 m, crack into prismatic or block-like lumps which often have shiny surfaces. During the dry season, and at the beginning of the rains, material from the surface falls into the cracks. When the soil is rewetted and swells, this extra material prevents the lower layers from returning to the original position. The result is that some of the soil is forced upwards and eventually, over many years, the whole soil profile is turned over and mixed by this movement. These soils are known as vertisols, or as self-mulching soils. Although often agriculturally valuable, they can be very difficult to work as they are extremely hard when dry, and very sticky when wet – the intermediate stage in which they are workable may not last more than a few days in each year.

The description of soil

A **soil profile** is a section through the soil from the surface to the underlying rock. In the tropics it may not be possible to achieve such a section, as unweathered rock may be found only at great depths – sometimes as much as 100 m. 'Rock' in this context means the parent material from which the soil is formed, so that unconsolidated wind- or water-borne sand counts as rock, just as does hard granite. The examination of a soil profile is best carried out in the walls of a soil pit dug as deep as necessary or possible (usually the latter in the tropics), but much valuable information may be obtained from holes bored with a soil auger.

Figure 2.9 Typical horizons in the surface layers of the soil. In certain tropical ecosystems, for example in rainforests, dead organic matter decomposes rapidly and these horizons cannot be recognized. In drier environments, or at drier times of the year, decomposition proceeds more slowly.

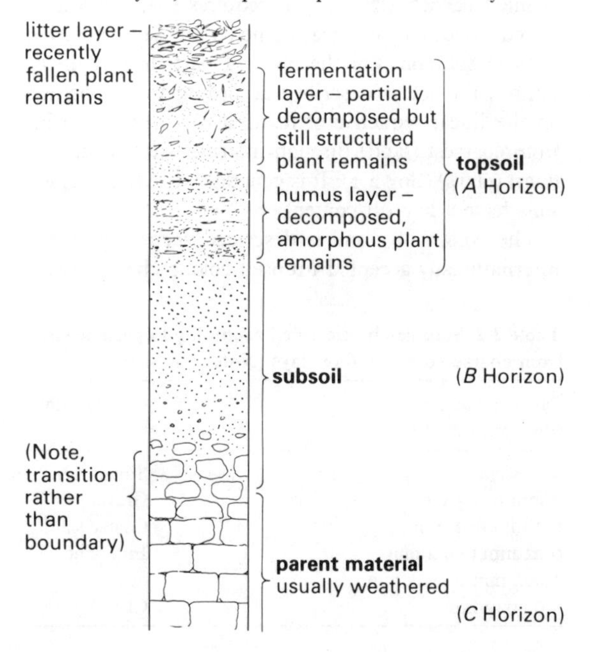

Within a soil profile it is usually possible to distinguish three main layers or horizons (see also Figure 2.9). Their colours are often noticeably different from each other. Near the surface is the '*A*' horizon which usually contains organic material, finely divided and well incorporated into the soil; it is the zone from which the downward percolation of rain-water has tended to remove finely divided or soluble materials. Below this is the '*B*' horizon, which is usually rather sharply delimited from the '*A*' horizon, and which contains little or no organic material. It is also the zone in which material washed out of the upper layers is redeposited, sometimes in distinct layers. The kinds of material which are removed from the '*A*' horizon and redeposited in the 'B' include clays, iron, and humus. Finally there is the '*C*' horizon which is, essentially, weathered parent material that has undergone neither depletion nor enrichment with the substances mentioned above. It often retains some of the structure of the parent material.

The main features that are taken into account when describing a soil profile are texture, colour, and structure. The texture of a soil is a product of its particle size distribution. The mineral particles which make up the inorganic skeleton of the soil are conventionally classified into size classes; these divisions are internationally accepted, with the exception of silt, where 'US silt' is from 2–50 μm, and 'International silt' is from 2–20 μm (see Table 2.2). In the field, a rough assessment of particle size distribution may be made by rubbing the soil moistened with saliva between the fingers to subjectively determine its grittiness. More detailed particle size analysis requires that the soil be sieved followed by measurements of sedimentation rates in solution. On the basis of particle size distribution, the soil is described according to its position on the linear sequence of textural classes which runs from coarsest (sand) through intermediates (loams) to finest (clays) (Box 2.1). It is generally true that tropical soils have a low silt content.

The colour of a soil is described according to the internationally accepted Munsell colour charts. These classify colour along three axes – hue (spectral colour), value (greyness) and chroma (strength of colour). This sounds complicated but in practice it is remarkably easy to allocate most soils to a colour value, and, other things being equal, there is reasonable agreement between workers as long as they have good colour vision. Most tropical soils tend to be reddish or yellowish in colour, in contrast to the rather brown soils which are common in temperate latitudes. These reddish and yellowish colours are caused by oxides and hydroxides of iron, the latter tending to be yellow and the former to be red.

It is unusual for a soil to be the same colour throughout the profile; typically the upper layers are darker and often brownish in undisturbed profiles, due to the incorporation of organic matter and sometimes, particularly in natural grassland soils, to the incorporation of finely divided carbon from grass fires. At depth, soils often become paler and more yellow, reflecting the usually greater moisture content. Soils which are seasonally waterlogged may be mottled, with patches of red–brown soil interspersed in a greyish matrix. These colour mottlings result from alternate oxidation and reduction of iron compounds – oxidized iron (Fe^{3+}) is reddish, while reduced iron (Fe^{2+}) is bluish or greenish. Lines of reddish soil often follow channels in the soil made by roots, which allow better aeration. This mottling of seasonally waterlogged soils is known as **gleying**. Waterlogging of soils produces various reduced ions as well as Fe^{+}; some of these, and also Fe^{2+} when in high concentrations, are toxic to plants, and species which can grow on such soils often show special features such as air-filled spaces in their roots and stems which improve oxygen supply to the cells.

The third important soil feature which is normally recorded in the field when describing a soil profile is structure. This is the way in which the soil particles are aggregated under natural conditions and is extremely important in assessing the agricultural potential of a soil. If structure is weak, the soil will break down into an impermeable mass, and root penetration will be difficult. Aeration will tend to be poor, and rain-water will tend to run off the soil surface rather than percolating easily. A well-structured soil has large spaces between the soil aggregates; indeed, as much as 60 per cent of the volume of the soil may be occupied by air- or water-filled space. In such a soil, water and air can both move freely, as can plant roots and soil fauna. The different horizons of the profile often differ greatly in their structure, and particular structures are strongly indicative of particular soil types. Thus soils with much sodium carbonate in the profile are characterized by

Table 2.2 Soils can be classified according to particle size from coarse stones to fine clays (Note: 1 mm ≡ 1000 μm)

Particle size (diameter)	Classification
> 7.5 cm	Stones
2 mm to 7.5 cm	Gravel
0.2 mm to 2 mm	Coarse sand
0.02 mm to 0.2 mm	Fine sand
0.002 mm to 0.02 mm	Silt
<0.002 mm	Clay

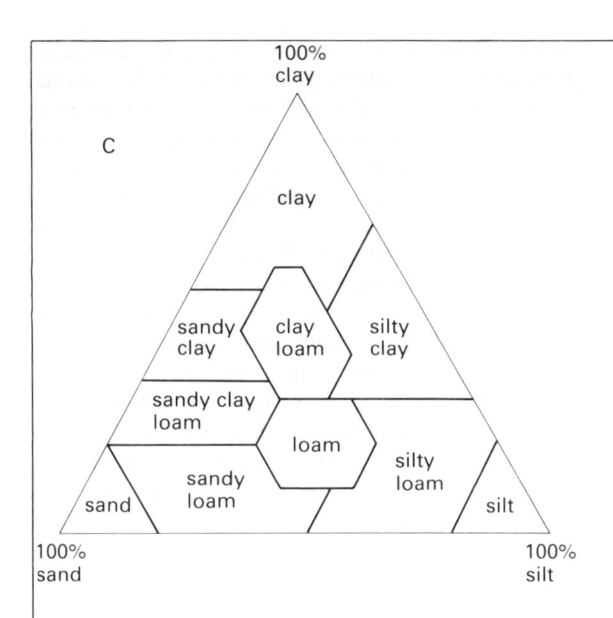

Box 2.1 Particle fractionation of soil

A Method of the analysis of particle size of soils. Soil should be thoroughly shaken or stirred into the water. Dispersal agents such as sodium hexametaphosphate or sodium oxalate are often added to break up aggregations of clay particles.

(a) Sieve soil through 2-mm sieves to remove coarse particles.

(b) Determine moisture content of soil weigh → oven-dry (80–105 °C) → weigh

(c) Thoroughly shake or stir 50 g in water, dilute to 1 litre. Take hydrometer readings (g $litre^{-1}$) after 5 minutes and after 5 hours.

(d) To separate coarse and fine sand if required. Pour off liquid, then wash and rub sediment through a 0.2-mm sieve. Weigh each fraction.

B Hydrometer readings and moisture content of the soil can be fitted to simple formulae to determine per cent clay, per cent silt and per cent sand. If a dispersal agent is used, make up a 5% w/v solution and subtract 1 from the first two formulae.

$$\text{Clay \%} = \frac{\text{5-hour hydrometer reading} \times 100}{50 - \text{moisture weight of soil (g)}}$$

$$\text{Silt + clay \%} = \frac{\text{5 minute hydrometer reading} \times 100}{50 - \text{moisture weight of soil (g)}}$$

$$\text{Silt \%} = (\text{Silt + Clay \%}) - \text{Clay \%}.$$

$$\text{Sand \%} = 100 - (\text{Silt + Clay \%})$$

C The results obtained from the formulae can be inserted into the triangle (which is based on the ISS particle fractionation) to obtain a soil classification. Methods follow Allen *et al.*, 1974.

'*A*' horizons with a thin platy or amorphous structure, and '*B*' horizons formed of columnar aggregates of soil with rounded tops to the columns. The structure also determines the extent of root penetration, and it is usual to record the presence or absence of roots in each soil horizon.

It is also usual when describing a soil to collect samples from each of the recognizable horizons, and to subject these to various analytical processes which give more information about the soil than can be obtained in the field. The most important of these is certainly the soil reaction; is it alkaline, neutral or acid? All but the most exceptional soils have pH values between 3.5 and 9.0; most tropical soils are weakly acid in reaction, with pH between 4.5 and 6.0. The pH of a soil is normally determined in a soil-water suspension; usually 1 part of soil to 2.5 parts of water, although conventions vary between countries and laboratories.

Two very important characteristics of soil are the **cation exchange capacity** and the **total exchangeable bases**. These two properties depend on the type and quantity of clay in the soil, and also on the content of organic matter in it. In order to understand these two parameters, it is necessary to know a little about the structure of clay minerals. All clay minerals are made up of two or three plate-like structural sub-units, themselves made up of silicon, oxygen, aluminium, magnesium and hydrogen. The surfaces of the clay minerals have unoccupied sites of negative charge, as do particles of organic matter. Positive ions, including hydrogen and metallic ions such as calcium and potassium, are loosely bound to these negative sites. The cation exchange capacity of a soil is a measure of the number of sites on clays and organic matter which are available. The number of total exchangeable bases is a measure of the extent to which these sites are occupied by metallic ions which are easily detached into solution. Different clay minerals, because of differences in their structure, have different numbers of sites available for binding of positive ions. Cation exchange capacity is measured in milli-equivalents per 100 g of soil. Kaolinite, a very common clay mineral, has a low cation exchange capacity (3–15 meq/100 g); illite has a rather higher one (10–40 meq/100 g); while montmorillonite, a frequent constituent of 'black cotton soils', has a very high capacity (80–150 meq/100 g). Organic matter in the soil also has a very high cation exchange capacity (100–350 meq/100 g), but since it forms only a very small proportion (usually less than 4 per cent) of lowland tropical soils it accounts for a small proportion of the total cation exchange capacity of the soil.

Soil as a rooting medium for plants

Plants obtain, from the soil, the water and mineral nutrients that are essential for growth and metabolism. Carbon, hydrogen and oxygen are the major constituents of plants, but are not normally considered as nutrients. Mineral elements that are required in large amounts by plants are referred to as **macronutrients** and those necessary in trace amounts are called **micronutrients** (Table 2.3). The absence of any one of these elements from soil has been found to have an adverse effect on some stage of the growth, development or metabolism of many plants.

Table 2.3 Chemical elements that are essential for healthy plant life. Absence of a single nutrient can adversely affect the growth and survival of plants

	Non-metals	Metals
Major constituents	Carbon, hydrogen, oxygen	–
Essential macronutrients	Nitrogen, phosphorus, sulphur	Potassium, calcium, magnesium
Essential micronutrients (trace elements)	Boron, silicon, chlorine, iodine	Iron, copper, manganese, zinc, molybdenum, cobalt, vanadium, sodium, gallium

The availability of nutrients in the soil is dependent on many factors including water status of the soil, soil pH and soil organic matter. Deficiencies or excesses of nutrients commonly occur and, in natural systems, a number of plant adaptations have evolved to allow a healthy existence under such circumstances (pages 78–79). In agricultural systems, fertilizer applications to crops provide a ready supply of available nutrients, thus increasing yields. The cycling of nutrients through the soil is discussed on pages 38, 39, 51–53.

The spaces in between mineral and organic particles of the soil are occupied by either air or water or both. Air spaces in the soil will tend to become saturated with carbon dioxide from the respiration of plant roots, microorganisms and soil fauna, and continued replenishment of oxygen by diffusion into these spaces is essential for a healthy soil. After rain, **gravitational water** fills the air spaces and gradually percolates downwards with gravity. When this gravitational water has drained away, **capillary water** remains as a thin film around soil particles and in capillary spaces.

Plant physiologists are of the opinion that water transfer should be considered as a continuum between the soil, the plant and the atmosphere, and refer to this as the **soil–plant–atmosphere continuum** (SPAC). As the soil dries after rain it becomes increasingly difficult for a plant to meet its water requirements, particularly on a sunny day when the plant is losing substantial amounts of water by transpiration. For this reason, plants tend to wilt during the day but to re-equilibrate with the soil and make up the losses during the night when stomata are closed.

If gravitational water remains in the soil for a long period it will cause waterlogging and a lack of oxygen in the soil, mainly because oxygen diffuses about 10 000 times slower through water than through air. This is used to advantage in rice paddies where rice can grow productively in waterlogged conditions but most weeds cannot. However, prolonged waterlogging leads to anaerobic conditions in the soil and leads to noxious concentrations of gases, such as hydrogen sulphide (H_2S) and methane (CH_4), and reduced metallic ions such as Fe^{2+} and Mn^{2+}. Plants adapted to waterlogged conditions often have well-developed air canals.

Soil has a complex structure and is not merely an inert substrate containing mineral nutrients in solution. In the next section, the biological processes that are involved in the development and maintenance of a healthy and productive soil will be considered.

Biological processes in the soil

'Soil is a function of many things' (Hans Jenny). Besides the effects of climate, parent material and the influence of time, the ecologist is often particularly concerned with biological processes in the soil. All plants and animals eventually die and contribute to the continuous cycling of nutrients through the soil (chapter 3) and also to the development of the soil.

Decomposition processes are complex and involve a succession of organisms which utilize dead organic matter as their energy source and release CO_2 in respiration. A typical succession, for example on a plant cellulose substrate, starts with colonization by fungal mycelia. After a period of days or weeks, bacteria start to grow profusely and support populations of nematodes and amoebae, and larger soil animals such as mites and springtails. Alternatively, termites may instigate the decomposition process. The rate at which decomposition proceeds is largely determined by moisture conditions, and leaching by rainwater may also accelerate the process.

The complexity of ecological interactions within the soil provides an active area of research. It has been found that 80–90 per cent of CO_2 released from the soil is from microbial respiration, but nevertheless soil animals also play an important role in decomposition. For example, oribatid mites can decompose organic matter directly, provide a larger surface area of substrate for microbial colonization, aid in the dispersal of microbial spores and also feed on microbes. Interestingly, this grazing on microbes can actually accelerate decomposition rates by maintaining microbial colonies in an active growth phase. A proportion of the nutrients released from dead organic matter become assimilated in the tissues of decomposer organisms (referred to as immobilization), whilst the remainder becomes available in the soil to plants (referred to as mineralization).

Some tissues decompose above the soil surface whilst others are first incorporated into the soil, often by the action of termites. In the latter case, the ratio of carbon to nitrogen (the C : N ratio) in the dead organic matter is of particular significance. Organic debris with a C : N ratio of less than 20 : 1 that is added to the soil will immediately provide available nitrogen for plant uptake. Freshly fallen tree leaves usually have a C : N ratio of between 25 : 1 and 54: 1 and must be acted on by decomposer organisms before they release nitrogen into the soil. One reason why small-scale farmers use compost heaps is to lower the ratio of carbon to plant nutrients and therefore provide readily available soil nutrients to their plants during the growing season.

The breakdown and incorporation of dead organic matter into the soil is sometimes referred to as humification. Humus is the organic fraction of the soil that consists of amorphous decomposed plant remains. Its incorporation into the soil improves soil structure and drainage and provides a continuous release of nutrients for plant uptake. Different zones in the upper layers of the soil can usually be recognized according to the extent to which dead organic matter has been incorporated (Figure 2.9).

Soil types

Examination of soil profiles in the field, and laboratory examination of samples from the profiles, gives detailed descriptions which can be compared with one another. Similar profiles can then be grouped together. Groups of similar soil profiles form a soil series, which can be used as a unit in preparing a soil map. Soil series are grouped into families; these into subgroups and greatgroups, and finally into suborders and orders, of which there are ten that include all world soils.

The classification of soils is a complex affair which has not been made easier for the beginner by the large number of systems in use in different parts of the world, and the complex nomenclatural systems used by some of them. A simplified version of one of the most widely used systems is shown in Table 2.4.

Two soil types in particular deserve further mention in relation to the African tropics, namely laterites and catenas which are examined in this section as special topics.

Table 2.4 Classification of soils in Africa following D'Hoor, often known as the Commission for Technical Co-operation in Africa (CCTA) System. A few local soil types, and some which only occur in the temperate regions of Africa, have been omitted here, as have most of the subdivisions within each class. In all, there are sixteen main groups and a total of 63 soil types

(a) Raw mineral soils
 Aa – Ac – on rock and rock debris
 An – Ar – on desert sands and detritus

(b) Weakly developed soils
 Ba – Bd – Lithosols (skeletal soils over rock)
 Bn – Br – Juvenile soils on recent deposits (volcanic ash, alluvium, etc.)

(c) Calcimorphic soils (calcareous: containing calcium carbonate)

(d) Vertisols and similar soils (high in swelling clays: crack widely upon drying)

(e) Podzolic soils (highly leached soils in humid areas)

(h) Eutrophic brown soils of tropical regions (containing optimal concentrations of nutrients for plant growth)

(j) Ferruginous tropical soils (weathered reddish-brown soils of savanna)

(k) Ferrisols (in small scattered areas: distinctive properties from parent material rather than weathering. In forest and forest–savanna zones)

(l) Ferrallitic soils (high degree of weathering: zonal soils of rain forest)
 La – Lc – dominant colour yellowish brown
 Ll – Ln – dominant colour red
 Ls – Lx – others

(m) Halomorphic soils (saline and alkaline soils)

(n) Hydromorphic soils (poor drainage conditions: marshes, swamps, seepage areas)

(o) Non-hydromorphic organic soils of mountains

Laterite

The red ferrallitic soils which are so widespread in Africa are often called latosols. This name should not be confused with laterite, a widely used but much misunderstood term. The term laterite was originally used in India to refer to material which is soft when quarried but which rapidly hardens to a brick-like consistency when exposed to the air (from the Latin: *later* – a brick). There has been a tendency to apply the term in Africa to almost any red soil material, soft or hard. Laterites most readily form on rocks rich in quartz, and oxides of iron and silica. When weathered intensively, the oxides decompose faster than the quartz which is therefore found in higher proportions in the resulting laterite. However, because the term 'laterite' has been applied to such a variety of different soils, many soil scientists prefer not to use it when describing soils.

Within iron-rich soils, small pea-like lumps of iron oxides are produced over a long period of time, precipitated from the soil solution. This is called pisolithic ironstone or plinthite. These ironstones are called 'secondary' because they are a product of weathering processes and not a part of the original rock. The exact mechanism of formation of these ironstones is very poorly understood, mainly because it takes place over very long periods within the soil, where it is difficult to make observations.

Soils in relation to topography – catenas

Much of the interior of Africa is covered by gently undulating country, with hills, sometimes crowned by

Figure 2.10 Typical soil–vegetation catena in Tanzania. These sequences are repeated throughout much of Tanzania. (After Vessey-Fitzgerald D., 1973, *East African Grasslands*, East African Publishing House: Nairobi, Kenya.)

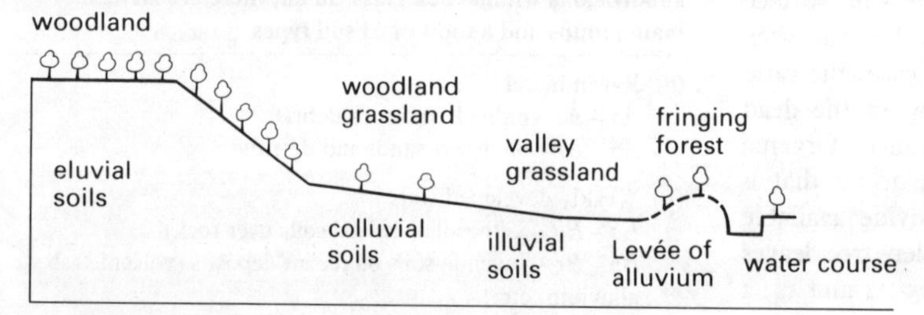

A Diagram of typical catena in valley with free drainage

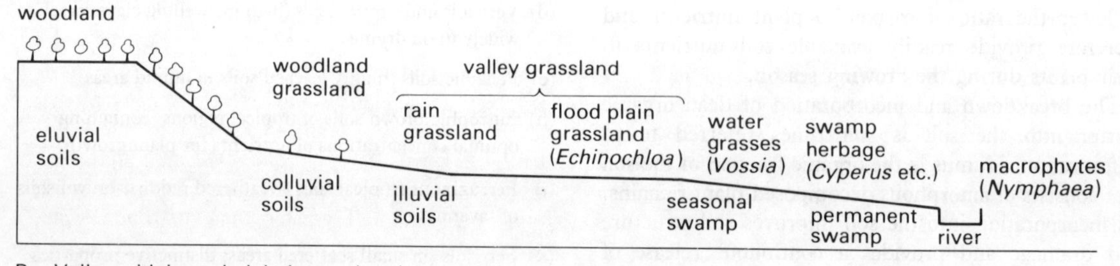

B Valley with impeded drainage showing extension of catena to include seasonal and permanent swamp

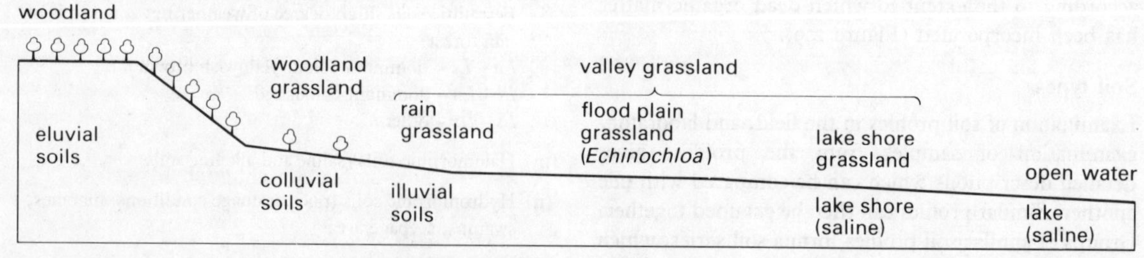

C Valley with closed drainage showing catena modified by incidence of lake shore grasslands surrounding shallow saline lake

rocks, separated by broad shallow valleys. Although both hills and valleys are underlain by the same kind of rock, they have very different soils, which occur in a regular and repeating sequence. These regular sequences of soil types were first recognized in Tanzania, and are known as **catenas**. The repeating pattern of soil types is usually marked by a correlated repeating sequence of vegetation types (Figure 2.10).

From hill and ridge tops, rain moves material downhill, either as solid particles in surface wash, or as dissolved material percolating through the soil. The residual material tends to be coarse and well oxidized, so that the predominant soil colour is red. In spite of this removal of material, including plant nutrients, these soils constantly receive new supplies of ions from rock weathering. These hilltop soils represent the **eluvial** part of the catena. On the hill slopes, material is constantly being moved downhill by wash and percolation, but new material is also constantly being added from further up the hill. This intermediate part of the catena is called **colluvial**. Soils here are generally deep, coloured predominantly red towards the top of the slope, and becoming browner or yellower towards the slope foot. In the valley bottom, material accumulates as a result of its downward transport from the hilltop and slope. This is the **illuvial** part of the catena. During the wet season, the valley floor is likely to be at least temporarily waterlogged, and under these conditions reduced forms of iron give grey and black colours to the soil. There is often also accumulation of salts in these soils, leading either to alkalinity and carbonate accumulation, or even to salinity.

Catenary soil patterns are very widespread in the tropics; their recognition provides an important framework which often helps to explain patterns of distribution of vegetation, and of the animals which are dependent upon it. It also simplifies mapping, the whole soil unit can be described and mapped as a single catena, instead of as a complex and recurring soil sequence.

Vegetation

The map (Figure 2.11) shows the distribution of the main types of natural vegetation in Africa. Comparison with the rainfall map (Figure 2.3) shows how many of the boundaries are similar to one another, emphasising the dependence of vegetation upon rainfall. Classification of African vegetation has caused much controversy in the past and the system used here is a very much simplified version of more complex schemes. It makes a first distinction between those vegetation types which are zonal – that is, which are related to a particular climate – and those which are azonal, occurring in areas with soil conditions which override the effects of climate. It should be noted that in most parts of Africa the natural vegetation has been considerably modified by the activities of man.

Zonal vegetation

Forest

Forest vegetation is predominantly woody. There are two or more layers of trees, with subordinate layers of shrubs and shade-tolerant herbs. Although any kind of woody vegetation is often called a forest, and many 'forest reserves' in Africa contain only a sparse woody growth with a dense undergrowth of grass. Ecologists generally restrict the term to vegetation in which the upper vegetation canopy is more or less continuous with few open spaces and where there is no substantial grass-dominated ground layer, although shade-tolerant grasses may be found there.

There are two major areas of forest in Africa. One occupies the Zaire River basin, and extends northwestwards into Nigeria. The second, much smaller area, lies along the West African coast from Ghana to Sierra Leone and extends up to 450 km inland. Both of these areas receive a high rainfall, although over most of them there is at least one month in each year with less than 100 mm of rain (Figure 2.2). In Sierra Leone, an intense dry period of at least three months is compensated for by very heavy rains during the wet part of the year. To the east of the Zaire Basin there are scattered patches of forest, but most of these are on high ground and will be considered under the heading of montane vegetation.

The most outstanding feature of forest vegetation is often its diversity (see also pages 90–92). The number of plant species per unit area is very great, and generally increases from the driest to the wettest forests. A survey of the forests of Ghana found a mean of 138 plant species in each 625-m^2 plot in the wettest forest type (rainfall greater than 1750 mm $year^{-1}$), but a mean of only twenty five species in the same area in the driest forest type (rainfall less than 1000 mm $year^{-1}$). Coupled with this diversity of species, and, indeed, probably a prerequisite of it, is the variety of different plant types, or **synusiae**. Synusiae are groups of plants occupying a similar habitat and making similar demands on their environment. The synusiae recognizable in most forests are:

A Autotrophic plants (with chlorophyll)
- (i) Mechanically independent plants
  - (a) trees
  - (b) shrubs
  - (c) herbs

(ii) Mechanically dependent plants (supported by others)
  (a) epiphytes
  (b) climbers
  (c) mistletoes

B Heterotrophic plants (lacking chlorophyll)
  (i) parasites (nourished by other living plants)
  (ii) saprophytes (living on dead plant or animal material).

Tropical forests have a more complex vertical structure than any other vegetation type (Figure 2.12); this contributes to their diversity. The tallest trees, occasionally reaching 60 m, extend above the general upper surface of the forest canopy, and are known as

Figure 2.11 Distribution of the main types of vegetation in Africa
(Based on Ewer & Hall, 1978, reference in chapter 1.)

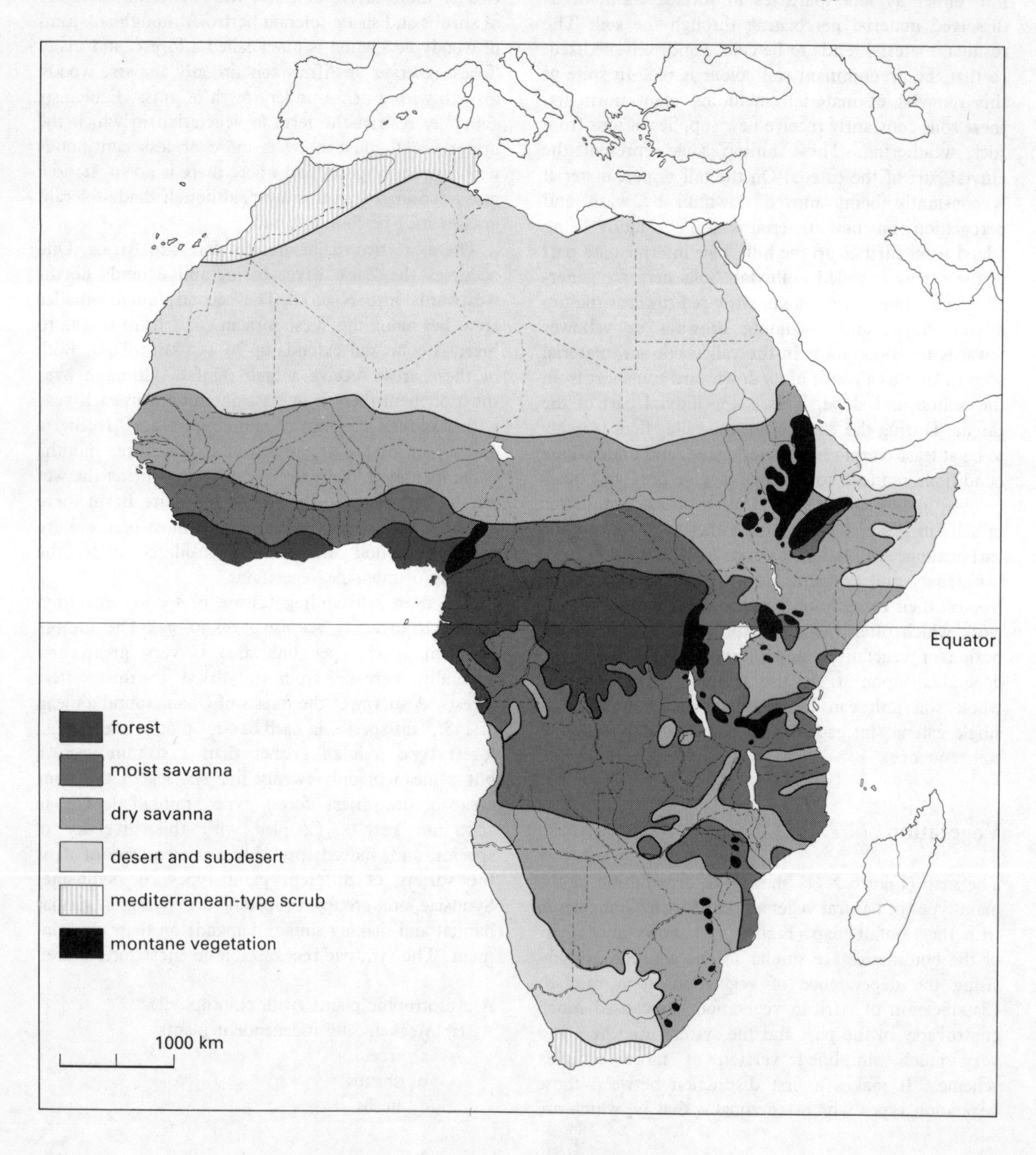

emergents. Below them, a second layer of large trees with rather flat crowns forms a continuous canopy, in which the individual trees are often linked by large woody climbers. Below this main canopy layer, whose component trees often carry many epiphytes, there is an understorey of smaller trees, often with vertically elongated crowns. Below this again is a layer of shrubs and even smaller trees, and finally a layer of herbs growing in the deep shade on the forest floor.

Temperate forests are usually made up of only a few tree species, and are therefore easy to classify and name on the basis of species composition. Tropical forests, on the other hand, are very much more diverse, and there may be forty or more large tree species making up the canopy in a small area. Although this diversity precludes the classification of tropical forests on the basis of a few tree species, it is sometimes possible to detect a predominance of particular families. In Africa, members of the subfamily Caesalpinioideae of the Leguminosae often form a significant part of the canopy. This phenomenon, known as family dominance, is much more strongly developed in south-east Asia where the family Dipterocarpaceae, scarcely represented in Africa, provides a high proportion of the canopy species in many of the forests. There are exceptional forests in Africa, with a canopy made up of only a few species, such as the *Cynometra alexandri* forests of western Uganda, and the very dry forests of southern Ghana dominated by *Talbotiella gentii*. The reasons for single-species dominance are unclear; *Talbotiella* forests are found on rocky hills at the extreme edge of the forest zone and thus occupy a peculiar habitat. The *Cynometra* forests, however, do not occupy any apparently unusual site and are, indeed, regarded as the climax forest stage in a succession of which the earlier states are represented by a much more mixed forest.

The plants of the forest floor are a peculiar assemblage of species which share the ability to survive in the extremely low light levels which are found there. Members of the families Rubiaceae, Commelinaceae, Acanthaceae and Zingiberaceae are often common, and often look extremely similar to one another, all tending to have broad thin leaves, irregular flowering, and often vigorous vegetative reproduction, reflecting the difficulty of seedling establishment in this poorly lit environment. Grasses and sedges (Gramineae and Cyperaceae), so characteristic of more open vegetation types, are very scarce in true forest, and those that do occur have broad thin leaves quite unlike the narrow linear leaves characteristic of species of the more open sites. *Leptaspis cochleata* is perhaps the most widespread of these unusual grasses. The members of these

Figure 2.12 Profile diagram of a wet evergreen primary mixed forest (Shasha forest reserve) in Nigeria. A detailed study of this forest showed that five vertical strata (A–E) could be recognized, although only the lowermost tree stratum (up to 15 m tall trees) was continuous (stratum C). Above this two further strata were identified: one of trees 37–46 m tall (stratum A) and the other of trees 15–37 m tall (stratum B). Stratum B contains trees which are widely dispersed and is not illustrated well in the small section of the forest shown in this profile. The shrub layer (stratum D) is not well developed and the forest is easy to penetrate. The ground layer (stratum E) of plants up to 1 m tall contains tree seedlings and herbs. Tropical forests also usually contain large numbers of lianes and epiphytes (see text). (From Richards, 1952.)

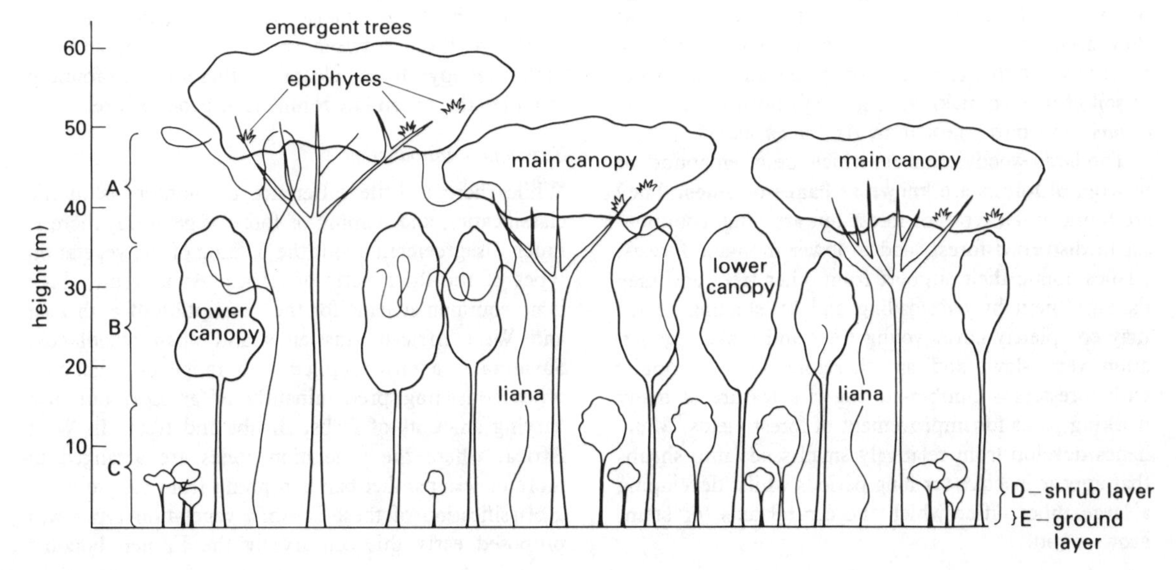

two families are normally wind-pollinated, but in the forest understorey there is extremely little air movement. It seems likely that at least some of the grasses and sedges of this habitat are pollinated by insects; this has been definitely observed in South American forest grasses but not, as yet, in African species.

Mention should also be made of two of the more specialized life forms found in forest, epiphytes and climbers. Epiphytes are plants which grow on others without being rooted in the soil. They obtain their nutrients from rainfall and surface runoff, or in some cases are carnivorous. Unlike parasites, they do not derive any water or nutrients from their 'host', only support. The range of habitats available to epiphytes in a tropical forest is very wide, ranging from the exposed upper twigs and branches of the canopy trees, where water is likely to be scarce and insolation strong, to the deeply shaded and usually moist bases of the trunks of the canopy trees. The most extreme habitats are often occupied only by mosses, with lichens also in the well lit sites. Intermediate sites are occupied by epiphytic flowering plants among which orchids are the most abundant both in numbers of individuals and species. Ferns, including the maidenhair fern, are also often prominent, and some ferns (e.g. the Hymenophyllaceae) tend to occur on the base of tree trunks in the wettest forests, near streams. They benefit from water running down the tree trunk and 'enjoy' a large catchment area. The surface of leaves of some trees growing in the wettest forests are often colonised by epiphyllous liverworts. Epiphytes have problems of acquisition and retention of water and nutrients. Many orchid roots are covered with a layer of dead cells which quickly fill with water during rain. This water is retained and absorbed by the plant. Many ferns have leaves arranged in such a way that they accumulate water and plant debris around the plant base, so that the plant comes to stand in a pocket of soil of its own making. (e.g. stag's horn fern, *Platycerium*, and bird's nest fern, *Asplenium nidus*).

The large woody climbers which are often abundant in tropical forests are known as **lianes** or **vines**. They are found in all types of forest but are often commonest in disturbed forest, and in lower montane forests. Lianes obtain their support from other trees and may damage them by overloading and by shading. They may completely cover young trees and make regeneration very slow, and are therefore very unpopular with foresters – climber-cutting is a feature of many working plans for improvement of forest yields. Many lianes develop from relatively small seeds into shrubs that appear inactive for long periods whilst developing a large tuber, after which the central axis (or stem) grows rapidly.

Other lianes start out as rapidly growing seedlings from large seeds. After a variable length of time lianes start to produce long shoots armed with hooks, spines or tendrils which attach themselves to understorey trees, and eventually reach the forest canopy. Once in the canopy, the liane produces a large crown which may extend over the crowns of several canopy trees. Since the trees support the weight of the liane, its stem has little supporting role, and acts only as a channel for the transport of water and mineral nutrients. Liane stems often contain very large and very long vessels which can be clearly seen with the naked eye, and the tension in the water columns in these vessels is such that a hissing sound can often be heard when the stem is cut, the water columns broken, and air drawn into the vessels. The wood of lianes is often peculiar, showing anomalous secondary thickening (i.e. the wood, instead of being deposited in regular cylinders as in a tree, is deposited in lobes or separate blocks, giving great flexibility to the stem).

The forest has a very well-defined layering of microclimates. The canopy and the emergent trees are in full sunlight, while light intensities at the forest floor may be as low as 2 per cent or even 0.2 per cent of this. Air movement, likewise, is profoundly reduced by the vegetation; air flow at the forest floor is about 1 per cent, or less, of that above the canopy. The maximum daily temperature is higher above the canopy, but there is often little difference between the minima in the various layers; this means that the daily temperature range is very much greater above the canopy than it is at the forest floor. Humidity shows a similar trend to temperature; fluctuations are greater above the canopy, while relative humidity near the ground rarely falls below 95 per cent. The differences between the forest interior and a clearing are very similar to those between the forest interior and the forest canopy; forest clearance therefore profoundly modifies the micro-environment at ground level.

Savanna – introduction

While there is little difference of opinion as to the classification and naming of forest vegetation, there is much disagreement about the naming of the vegetation types of the drier parts of Africa. As a result, there is no common scheme for the description of both East and West African grasslands and open woodlands. **Savanna** is a term applied to a range of vegetation types consisting predominantly of grasses, but also varying amounts of forbs, shrubs and trees. In West Africa, where the vegetation zones are arranged in more or less parallel bands running from east to west, a classification of these savanna vegetation types was proposed early this century by the French botanist

Auguste Chevalier. The classification into three main savanna types (Guinea, Sudan and Sahel) has proved very useful and there is no reason to abandon it now. In East Africa, on the other hand, several classifications have been used but there is still no general agreement. The problem has probably arisen because East Africa is an exceptional area; nowhere else in the world is there such an extensive area with low rainfall spanning the equator. The vegetation types do not have counterparts in any other part of the world with which they can be compared. Attempts to include them in pre-existing schemes, developed in the seasonal tropics, have not been very satisfactory. The tendency now is to give the savanna-type vegetation of East Africa descriptive names which, if correctly applied, should be unambiguous.

East African savanna types

East African savannas, particularly those in Kenya, Uganda, southern Somalia and northern Tanzania, lie close to the equator and tend to have two rather unreliable wet seasons each year. Furthermore, large areas in Kenya and northern Tanzania, and to a lesser extent in western Uganda, have been scattered with volcanic ash in the relatively recent past. This has produced juvenile and often highly fertile soils which still influence the vegetation. Finally, much of East Africa is high enough to produce a climate very different from that of the West African savanna zone.

Table 2.5 A simple classification of East African rangelands based on the amount and height of woody vegetation adapted from Pratt, D. J. and Gwynne, M. D., 1977, *Rangeland Management and Ecology in East Africa*, Hodder and Stoughton; London.

Woodland – Tree canopy cover >20%. Ground cover mainly grasses (see Figure 2.13).

Wooded grassland – Scattered trees covering 2–20%. Ground cover mainly grasses (see Figure 2.13).

Bushland – Woody plants <6 m tall and usually multistemmed at the base, providing 20–80% woody canopy cover (see Figure 2.13).

Bushed grassland – Dominantly woody shrubs and tall grasses. Height of vegetation <6 m with woody canopy cover of 2–20%.

Grassland – Dominantly tall or short grasses with scattered trees or shrubs. Total woody canopy cover <2%.

A simple classification of East African rangelands (a general term for land which can support grazing animals) shown in Table 2.5 divides them according to the amounts of woody vegetation, following a scheme originally proposed by Greenway. Examples are illustrated in Figure 2.13.

In north-west Uganda and south-west Sudan the most widespread vegetation is a broad-leaved woodland, in which *Isoberlinia doka* is abundant. (This tree is a diagnostic feature of northern savanna in West Africa). Further south in Uganda this is replaced by woodland, also broad-leaved, but in which species of *Terminalia* and *Combretum* are the commonest trees. In both these areas the grass is very tall, up to 4 m by the end of the rainy season; the most abundant grasses are species of *Hyparrhenia*. These areas tend to have a single wet season. South of the equator, in regions which also have a single long wet season, (e.g. Kasama, Figure 2.2), vast areas of Tanzania, Zambia, Mozambique, Zimbabwe, southern Zaire and Angola are covered by woodland in which species of *Brachystegia* and *Julbernardia* are the commonest trees. This vegetation is often given the Tanzanian (strictly Kinyamwesi) name 'miombo'. Here 900–1300 mm of rain falls in the single wet season which lasts 4–6 months. The trees are fire-resistant, and the ground cover of tall grass is burnt almost every dry season. This is a strikingly unproductive vegetation type because tsetse fly infestation prevents cattle ranching and there is only one important timber species (mninga – *Pterocarpus angolensis*). The only other products of any importance are honey and beeswax produced by bees from both natural and artificial hives. The vegetation supports quite large game populations, but they are difficult to see in the woodlands and this makes tourist game-viewing difficult – a problem shared by the West African Guinea savannas.

Large areas of western Uganda, central Kenya, and northern Tanzania are occupied by open plains of grassland or wooded grassland. The main grass is *Themeda triandra* – red oat-grass, and the trees are usually species of *Acacia*, especially *A. tortilis*. This vegetation is, or was, occupied by large populations of wild ungulates, and is still the most familiar background to tourist photographs of game animals in national parks: see for example Figure 5.20. *Themeda* is scarce in West Africa although why this should be is not clear. These grasslands and wooded grasslands are often occupied by pastoral peoples who have a tradition of grassland management by burning; at least in the higher rainfall areas they have doubtless profoundly modified the composition of their habitat.

Much of northern Kenya, north-eastern Uganda, and parts of northern Tanzania have a comparatively dry climate with two ill-defined and unreliable wet seasons, rather than one. Often grasses are not a prominent part of the vegetation; various small species of *Commiphora* and *Acacia*, in particular *A. mellifera* and *A. nubica*, are the most abundant woody plants, forming a bushland or bushed grassland. Most of the

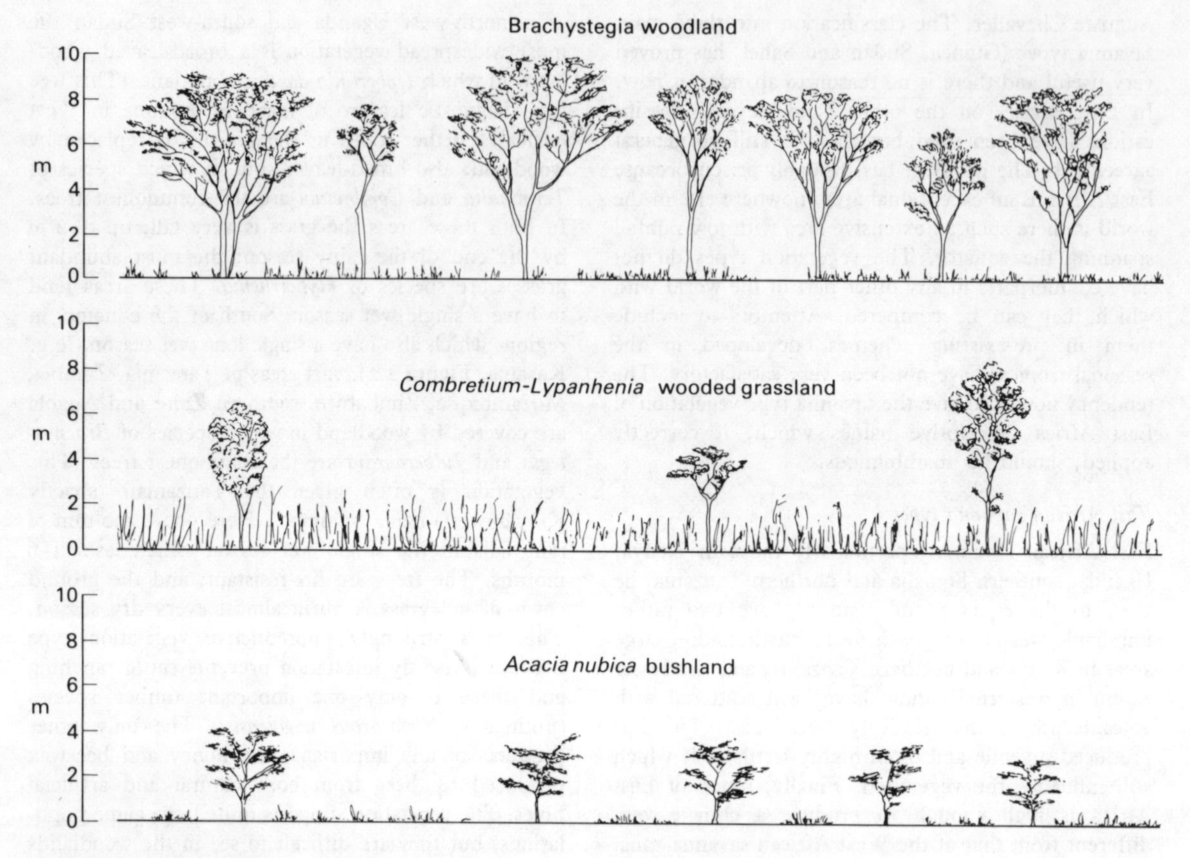

Figure 2.13 Profiles of some East African savanna types, corresponding to Table 2.5. This is a useful classification scheme to illustrate community structure based on ground cover and vegetation height.

Areas which appear to be uniformly *Acacia-Commiphora* bushland during dry periods can quickly become amazingly varied if sufficient rain falls. grasses are short-lived, and exist during the dry seasons as seeds, which germinate when rain falls.

West African savanna types

In West Africa the vegetation belts run east–west, more or less parallel with the coast (Figure 2.14) and they can be dealt with conveniently in a series extending from the forest edge northwards to the desert. The transition from forest to savanna is often very abrupt, taking place within a few metres. In many places, the precise position of the forest–savanna boundary has certainly been modified by man, but the extent of this modification is still disputed. Large areas of land regarded as derived savanna were once forests. In many places, however, rather dry types of forest are bordered directly by species-rich woodland or wooded grassland, rather than by a species-poor grassland which might be expected if forest had recently been destroyed.

The southernmost West African savanna type is the Guinea savanna comprising mainly broad leaved trees, though some are pinnate or bipinnate. Tree density is variable. There is an understorey of tall grass, mostly composed of species belonging to the Andropogoneae (a tribe within the family Gramineae). These reach 3–4 m by the end of the wet season. The annual rainfall is between 850 and 1500 mm and the dry season lasts 4–5 months. The Guinea savanna is often divided into two subzones, a southern one in which the palm *Borassus aethiopum* is common, and the trees *Lophira alata* and *Daniella oliveri* are prominent, and a northern one in which the most abundant tree is *Isoberlinia doka*, which is often present as a shrub form. This northern Guinea savanna extends eastwards to southern Sudan and north-east Uganda.

The next zone encountered as one travels north in West Africa is the Sudan zone. This is an intermediate zone, now very heavily settled and cultivated. The annual rainfall is from 600–850 mm and the dry season lasts about six months (see e.g. Kano, Figure 2.2). Some of the broad-leaved trees of the northern Guinea savanna extend into this zone, and there are

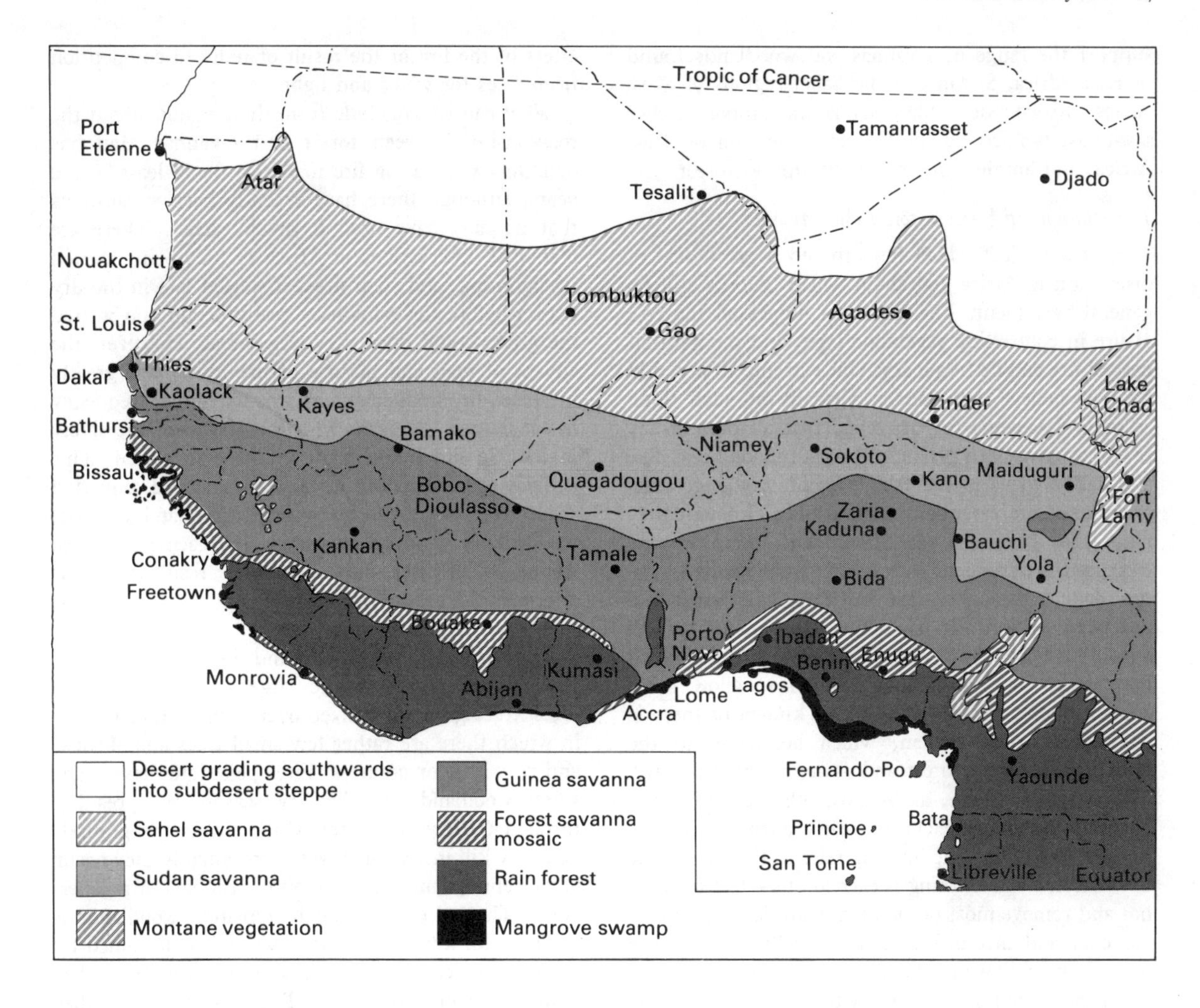

Figure 2.14 Distribution of West African vegetation showing the different types of savanna: (see rainfall data in Figure 2.3). (From Lawson, G. W., 1966, *Plant Life in West Africa*, Oxford University Press.)

also some of the *Acacia* species which are such a prominent feature of the Sahel zone. Perhaps the most distinctive tree of the Sudan savanna is the Doum Palm, *Hyphaene thebaica*.

The final zone crossed in this transect from forest to desert is the Sahel savanna, where rainfall is less than 600 mm in a year and the dry season lasts for more than seven months (see e.g. Gao, Figure 2.2). Perennial grasses tend to give way to annuals, and the trees are mostly compound-leaved, with species of *Acacia* being particularly abundant. The grass layer tends to be too scanty and discontinuous to support the regular fires which are such a feature of the more southerly vegetation zones. A further feature is the unreliability of the rainfall. It is generally true that low mean annual rainfalls are associated with variability and this tendency seems to be at its most marked in the Sahel zone. This zone is similar in many ways to the bushlands and bushed grasslands of northern Kenya.

The northern edge of the Sahel, and the driest parts of the East African bushlands, merge into deserts. It is extremely difficult to draw a precise line delimiting a desert; very little even of the Sahara is devoid of plant life, even if it is confined to the beds of seasonal watercourses. Deep-rooted trees and shrubs, with a few drought-resistant perennial grasses, grow here, and after heavy rains there is often a diverse flora of short-lived species.

Other terms that are commonly used in relation to a zonal classification are Arid, Semi-arid, Moist and Humid. **Arid** lands include deserts, which contain vegetation only along watercourses and semi-deserts, which contain scattered vegetation away from watercourses but mostly on valley sides. **Semi-arid** lands support the range of savannas and woodlands found in East Africa, Sudan and the Sahel regions of West Africa. **Moist** (or subhumid) zones support richer savannas, typified by the Guinea vegetation of West Africa, and **humid** zones naturally support forest.

The relationship between forest and savanna

This is a subject which has aroused a great deal of discussion in Africa and about which there is still no general agreement. The argument hinges on the role of fire in controlling vegetation type, and the naturalness of fire and its frequency. One extreme view would have us believe that there is virtually no such thing as natural savanna in Africa and that all present-day savannas have been produced by man-made fires. The alternative view is that the present day vegetation boundaries are essentially natural ones although they may have been sharpened here and there by the increased fire frequencies resulting from man's activities. As is common in such controversies, a great deal has been written but few actual experiments carried out. There are, however, a few experiments which have been kept going long enough to allow some conclusions to be drawn. The best known of these is in Nigeria, near Ibadan, which lies close to the boundary between forest and savanna (Figure 2.14). The treatments here, as in most other experiments, have compared fire exclusion, early burning and late burning (early and late burning refer to the part of the dry season when burning is carried out – late fires are hot and remove most of the plant material – early fires are cool and incomplete, but generally prevent late fires from following them). An essentially similar experiment, but in an area with a long dry season which supports *Brachystegia* woodland, has also run for a long period in Zambia. In addition there are other experiments in Ivory Coast and northern Ghana which have both run for a reasonable length of time.

The general conclusions from these experiments are that late burning is very detrimental to tree growth, probably because the trees themselves are usually severely damaged, and regeneration from seed is effectively prevented, so that deaths among the adult trees are not replaced. Early burning appears to encourage tree growth, or, in the driest sites, discourage it much less than late burning. Complete protection usually leads to increased tree densities and sometimes to the appearance of tree species normally considered to be characteristic of forest and susceptible to fire. Not surprisingly, the forest species appear in greatest numbers if there are seed sources nearby. The effect on the grass is the reverse of that on the trees; grass growth is more vigorous in late-burned plots, although it is difficult to know if this is a direct effect of the fire or the result of reduced competition from trees for water and light.

What can we conclude from these results about the relationship between forest and savanna? Man has definitely been using fire in Africa for at least 60 000 years, although there have recently been suggestions that he may have used it much earlier. There are plenty of observations in Africa of fires being started by lightning, and such fires, if started late in the dry season before the rains have really begun and wetted the grass, could burn very large areas. However, the advent of man and his use of fire must have greatly increased fire frequency, and it is likely that frequency has increased yet again with the arrival of the safety match. So fire frequency must have increased. This is likely to have opened up the savannas and converted them from woodland to wooded grassland or from wooded grassland to grassland. In places, increased frequency of fires may also have resulted in the altering of the position of the boundary between the forest and the savanna.

At present, it is possible to find sites in Nigeria (and probably elsewhere) where large savanna trees and forest trees grow intermixed over a sparse understorey in which there are rather few small trees and shrubs, and very little or no grass. Fires penetrate these 'transition woodlands' in the dry season by spreading through the dry leaf litter which covers the ground, and they kill many small trees. Presumably they occur at long enough intervals to allow occasional tree regeneration. The existence of such transition woodlands is evidence for a fixed boundary which is determined by some ecological factor such as water supply, rather than by man-made fires. The situation is rather different when forest is cleared and cultivated and then allowed to return to fallow. If grass becomes established on a former forest site thus cleared, fire will maintain the grassland and prevent the return of forest. Most grasslands of this type are rather different from natural savannas and do not contain a wide range of typical savanna plants.

Azonal vegetation

All the vegetation types so far mentioned are determined, in general, by the climate of the region in which they occur. In contrast, other vegetation types are determined by local factors which are strong enough to override regional climatic effects. Such vegetation types include swamp and aquatic communities where the overriding local factor is flooding or waterlogging of the soil; maritime vegetation, where the factor is the salt water of the sea; and montane vegetation, where the peculiar local climate characteristic of high altitudes in the tropics overrides all else.

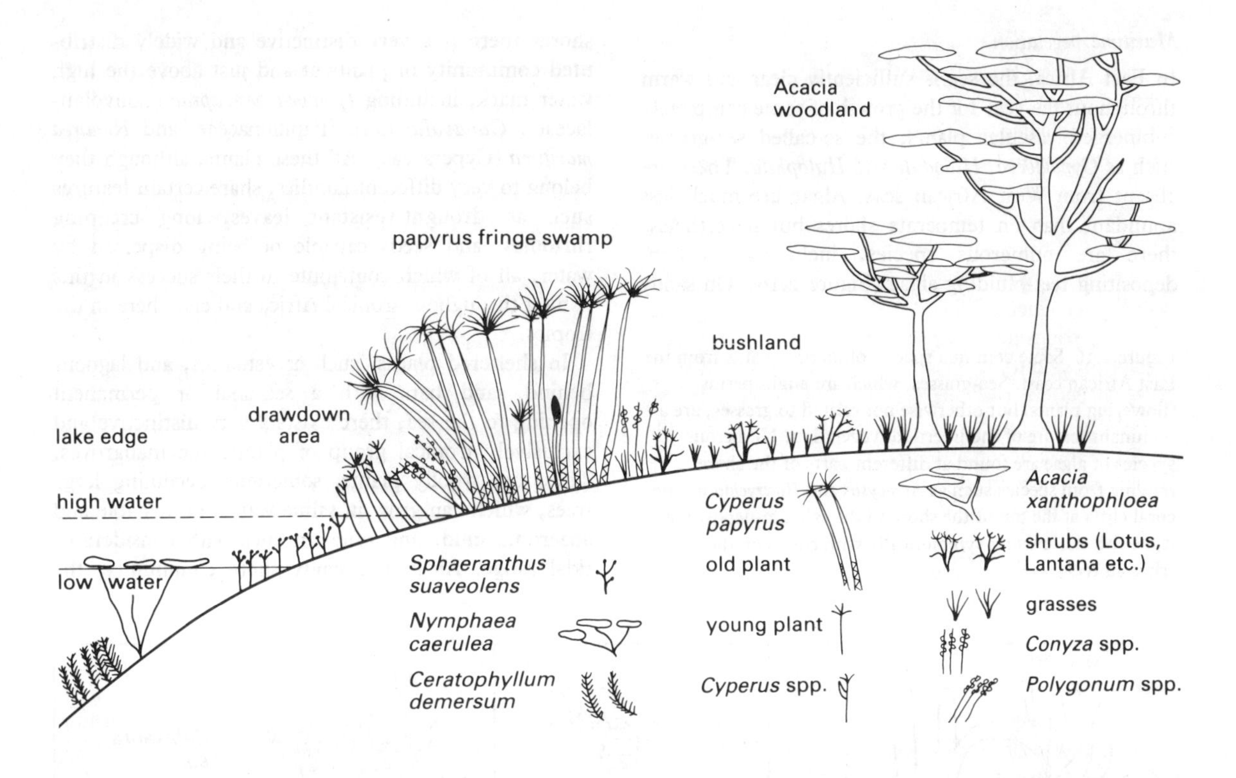

Figure 2.15 An example from Lake Naivasha, Kenya, of lakeside communities showing a transect from dry ground with trees through a papyrus swamp, to aquatic free-floating and submerged plants such as *Ceratophyllum* (From Gaudet, J. J. 1977, *Aquatic Botany* 3 1–47).

Swamp and aquatic vegetation

This vegetation can conveniently be described by considering a series of sites along an imaginary transect running from deep water to marginal soil which is permanently wet (Figure 2.15). In the deep water, if it is sufficiently clear, there are totally submerged plants such as species of *Ceratophyllum*, *Hydrilla*, *Utricularia* and *Vallisneria*. Some of these flower underwater; others, such as *Vallisneria*, produce their flowers on long stalks at the water surface and are therefore limited to certain maximum water depths – at least if they are to flower successfully. On the surface of open water there may also be free floating plants such as *Lemna* and *Pistia stratiotes*; these have roots which hang down in the water and absorb all the nutrients that they require. They are completely unattached and can therefore be blown about by wind or carried by water currents.

In shallower water there are plant species which are rooted under the water but which have their leaves floating on the surface. Such are the species of *Nymphaea* – the water-lilies, and some of the species of *Potamogeton*. Other species are rooted in the bottom but project upwards into the air, such as *Typha* and some species of *Scirpus*.

At the water margin there are plants rooted in the mud. Some of these, such as *Cyperus papyrus*, have such extensive root and rhizome systems that they form a mat which floats on the water surface and can extend out over open water (Figure 2.15(b)). Swamps dominated by this species cover vast areas in Uganda, and are also found in Sudan where they are known as Sudd, and at the southern edges of Lake Chad.

All aquatic plants which are rooted in mud face considerable problems of oxygen supply to the root system. The diffusion rate of gases in water is 10 000 times slower than in air, and soils which are waterlogged (with the spaces between the soil particles filled with water and not with air) very quickly become anaerobic because the oxygen within them is quickly used up by the respiration of roots and bacteria and cannot be replaced fast enough by diffusion. In aquatic and marsh plants, the oxygen needed by the roots, which in most plants is usually obtained from the soil, is obtained instead through systems of air canals which permeate the stems and roots, and connect with the outside air via the stomata in the leaves.

Maritime vegetation

In East Africa, the sea is sufficiently clear and warm throughout the year for the growth of some completely submerged vascular plants, the so-called sea-grasses such as *Cymodocea, Halodule* and *Halophila*. These are absent from West African seas. Algae are much less abundant than on temperate shores but nevertheless there are numerous species, including calcium-depositing reef-building algae (Figure 2.16). On sandy shores there is a very distinctive and widely distributed community of plants at and just above the high water mark, including *Ipomoea pes-caprae* (Convolvulaceae), *Canavalia rosea* (Papilionaceae) and *Remirea maritima* (Cyperaceae). All these plants, although they belong to very different families, share certain features such as drought-resistant leaves, long creeping rhizomes, and seeds capable of being dispersed by water, all of which contribute to their success in this habitat throughout tropical Africa and elsewhere in the tropics.

In sheltered places, such as estuaries, and lagoons behind sand bars with a seasonal or permanent opening to the sea, there exists a very distinctive and exclusively tropical group of plants, the mangroves. These are woody plants, sometimes becoming large trees, which can grow in saline water, in unstable and anaerobic mud, and often in sites with considerable tidal range. As in the sandy shore community, the

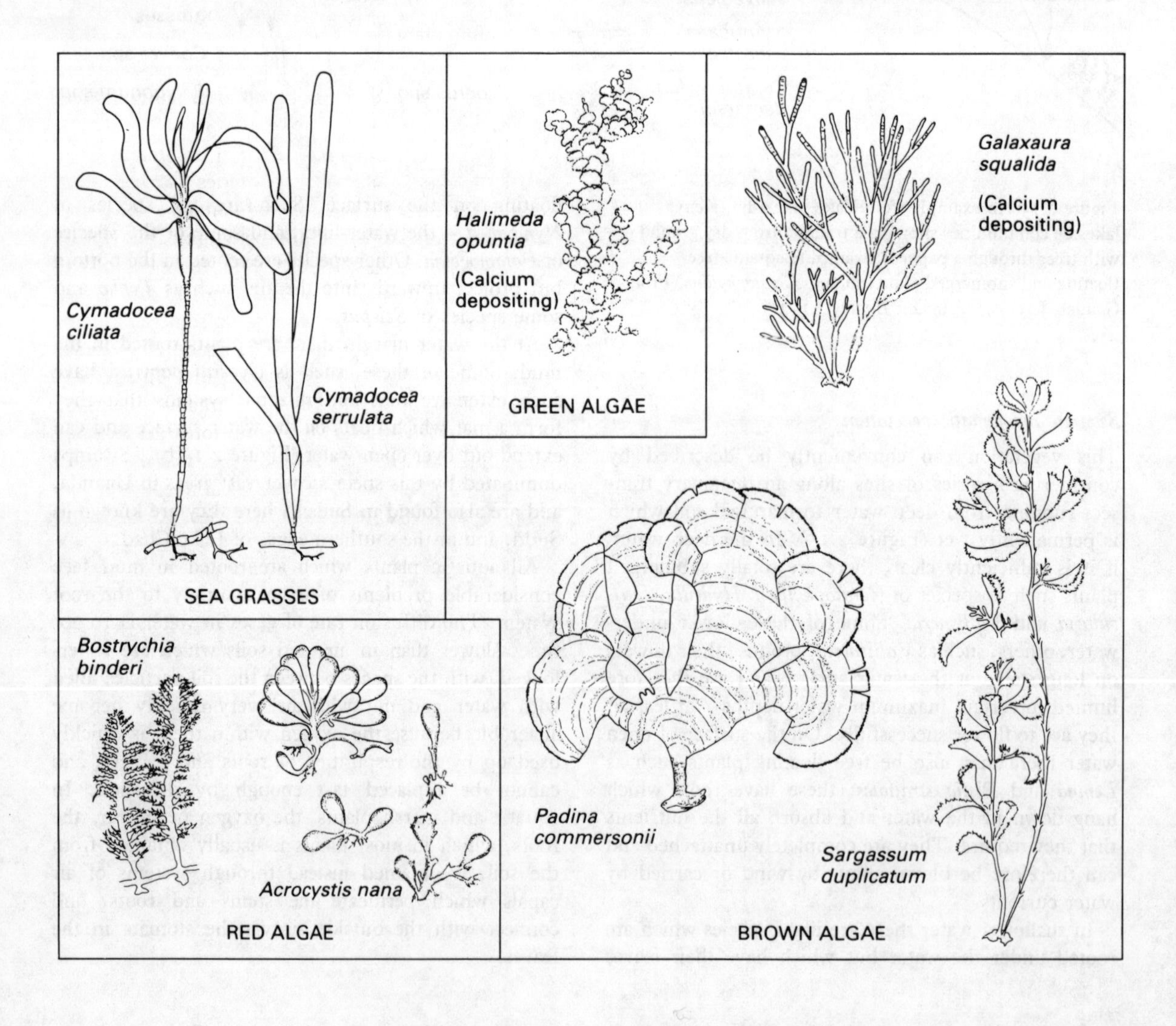

Figure 2.16 Some common species of marine plants from the East African coast. Sea-grasses, which are angiosperms (flowering plants) but otherwise not related to grasses, are a dominant feature of the intertidal vegetation. Numerous species of algae are found at different parts of the shore, ranging from species such as *Acrocystis* and *Bostrychia* on the coral cliffs at the top of the shore to the calcium-depositing algae and those living symbiotically with corals on the fringing reef.

Figure 2.17 A mangrove swamp on the coast of East Africa. Note the supporting stilt roots.

members of this group belong to several families; examples are *Rhizophora* (Rhizophoraceae), *Avicennia* (Avicenniaceae), and *Heritiera littoralis* (Sterculiaceae). The first two occur on both the west and east coasts of Africa; the last, only on the east coast. The problems of oxygen supply to the roots are extremely severe in the mangrove habitat, and all the species have specialized aerial roots (pneumatophores) which assist in aeration, and sometimes also in support (Figure 2.17).

Figure 2.18 Vegetation zonation on some East African mountains. The left hand side of each figure is the wettest side of the mountain. Note the absence of bamboo from Kilimanjaro, and the absence of moorland from the wet Rwenzori. Note also the very marked contrast between the two sides of Mt Kenya which have very different rainfall regimes.
(From Hedberg, O., 1951. *Svensk Botanisk Tidskrift*, **45**, 140–202)

Montane vegetation

East Africa has a number of scattered mountains and mountain ranges; West Africa has almost none, except for Mt Cameroun and the much lower summits of the Fouta Djalon in Guinea. These tropical mountains have a very peculiar local climate which has been described as 'summer every day and winter every night'. In the thin atmosphere of the high altitudes, incoming radiation is strong, and heats up the environment substantially during the day, but this same thin atmosphere also allows very rapid outflow of radiation, so that above about 3500 m temperatures fall below freezing every night. Up to about 3000 m, rainfall tends to increase, and total precipitation is increased by condensation from mist. Persistent cloud and low temperatures also lead to reduced evaporation. At about 3000 m there tends to be a decrease in rainfall, but measurements are scarce. It is certain, however, that on the saddle between the two peaks of Kilimanjaro, at an altitude of over 5000 m, the rainfall averages only about 150 mm per year, and the vegetation is desert-like with very few and scattered plants. On other mountains the drop in rainfall is less pronounced.

Travelling up a tropical African mountain, one passes through a characteristic series of vegetation zones (Figure 2.18). The lowest montane zone is forest, rather similar to tropical lowland forest, but usually with shorter canopy trees, many more epiphytes including long strands of pendant mosses and *Usnea* lichens, and different tree species. On some mountains this gives way to a zone of bamboo (*Arundinaria alpina*), which forms almost pure stands. This zone is strangely absent from some mountains including Kilimanjaro. Above this is the heath forest, dominated by members of the family Ericaceae, mainly species of *Erica* and *Philippia*. These are small trees, with very bent and distorted trunks and branches, covered with

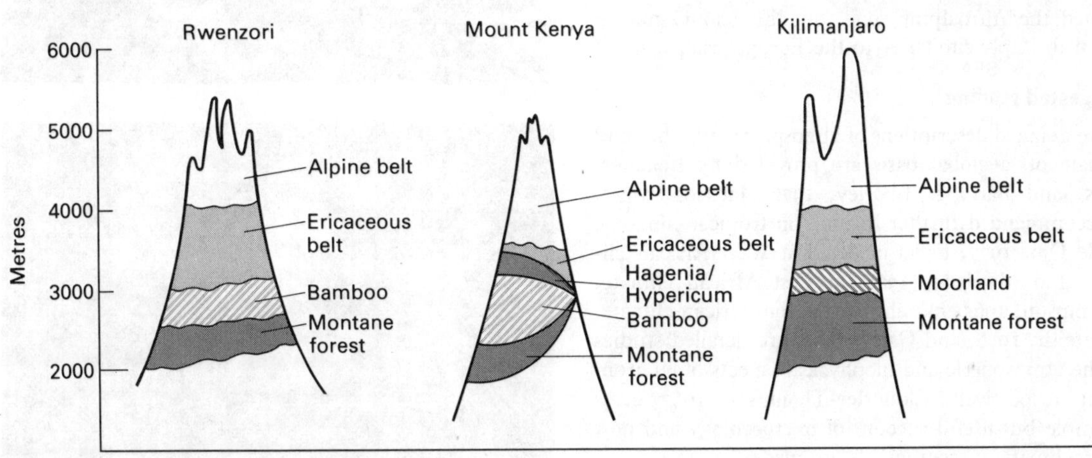

Figure 2.19 Trees with a very heavy growth of mosses and other bryophytes growing on Mt Kenya. (M. W. Service.)

a very heavy growth of lichens and bryophytes (Figure 2.19). Within this zone are scattered individuals of the plants most often associated with high altitudes on African mountains such as the giant groundsels (*Dendrosenecio* spp.) and giant lobelias (*Lobelia* spp.). These form dense woodlands at higher altitudes. Their peculiar structure, with a thick trunk usually covered with a dense layer of dead leaves, and an apical rosette of huge leaves, is best regarded as an adaptation to the diurnal temperature fluctuations. During the night the rosette leaves of some species fold inwards to cover the terminal bud and prevent it from freezing (Figure 2.20). Likewise the dead leaves covering the stems protect the delicate layers of the cambium from extreme cold. Above this zone, scrub and sparse grassland, with many small species of *Helichrysum*, and grasses belonging to otherwise temperate genera such as *Festuca*, extend to the upper limit of plant growth at about 5000 m. This unique environment has been termed the Alfroalpine zone, but this should not be taken to imply similarity to the European alps.

Suggested reading

More detailed descriptions of atmosphere, weather and climate on a global basis are provided by Boucher, 1975, and Barry & Chorley, 1976. Jackson, 1977, is recommended further reading on tropical climates, while Ojo, 1977, looks in detail at West African climate and Mwebesa, 1978, at East African climate; Thompson, 1965, has also written on African climate. Monteith, 1976, and Gates, 1980, are detailed studies of the atmospheric and biophysical aspects of environments respectively. Cloudsley-Thompson, 1967, gives a simple but useful account of microecology and how microclimatic measurements are made.

Figure 2.20 Afroalpine vegetation. Many plants have evolved into giant forms as an adaptation to harsh subzero temperatures every night and tropical sunshine every day. The picture above shows *Dendrosenecio adnivalis*. The two pictures below show *Senecio brassicae* during the day (above) and at night (below). Water is stored inside the rosette of leaves, which fold into a ball at night. The water does not freeze during the night and is thought to be used as a water supply for the plant in the morning so that it can photosynthesize in the bright sunshine even though the ground is still frozen. (N. Dickinson.)

Young, 1976, is a useful reference book on tropical soils, together with Sanchez, 1976, which also covers soil management in relation to agricultural crops. Etherington, 1982, is a good introduction to the chemical and physical environment of plants and includes soils, water relations and mineral nutrition. Allen *et al.*, 1974, is one of the better standard texts dealing with methods of soil analysis.

Three introductory texts by Money, 1982, examine the environment, characteristics and development of arid lands, tropical savannas and tropical rain forests. More detailed examination of the environment and vegetation of the tropics is provided by Walter, 1971, and, dealing specifically with tropical forests, Richards, 1952, is still much to be recommended. White, 1983, provides excellent maps of Africa showing the main vegetation types, and there are accompanying descriptions, while the ecology of the vegetation of West Africa is the subject of two good introductory texts by Longman & Jenik, 1974, and Hopkins, 1979. Lind & Morrison, 1974, is essential background reading on the vegetation types in East Africa. Excellent recent reviews of tropical forests are to be found in Sutton *et al.*, 1983, Mabberley, 1983, and Unesco, 1978, and of savannas in Bourlière, 1983, and Unesco, 1979.

Essays and problems

1 How and why does the periodicity of rainfall vary throughout tropical Africa, and which factors influence it?

2 In terms of the relative importance of convection, evaporation and reradiation, suggest reasons for the following:
(a) reptiles basking in the sun
(b) amphibians staying close to water
(c) leaf hairs on plants
(d) reddish-coloured leaves on plants?

3 What is the importance of organic matter in soils? How is organic matter affected by agricultural practices?

4 How widespread are soil catenas in your country? How do they influence the vegetation of the region and the agricultural crops that are planted. Consult a soil map to find the soil types and investigate their characteristics.

5 What do you consider to be the natural vegetation of your home region? How much has it been influenced by the activities of man? Do you consider that the changes that have taken place are an improvement, or something to worry about? What changes do you foresee in the future?

6 How important is fire in determining the vegetation of Africa? Do you consider fire to be a natural event and in which instances is it useful in management of the vegetation?

7 The environment of the Afroalpine region of equatorial mountains has been described as, 'summer every day and winter every night'. How does this affect the plants that grow there?

References to suggested reading

Allen, S. E., Grimshaw, H. M., Parkinson, J. A. & Quarmby, C. 1974. *Chemical Analysis of Ecological Materials*, Blackwell: Oxford.
Barry, R. G. & Chorley, R. J. 1976. *Atmosphere, Weather and Climate*, (3rd edn). Methuen: London.
Boucher, K. 1975. *Global Climates*, English University Press: London.
Bourlière, F. (ed). 1983. *Tropical Savannas*, Elsevier; Amsterdam.
Cloudsley-Thompson, J. L. 1967. *Microecology*, The Institute of Biology's Studies in Biology, no. 6, Edward Arnold: London.
Etherington, J. R. 1982. *Environment and Plant Ecology*, (2nd edn). Wiley: London.
Gates, D. M. 1980. *Biophysical Ecology*, Springer Verlag: New York.
Hopkins, B. 1979. *Forest and Savanna*, (2nd edn). Heinemann: London.
Jackson, I. J. 1977. *Climate, Water and Agriculture in the Tropics*, Longman: London.
Lind, C. M. & Morrison, M. E. S. 1974. *East African Vegetation*, Longman: London.
Longman, K. A. & Jenik, J. 1974. *Tropical Forest and its Environment*, Longman: London.
Mabberley, D. J. 1983. *Tropical Rain Forest Ecology*, Blackie: Glasgow; in USA distributed by Chapman & Hall: New York.
Money, D. C. 1982. 1. *Arid Lands*; 2. *Tropical Savannas*; 3. *Tropical Rainforests*, Evans Brothers Ltd.: London & Ibadan.
Monteith J. L. (ed.) 1976. *Vegetation and the Atmosphere*, (2 volumes) Academic Press: London.
Mwebesa, M. 1978. *Basic Meteorology*, East African Publishing House: Nairobi.
Ojo, O. 1977. *The Climates of West Africa*, Heinemann: London.
Richards, P. W. 1952. *The Tropical Rain Forest. An Ecological Study*, Cambridge University Press: Cambridge.
Sanchez, P. A. 1976. *Properties and Management of Soils in the Tropics*, Wiley: New York.
Sutton, S. L., Whitmore, T. C. & Chadwick, A. C. 1983. *Tropical Rain Forest: Ecology and Management*, Blackwell: Oxford.
Thompson, B. W. 1965. *Climate of Africa*, Oxford University Press: Oxford.
UNESCO 1978. *Tropical Forest Ecosystems*, A state of knowledge report prepared by UNESCO/UNEP/FAO, UNESCO, Paris.
UNESCO 1979. *Tropical Grazing Land Ecosystems*, UNESCO, Paris.
Walter, H. 1971. *Ecology of Tropical and Subtropical Vegetation*, Oliver & Boyd: Edinburgh.
Walter, H. 1971. *Ecology of Tropical and Subtropical Vegetation*, Oliver & Boyd: Edinburgh.
White, F. 1983. *The Vegetation of Africa. Descriptive Memoir and Maps*, UNESCO, Paris.
Young, A. 1976. *Tropical Soils and Soil Survey*, Cambridge University Press: Cambridge.

3 Communities and ecosystems

From our general review of African environments, we now proceed to examine the nature of ecosystems. These are of many kinds: some are aquatic whilst others are terrestrial; and they may be natural or man-made. Despite this variety, all ecosystems share some important common properties. An energy source is a necessity and this is usually sunlight, whose energy is trapped by green plants. Living tissues are composed of various materials, derived from water, minerals and gases. The minerals circulate mainly within ecosystems, but the cycles of water and gases involve the biosphere as a whole. In each ecosystem the energy flows and the nutrients cycle through a series of organisms which typically consist of producers, consumers and decomposers, constituting various trophic levels.

After considering these topics in general terms, we take three actual ecosystems that have been studied in considerable detail, and for which the productivities of the various trophic levels have been estimated.

Following this review of ecological energetics, we conclude the chapter by looking at how individual species fit into their communities, each of which is part of an ecosystem. We find that for any particular community, such as birds, some species are common but the majority are rare. Communities have other characteristic properties, including diversity, and we observe that this is greatest in the tropics. The number of species found in an area also depends, not surprisingly, on its size.

A special case, with implications in the field of conservation, is that of island communities, with their problems of colonization and apparently high extinction rates.

Introduction: the concept of ecosystems

In chapter 1 we briefly introduced the terms biome and ecosystem. Each of the various types of vegetation that was described on pages 21–32 (and those shown in Figure 2.11) is characteristic of a particular climatic zone. These major vegetation groups, together with their characteristic animals, are called **biomes**. It is convenient to describe the different biomes, such as savanna, forest or montane, in terms of their most conspicuous plant-types; but in fact many other organisms, both plant and animal, live there too.

When we consider smaller areas, we find that the vegetation within a biome is not uniform. Within the dry savanna, for instance, there are bushed grasslands and wooded grasslands, whilst along the water courses running through the savanna there may be quite dense riverine forest. Each of these different types of vegetation, together with all the other living organisms that inhabit the same place, and the non-living things upon which they depend, is an **ecosystem**. The word 'ecosystem' was introduced in 1935 by one of the first modern ecologists, A. G. Tansley. He particularly stressed that they were **systems**. The living organisms in an ecosystem (the biotic components) interact with one another, and with their physical (or abiotic) environment. So an ecosystem has a wholeness of its own; it is a largely self-contained segment of the biosphere which can be of varying kind and size.

Ecosystems vary in the extent to which they are self-contained. In some, such as rainforest, there is very little exchange of organisms or materials with other ecosystems. Exchanges in aquatic ecosystems are usually more significant. A pond, for instance, gains nutrients from leaves that fall into it, and many of its animal inhabitants (such as frogs and dragonflies) spend part of their lives in other ecosystems.

Originally there were only **natural ecosystems**, assemblages of plants and animals which lived together, interacting with each other as a consequence of millions of years of co-existence. Such processes of **co-evolution** brought about the interdependence of the various species so that, to a greater or lesser extent, they all need each other. Natural ecosystems are sometimes spoken of as being 'balanced'. They are also considered to be stable; if disturbed they tend to revert to their original condition. Thus when a forest tree dies, and falls, a clearing is created: but, through the processes of succession (pages 87–90), it is colonized by a sequence of plant species of increasing size until, finally, the gap is filled again by a young forest tree.

It is sometimes difficult to see where one ecosystem ends and another starts. It is unusual for the edges of natural woodlands, for example, to be clearly defined. Rather, the size of the trees decreases gradually towards the edge, and they become further apart. This zone merges into wooded bush or grassland, where the trees are scattered, singly or in clumps. Often they are of different species from those within the woodland. Such transitional zones are known as **ecotones**.

Nowadays, many of the natural terrestrial ecosystems of Africa are fast disappearing, the main exceptions being in National Parks and also in some of the drier parts of the continent. In most places, human activities have resulted in modified or man-made ecosystems, although the degree of modification varies greatly. An example of a **modified ecosystem** is the conversion from dry savanna to cattle ranching. The original large mammals, such as antelopes and perhaps elephants, have been replaced by domestic species. There are now waterholes, trees are cut, fire is much more frequent, and other management processes modify the original flora and fauna. In some cases, the degree of modification is much greater, as with arable farmland. Such **man-made ecosystems** are ones where the original vegetation of vascular plants has been removed completely, and replaced by (for example) crops, or buildings.

Modified and man-made ecosystems tend to be ecologically unstable (pages 87–89): when management stops, they are likely to revert, through succession, to the original natural ecosystem (although the more they have been changed, the longer this would take). One objective of management is to prevent this reversion.

Ecosystems

The characteristics of ecosystems

Despite their great variety, there are certain feaures that are common to all ecosystems – natural and

Box 3.1 Feeding relationships amongst living organisms, and some of the names applied to them

Primary producers (or autotrophs)	Green plants deriving energy directly from sunlight by photosynthesis	
	Herbivores – eat parts of green plants	grazers – predominant food is grass browsers – mainly eat leaves and twigs of trees and shrubs frugivores – fruit-eaters granivores – seed-eaters, especially of grasses nectar-feeders (nectivores) sap-suckers wood-eaters – etc.
Secondary producers (or heterotrophs, or consumers)	**Carnivores*** (or predators) – feed on other animals	free-living insectivores* – eat insects free-living piscivores – eat fish free-living predators of birds, of mammals, etc. do not have special names parasites – belong to the same trophic level
	Decomposers and related groups, feeding on dead plant and animal material (see also Figure 3.10.)	primary decomposers (or reducers) – i.e. bacteria and fungi (microflora) feeding directly on dead organic matter and reducing it, as biproducts, to simple constituents detritivores+ – feed on dead, macroscopic matter mostly of plant origin scavengers+ – feed on dead, macroscopic animal matter coprophages – feed on dung herbivores and carnivores of the detritus food chain – feeding on the first and remaining groups above, respectively

* the terms 'carnivore' and 'insectivore' are sometimes restricted to members of the mammalian orders Carnivora and Insectivora

+ These terms can have various meanings. Detritus sometimes includes small-sized material of either plant or animal origin.

otherwise. Firstly, they all contain water, either as a medium (as in lakes, rivers and seas), or in the substrate (soil or rocks) in the case of terrestrial ecosystems. In these latter, air is the universal medium, and this too contains water, as a vapour. Secondly, all ecosystems contain inorganic substances, those in solution in the water being of special importance. Finally, ecosystems contain living organisms. Typically, these comprise green plants, with herbivorous animals that feed from them, and usually carnivores which depend upon the herbivores. In addition, there are microorganisms which decompose dead plant and animal matter. Thus the organisms in an ecosystem can be characterized by their mode of nutrition; and the term **trophic level** is used to express the idea that feeding relationships are a basic attribute of ecosystems. Green plants, which are **autotrophs**, are the first trophic level; herbivores are the second, and so on for other **heterotrophs**. Few ecosystems have less than three trophic levels, whilst some have four or more. Box 3.1 summarizes the relationships, and introduces a number of terms that are commonly used to describe the specialists in the various trophic levels.

Most organisms belong clearly to a particular trophic level, but there are exceptions. Some, like man or baboons, are omnivores, belonging to both second and third trophic levels. Some beetles and hoverflies have larvae which are predatory, while their adults feed on nectar and pollen. In blackflies, *Simulium* spp., the aquatic larvae are detritivores, whilst the adults have piercing mouthparts, used by the females to suck blood from vertebrates. Most weaver-birds eat seeds as adults but feed their young on insects. Adult male mosquitoes feed on plant juices, and so belong to the second trophic level. However, the females, as with blackflies, suck blood, and belong to the third level. Cows undoubtedly prefer living grass, but if there is no alternative they will eat dead grass and at such times are (strictly speaking) detritivores. Some plants are heterotrophic; for instance, sundews (species of *Drosera*) are insectivorous, while bladderworts (species of *Utricularia*) catch small aquatic organisms, although they photosynthesize too. These examples illustrate a point which applies to most generalizations in ecology: namely, that any set of categories that we invent is mainly for our own convenience. Nature is less simple, and almost every rule has its exceptions.

In studying any particular ecosystem, the first step is to identify the main species of plants and animals living there, and then to determine what the animals eat. This information can be used to construct a food-web, of which Figure 3.1 is an example. Some animals will be found to have only a single food source, or at least one that forms the greater part of their diet, like grass for cows or blood for tsetse flies. Others, such as the shrike or termite, both incorporated in this figure, take a variety of foods – they are said to have a wide or **catholic** diet, compared to the **restricted** diets of animals like the cow.

A number of other general terms are widely used in ecology. For example, it is sometimes convenient to refer collectively to the organisms living in a particular place as a **community** (pages 61–71).

Figure 3.1 A food web illustrating the feeding relationships of two plant and seven animal species. The arrows point towards the consumers. The shrike is a carnivore; if in turn it is eaten by the hawk, the latter would be termed a 'top carnivore'. Most ecosystems, natural and otherwise, contain many more species than this.

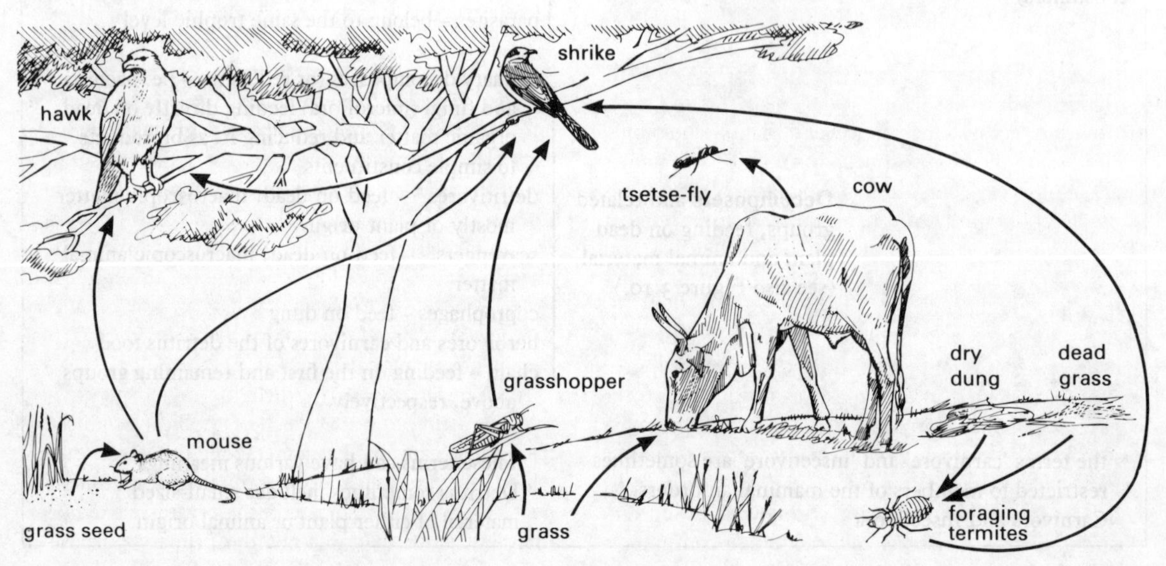

Usually the term is restricted to a single taxonomic group. Thus we could have the 'plant communities of Kivu National Park' or the 'insect community of a Ghanaian cocoa farm'. The word *habitat* appears frequently in the zoological literature. Its meaning is not very precise, but it refers mainly to the structural aspects of an ecosystem, such as ponds, rocks, short grass, or trees. It often implies the requirements of a particular species, as in 'the elephant's preferred habitat is woodland and forest' or 'tsetse flies can be controlled by modifying their habitat'. A largely conceptual term (in the sense that, unlike a habitat, one cannot go out and look at it) is **niche**. It is usually taken to mean the sum of an organism's requirements, together with its position, or 'role', within the ecosystem. Many ecologists studying communities find the concept of the niche to be useful, and we shall return to it on page 61.

Biomass and energy flow

Ecologists, like other scientists, gain insight into their subject by making comparisons. The simplest units for comparing ecosystems would be the total numbers of organisms, or the numbers per unit area, but a moment's thought shows that such units are unrealistic; organisms vary enormously in size. This problem can be overcome by taking the weights of the organisms instead. Comparing weights per square metre of vegetation from different ecosystems seems logical because weights are independent of whether the individuals are large like trees or small, like grass plants. The weight (more strictly the mass) of living organisms is referred to as their **biomass**. For plants, the phrase 'standing crop' is sometimes used; it has much the same meaning, except that biomass may also include non-living material, such as dead leaves and other litter.

In terms of biomass, more than 99 per cent of life on land and in most inland waters, is plant material. In marine ecosystems, however, plants contribute rather less to the biomass, usually around 75 to 80 per cent. In all ecosystems, the total biomass of carnivores is much less than that of the herbivores, and of higher level carnivores it is very small indeed. The biomass of **decomposers** is generally between that of the autotrophs and the herbivores (Figure 3.2).

Measurements of biomass are widely used in ecology, for example, recording yields of crops or carrying capacities of livestock or wildlife. But the concept of biomass implies a static system whereas in reality ecosystems are dynamic: plants use the sun's energy to produce protoplasm, which is eaten by animals, and ultimately death is followed by decomposition. These processes, although not continuous, happen every day and, like all processes within an ecosystem, they involve transfers of energy. Thus, if we could measure the amounts of energy and rates of transfer from one organism to another, we could begin to build a dynamic picture of an ecosystem.

trophic levels
4 top carnivores
3 carnivores
2 herbivores
1 primary producers
decomposers

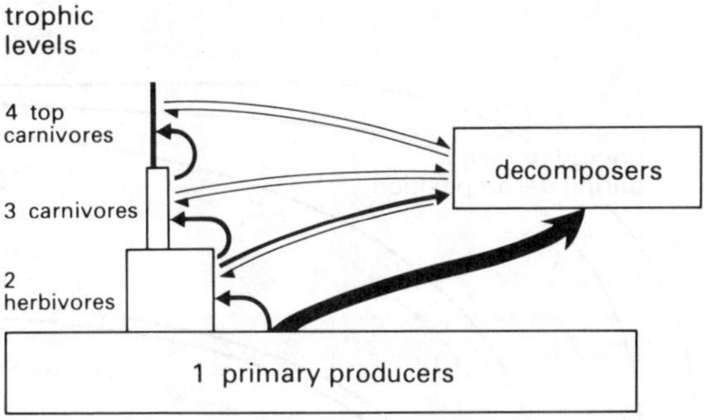

Figure 3.2 The 'pyramid of biomass' – a generalized case. The sizes of the boxes are proportional to biomass. Arrows indicate feeding by organisms of one trophic level on those of another, and their thickness reflects the quantities involved. The exact proportions of the organisms' biomasses in the various trophic levels vary between ecosystems, but are roughly similar for most non-marine ecosystems. In marine ecosystems, the primary producers have a proportionately smaller biomass. Some ecosystems such as streams have almost no primary producers; their detritus-based food-chains receive a supply of materials such as dead leaves from other ecosystems.

Figure 3.3 illustrates the pattern of energy flow in a generalized ecosystem, in which all the species of plants and animals have been grouped into their respective trophic levels. Physicists define energy as 'the capacity to do work' (and work as 'the rate of using energy'). Energy itself can take several forms, such as chemical, mechanical and heat. For all practical purposes, energy flow in ecosystems obeys the First Law of Thermodynamics, which states that **energy can be transformed from one kind to another, but it cannot be created or destroyed**. An example of a transformation is the fixing of sunlight by green plants, although they utilize only a small proportion of the incident energy. Plants convert energy in the form of light into energy in the form of chemical bonds, plus low-grade energy in the form of heat. Another example is the conversion of energy from chemical bonds into mechanical energy, plus heat, when a muscle contracts.

All transfers of energy in living organisms involve the production of heat. Much of the heat produced is from the organisms' basic metabolic processes, during which they release energy through respiration. The heat produced causes a rise in the organism's temperature, but since that was not the primary purpose in

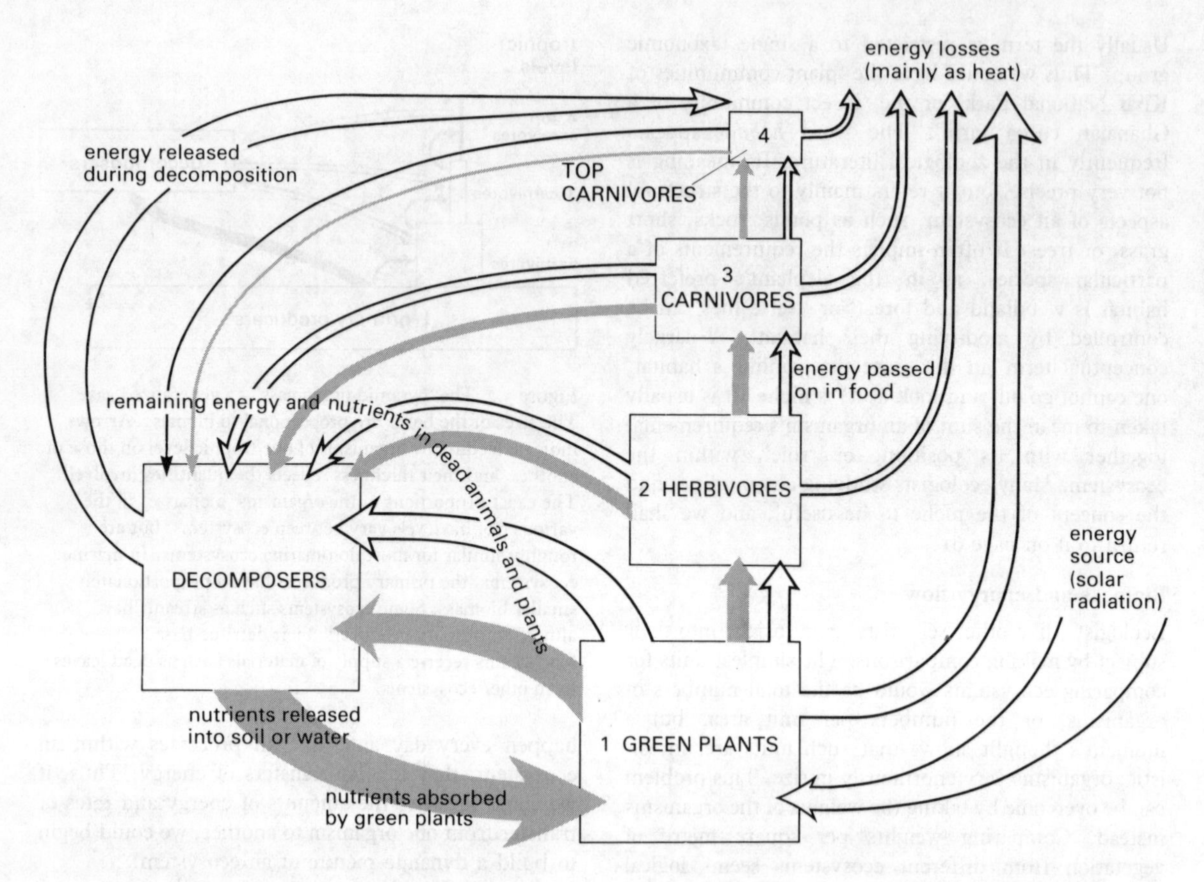

Figure 3.3 A model of a closed ecosystem. Boxes represent organisms of a particular trophic level, numbered as in Figure 3.2. White arrows indicate the **one-way flow of energy** whilst dark arrows represent the **cycling of nutrients**. Wider arrows indicate greater flows. All energy is finally converted to heat and lost from the ecosystem, whose continued functioning therefore depends upon the energy received daily from the sun. Nutrients, on the other hand, are not expended, and can be recycled indefinitely, usually taking part in chemical changes as they do so.

releasing the energy, heat produced is usually considered as a by-product. This is an example of the Second Law of Thermodynamics: **energy transfers are never 100 per cent efficient; a proportion (often a large one) always dissipates as heat.**

Since the organisms of each trophic level lose most of their intake of energy as metabolic heat, it follows that only a small proportion is available for forming chemical bonds in their tissues. These bonds, however, are the stored energy that will ultimately be available to the next trophic level as its source of energy. The proportion of the intake of energy of one trophic level that is passed on to the next is around 10 per cent, but it varies widely.

The fate of nutrients in an ecosystem contrasts sharply with that of the energy: for whilst energy has a one-way flow and is progressively lost to space – and thus needs to be supplied continually – nutrients remain and circulate indefinitely. In the broadest sense, nutrients are those chemical constituents of the ecosystem which form the living tissues: hydrogen, carbon, oxygen, phosphorus, nitrogen and many others.

If we go back for a moment and consider the biosphere as a whole, then during the course of a year (i.e. a complete cycle of seasons), the energy gained from the sun will equal the amount lost to space. Were this not so, either the earth's temperature, or the biomass of the biosphere, would change. At the same time, water and nutrients cycle in various ways within the biosphere. A similar pattern of energy gained and lost, and of nutrients cycling, also applies to biomes and ecosystems, but there is a difference. These smaller units are less self-contained than the whole biosphere, and exchanges of energy and nutrients takes place between biomes, and between ecosystems. On the whole, however, any particular biome or ecosystem will gain about as much as it loses, so that the overall result is one of dynamic equilibrium. In the simple ecosystem model of Figure 3.3, the nutrients are represented as cycling entirely within the ecosystem. This is an example of a **closed ecosystem**, contrasting with more **open ecosystems** such as rivers. In a few ecosystems there is an accumulation of stored energy, which eventually forms oil or coal.

The study of energy flow within and between ecosystems is called ecological energetics. It has led to a better understanding of many ecological processes, although like any approach it has its limitations.

Before proceeding to more specific aspects of the subject, it will be useful to mention the units which ecologists use. They are based upon an internationally agreed set of units known as 'SI', from the French Système International Unités. This system is sufficiently flexible to allow the ecologist a wide choice. Box 3.2 summarizes a number of the commoner ones, including those used in this book.

Primary production

The process

The basis of primary production is photosynthesis, a process which has been studied in great detail for more than a century and is now very well understood. It can, of course, be represented by the simple summary equation:

$$6CO_2 + 6H_2O \xrightarrow{\text{energy}} C_6H_{12}O_6 + 6O_2$$

The carbon that is incorporated into the glucose molecule ($C_6H_{12}O_6$) is said to have been fixed, but it does not usually remain in this form for long. Instead, the fixed carbon is released by the plant during respiration, or stored, or incorporated into other compounds, such as cellulose and proteins (the latter involving the addition of nitrogen and sometimes other elements). The complex biochemical aspects of photosynthesis need not concern us; what is important ecologically is that all biological activity, including our own, is ultimately dependent on the fixation of carbon by green plants. So the term **primary production** is clearly appropriate.

The energy reaching the earth's atmosphere after its journey of 150 million km through space is of various wavelengths. Radiations of some wavelengths penetrate the earth's atmosphere and reach the ground. The remainder are reflected back into space, or are absorbed by the atmosphere and re-radiated as heat. The photosynthetically active radiation is almost entirely within the range of wavelengths of visible light and especially in the blue and red parts of the spectrum (about 380 and 700 μm respectively). Light of intermediate wavelengths is absorbed less efficiently by the major plant pigments, chlorophyll *a* and *b*: the reflection and transmission of this light causes plants to appear green.

All but a few living organisms respire aerobically, using the stored chemical energy of the glucose molecule to release energy for their numerous activities. The overall equation for respiration:

$$C_6H_{12}O_6 + 6O_2 \xrightarrow{\text{energy}} 6CO_2 + 6H_2O$$

is the opposite of that for photosynthesis, although the many intermediate steps are quite different.

In plants, the total amount of glucose produced by photosynthesis is termed **gross primary production**.

Box 3.2 Examples of SI units used in measuring biomass and energy in ecosystems. They are not the only units possible; but illustrate the general pattern which is followed in this book

Units convenient for areas or organisms which are:	Biomass (or standing crop)		Amounts of energy	
	Fresh or live weights*	Dry weights†	Fixed per unit time,‡ normally as a dry weight	Flow per unit time‡ per unit area §
– small	wet g m^{-2}	dry g m^{-2}	dry g m^{-2} day^{-1}	kJ m^{-2} day^{-1}
– medium	wet kg ha^{-1}	dry kg ha^{-1}	dry kg ha^{-1} day^{-1}	kJ ha^{-1} day^{-1}
– large	wet t km^{-2}	dry t km^{-2}	dry t km^{-2} day^{-1}	kJ km^{-2} day^{-1}

1000 g ≡ 1 kg 1000 kg ≡ 1 metric tonne (t) 10 000 m^2 ≡ 1 hectare (ha) 100 ha ≡ 1 km^2

* Strictly speaking, one should use mass in this context, not weight

† After drying at 100 to 105 °C to remove free water

‡ Here as per day, but sometimes per year is more appropriate, as it averages out seasonality

‡ **Rates of production** such as these, are termed **productivities**.

§ The SI unit of energy, the joule, has been slow to replace the traditional calorie. Its use is, however, to be encouraged (1 kcal ≈ 4.2 kilojoules (kJ))

Some of this is used up in the plants' own respiration leaving a smaller amount as the **net primary production** (NPP), which is the total amount turned into new tissues or stored. We can estimate the NPP by determining the increase in dry weight of plants over a period of time. (This figure will also include the weight of the minerals taken up by the roots and incorporated into the plant, which usually amounts to about 5 per cent of the total dry weight.) So long as the conditions for photosynthesis remain favourable, the plant adds new tissues, or stores the excess glucose (often as starch), and NPP has a positive value. But when conditions are unfavourable, the rate of NPP may fall to zero, or even become negative if respiratory losses exceed photosynthetic gains (Figure 3.4). It is common for plants to be shedding some tissues, such as old leaves, at the same time as others, such as new leaves, are being produced. NPP includes all the material produced within a given time, including that lost.

Figure 3.4 A diagram illustrating the effects of rainfall on above-ground primary production in an area with one rainy season per year. In an ecosystem with a warm climate, the rate of primary production mainly depends upon the availability of soil moisture. Primary production continues throughout the year in the evergreen perennial plant, *A*, but is more rapid in the rainy season. In the dry season the rate of photosynthesis declines and is exceeded by respiration. Nevertheless, at the end of the year, there has been an overall increase in the plant's size (and therefore weight); this is the most obvious component of its annual NPP. However, the plant's total NPP is greater than the increase in biomass, because some parts will have been lost during the year, such as leaves that drop or are eaten by herbivores, as well as fruits that are eaten or dispersed. Plant *B* is a short-lived species, surviving the dry season as a seed. It germinates with the rains and grows rapidly. After seeding, it begins to die back; photosynthesis ceases and the plant's weight decreases until it dies. The weight of the seeds and other structures that are shed or eaten, plus the dead tissues, represent the NPP for the year.

Factors affecting primary production

Classical laboratory experiments have shown how the rate of photosynthesis is affected by light, temperature, water, and the availabiliy of carbon dioxide. In the case of light, both intensity and wavelength are important.

(a) The availability of *carbon dioxide* (CO_2) differs between ecosystems. It is generally less in aquatic than in terrestrial ones, but for any particular ecosystem it is fairly constant. This relative constancy means that it is of only marginal importance as an ecological factor in the field. However, its concentration can be raised artificially, as in a greenhouse, thus promoting carbon fixation.

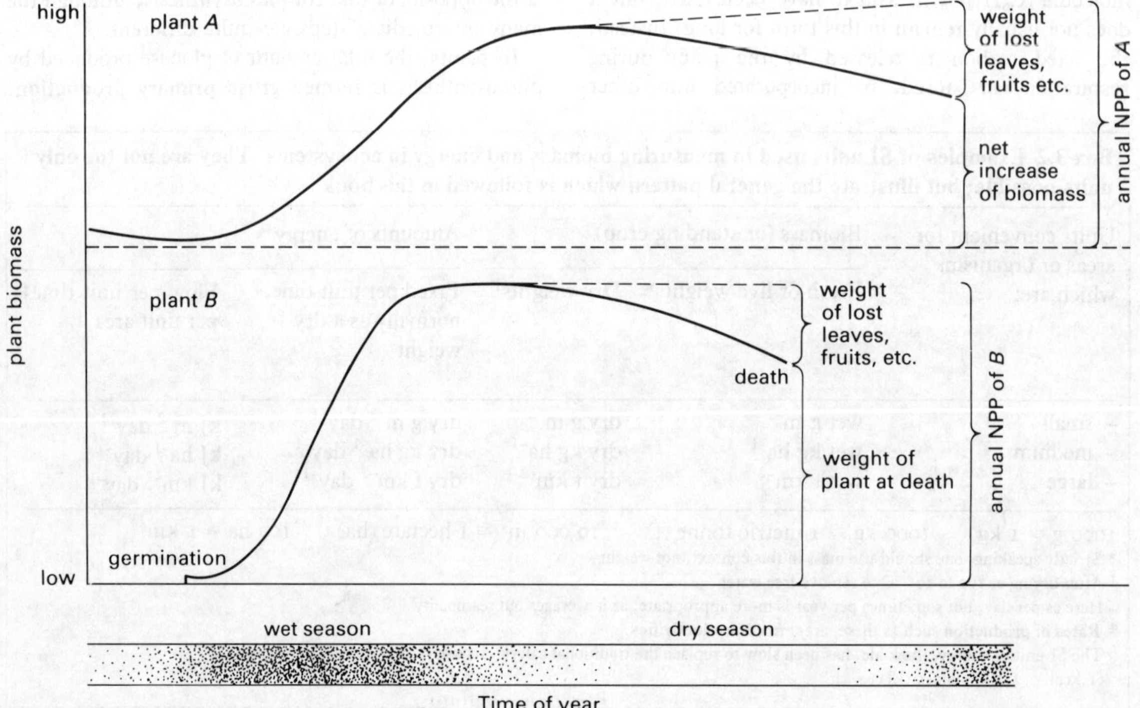

(b) In tropical Africa, the availability of *water* is the major variable in most terrestrial ecosystems. In turn, this depends upon rainfall. The overriding significance of rainfall for plant production is clearly evident in Figure 3.5 which includes data from many of the world's climatic zones, including cold ones. There is a strong correlation between NPP and rainfall, at least up to about 2000 mm yr^{-1}; only about 5 per cent of Africa receives more than that (Figure 2.4). In some areas, local rainfall is supplemented by groundwater which originates from rain that falls on higher ground, sometimes far away. Nowadays, too, man is increasingly supplementing groundwater by various forms of irrigation.

Water is a basic constituent of all living cells, but apart from that it affects photosynthesis in two ways. Firstly, it is an ingredient in the chemical transformation of CO_2 to organic molecules, and secondly, its shortage leads to stomatal closure. This in turn can reduce the intake of CO_2 by the photosynthetic tissues (see pages 74–77).

(c) NPP increases with *temperature* in the cooler parts of the world, but the rate of increase declines and for mean annual temperatures above about 25 °C there is little increase in NPP (Figure 3.6). In the tropics, temperature is not a major factor in plant production except at high altitudes. However, it limits NPP at intermediate altitudes, such as in the highlands of eastern Africa, where mean temperatures are about 15 °C and maxima about 20 °C. High temperatures also reduce photosynthetic rates, so that on hot days the rate is highest at mid-morning and again at mid-afternoon, and lower at midday.

(d) Seasonal variations in *light* (particularly day length) increase in importance with latitude. There are also major differences within ecosystems. In aquatic environments, light decreases rapidly with depth, and even in the clearest waters, photosynthesis ceases completely below a depth of about 50 m. The **euphotic zone** (beneath which light is insufficient for photosynthesis to exceed respiratory loss) is usually much less than 50 m, and in rivers with a high silt load, it is frequently less than one metre. Amongst

Figure 3.5 The influence of rainfall on net primary production (NPP). Data include both above- and below-ground production, and were obtained from many parts of the world, with a wide range of temperatures and soils, but despite that there is a clear pattern. This is described by the fitted line, whose equation is also shown (Y being the NPP and x the rainfall; e is the constant, 2.718). Most of tropical Africa has an annual rainfall of less than 2000 mm (Figure 2.3).
(From Leith, H., 1973 *Human Ecology*, **1**, 303–32.)

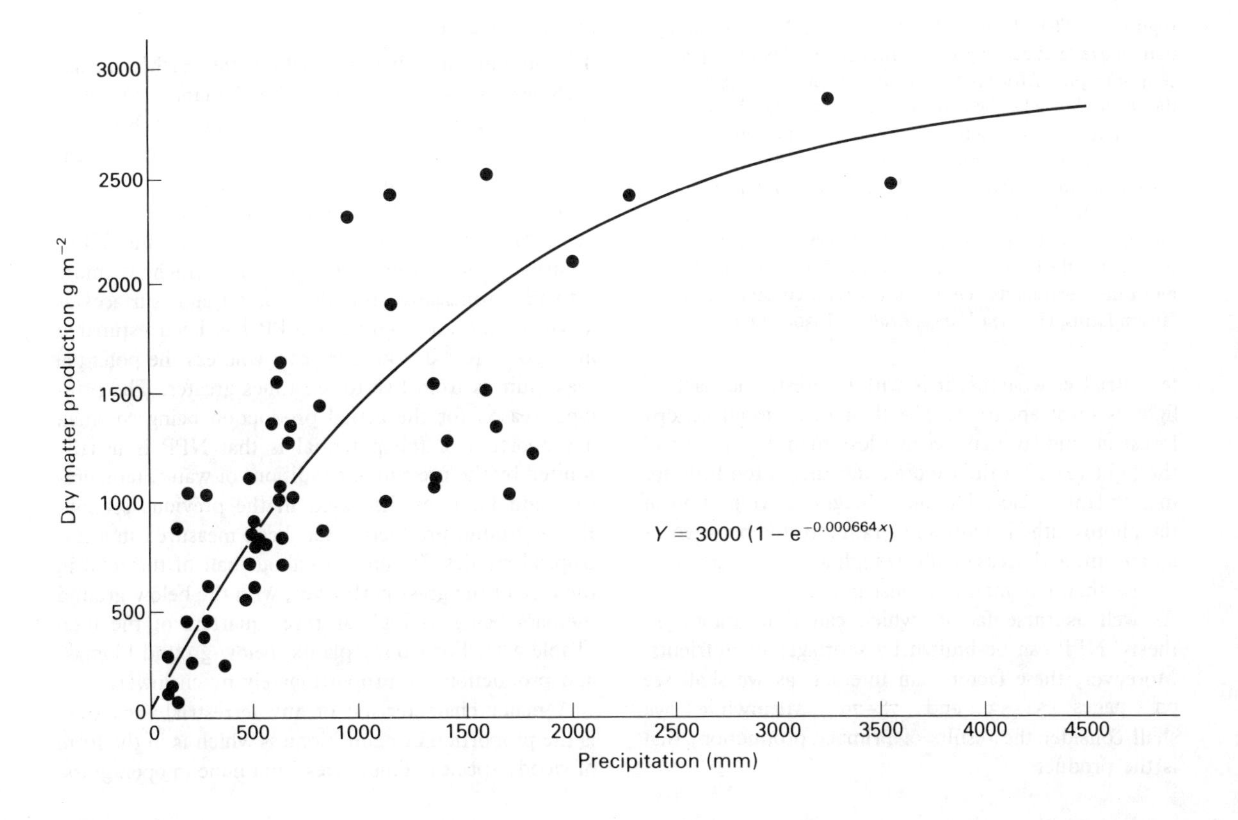

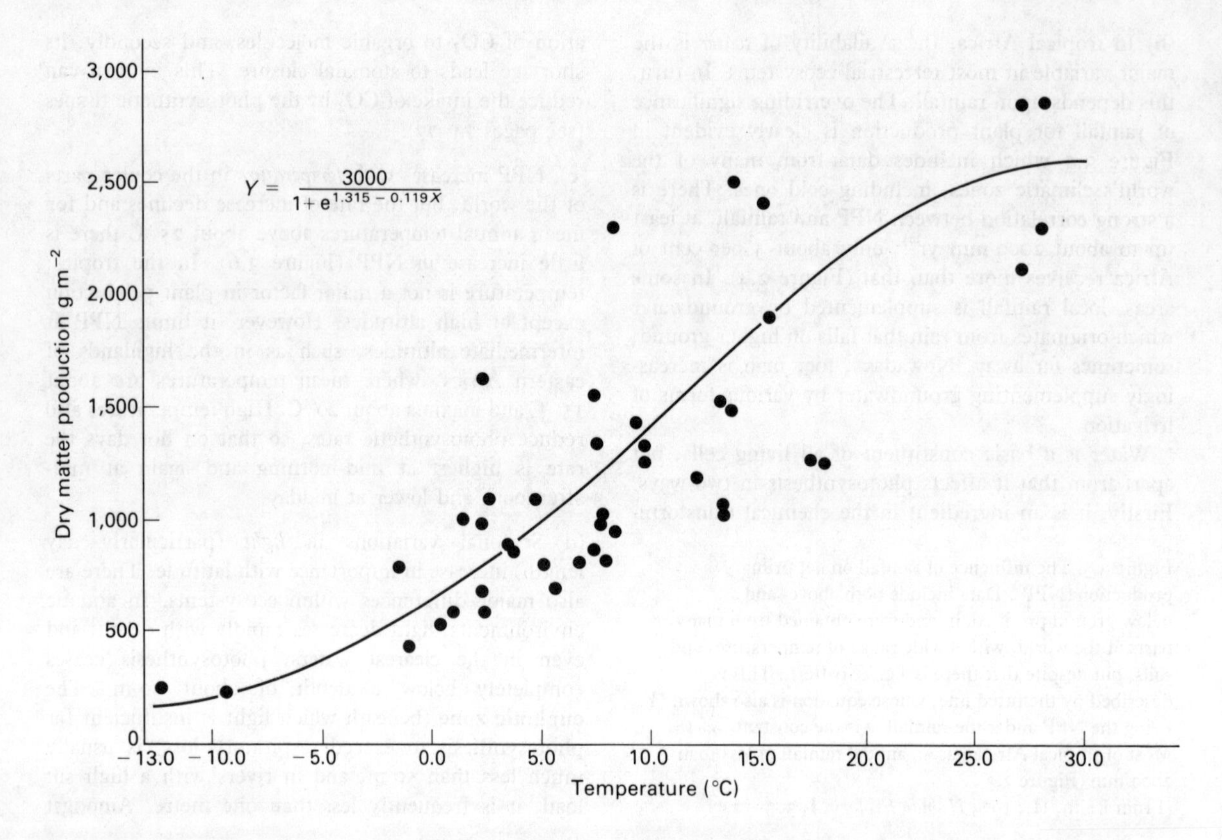

Figure 3.6 The influence of temperature on NPP (including both above- and below-ground production). About 90 per cent of tropical Africa has mean annual temperatures within the range shown by the dense clustering of dots (the main exceptions being the montane zones). From the graph, it appears that low temperatures are comparatively unimportant in determining NPP in tropical Africa, except at higher altitudes. In this diagram the effects of rainfall and soil type are ignored, but again a striking correlation is apparent. The equation of the fitted curve is shown, with Y being the NPP, and x the mean annual temperature; e is the constant, 2.718. (From Leith, H., 1973 *Human Ecology*, **1**, 303–32.)

terrestrial ecosystems, it is within forests that lack of light is most apparent. The floor of a closed canopy forest at midday may receive less than 1 per cent of the light reaching the canopy, and this is too little for many plant species. The mean height above ground of the photosynthetic tissues in a rain forest is between 20 and 30 m, and occasionally as high as 40 m, compared to less than one metre in most grassland.

As well as these factors which can limit photosynthesis, NPP can be limited by shortages of nutrients. Moreover, these factors can interact, as we shall see on pages 51–53 and 78–79. Meanwhile we shall consider the results of primary production, that is the products.

Primary products

The amount of radiation reaching the earth's surface each year is very large indeed; for the land areas alone it is approximately 3.5×10^{23} joules per year. But even in the most productive ecosystems, only 3–4 per cent of the solar radiation is fixed in plant tissues, and on a world-wide basis the amount is probably only 0.2 per cent. Most studies measure the above-ground NPP, ignoring below-ground production, which is more difficult to measure. For the world's land surfaces as a whole, the above-ground NPP has been estimated at $5\text{–}10 \times 10^{13}$ dry kg per year; whereas the potential maximum is from five to ten times greater. The principal reason for the actual production being so small a proportion of the potential is that NPP is usually limited by the prevailing conditions of water, temperature and light, as discussed in the previous section. Below-ground production has been measured in a few tropical studies. It comes to about half of the total in the case of the grass/herb layer, with the below-ground biomass being as high as three-quarters of the total (Table 3.1). For woody plants, below-ground biomass and production are proportionately much lower.

A major characteristic of any terrestrial ecosystem is the proportion of plant biomass which is in the form of woody species. This varies from none in open grass-

Table 3.1 Estimates of above- and below-ground biomass (in dry g m^{-2}) for several ecosystems. Below-ground primary production, as a proportion of the total, is also given

	American forests		Tropical forests*	African savannas				
	Mature deciduous forest (Tennessee)	Young mixed forest (New York)		Lamto, Ivory Coast Savanna woodland, consisting of: woody plants	grass layer	open grassland	Uganda Grasslands: grazed	ungrazed
Wood and bark	50 250	5 950	38 000					
Leaf	350	405	4 000					
Fruit and flower	20	20	–					
TOTAL ABOVE-GROUND	50 620	6 375	42 000	5 820	460	480	200	440
Roots	7 880	3 320	8 000	2 600	1 050	1 640	1 040	1 170
Roots as % of total biomass	14	34	16	31	70	77	84	73
% below-ground production	19	25	20 approx.	7	46	57	82	76

* Data for African forests are incomplete; those listed here are approximate, and apply to the tropics in general.

(Based on various sources including Strugnell, R. G. & Pigott, C. D. 1978. *Journal of Ecology*, **66**, 73–96; Swift, *et al.*, 1979; Whittaker, 1975 – reference in chapter 1.)

lands to nearly 100 per cent in rain forests. As we have seen (Figure 2.11), forests are characteristic of regions at medium and low altitudes receiving at least 1400 mm of rain a year, and without severe dry seasons. (At higher altitudes forests give way to montane vegetation.) Drier areas, most of which have a markedly seasonal pattern of rainfall (Figure 2.2), support woodlands and grasslands. For a given level of rainfall, the proportion of woody plants is strongly influenced by the availability of nutrients.

Much of tropical Africa is underlain by ancient granitic rocks or metamorphic rocks of the basement system. These generally support soils with a low nutrient status. The major exceptions are in eastern Africa, from Somalia to Zimbabwe, where extensive deposits of sedimentary rocks occur, and where there are large areas of volcanic rocks, mainly associated with the rift valleys. Soils overlying these rocks have a higher nutrient status, and typically support wooded and bushed grasslands. The predominant grasses of these areas are of short or medium height, such as species of *Themeda, Pennisetum* and *Andropogon*, which are generally palatable to mammalian herbivores. By contrast, on the nutrient-poor soils, the dominant woodland grasses are taller and usually tough, such as species of *Hyparrhenia* and *Hyperthelia*, which are only palatable when young.

Secondary production and products

Secondary producers

Secondary producers are heterotrophs; they depend upon the autotrophic producers, and are sometimes called **consumers** (Box 3.1). They consist of the herbivores feeding on live plants, and the carnivores that feed on the herbivores.

The rate of secondary production by herbivores is determined by a variety of factors, of which we shall consider four.

(a) *The amount of food that an animal can find*, and the ease or difficulty of finding it, can have a considerable effect on the rate of secondary production (pages 156–160).

(b) More than 99 per cent of the biomass in most terrestrial biomes consists of plant material (Table 3.2). The *quality of this vegetation* is of great importance to herbivores, and the most significant aspects of quality are digestibility and nutritive value. The ratio of soluble to structural carbohydrates and the proportion of proteins are useful indicators of food quality. Herbivores will often select the most nutritious parts of plants; thus many insect herbivores in trees feed only on young leaves. The process of selection by herbivores is well illustrated by the studies on a forest monkey described in Box 3.3.

Different types of vegetation vary in their quality. Forest trees provide few tissues that are edible to most herbivores, in comparison to the rapidly growing

Box 3.3 An example of how animals select food on the basis of quality

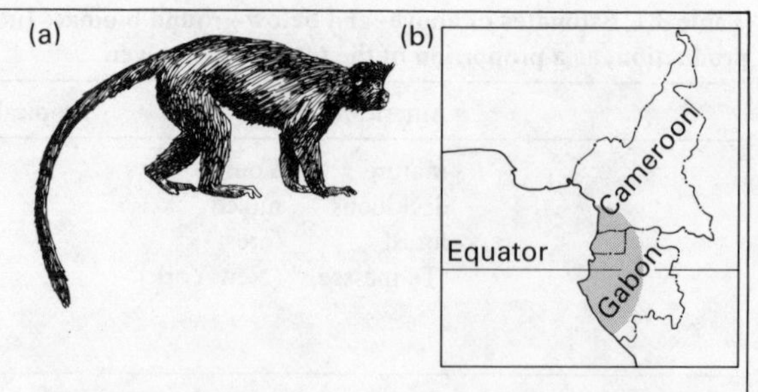

In recent years ecologists have become increasingly aware of the degree to which animals are selective in the food they take (see also pages 107–109). One of the most detailed studies that has yet been made is that of the black colobus monkey. There are seven species of colobid monkeys in Africa, all being leaf-eaters living in the forest canopy. The black colobus (a) has a rather restricted distribution, within which it is confined (b) to coastal rain forests. The monkeys live in groups, and one particular group became so used to being followed by scientists that they could observe the monkeys' behaviour freely and in detail. More than fifty species of trees occurred within the monkeys' home range; records were kept of the food selected, which consisted of young leaves, mature leaves or fruits. Only mature leaves of evergreen trees were available throughout the year. The scientists also analysed the leaves and fruits chemically. Their main conclusions were these:

(a) More than 99 per cent of the monkeys' food was provided by about one-quarter of the tree species, thus demonstrating a very high degree of selectivity. Some of the more important trees to the monkeys were amongst the rarest in the forest.

(b) The food items chosen were found to be those richest in energy, nitrogen and mineral nutrients; and low in substances that inhibit digestion, such as tannin and lignin.

(c) Young leaves were preferred to mature ones, but when mature leaves were eaten, they were from the most nutritious species.

(d) Seeds and fruits are rich in energy, but many contain toxins, and most of these were avoided. However, monkeys were able to eat some seeds containing toxic substances probably because their gut microflora can detoxify them.

Overall, there was a very close correlation between the quality of the food and the quantity of it in the monkeys' diets. Such detailed studies will greatly help in understanding the factors which limit the distribution and abundance of a species (chapter 5); almost certainly these factors will be found to differ between species. (Data from McKey D. B. *et al.*, 1981. *Biological Journal of the Linnaean Society*, **16**, 115–46.)

herbs and grasses which provide much higher proportions of edible material. Thus the biomass of animals is proportionately much higher in grasslands than in bushlands or forest (Table 3.2). Plant quality, from the herbivore's point of view, also depends upon the nutrient status of the soil (Figure 3.7).

Aquatic biomes have proportionately higher biomasses of animals than most terrestrial ones (Tables 3.2). Vascular plants form a small part of the vegetation in oceans and most rivers. In the oceans, there is no suitable substrate for them, whilst in coastal waters and fast flowing rivers only a few species of vascular plants have evolved structures capable of withstanding the current. Plants in these ecosystems are mainly algae, which are generally more digestible than vascular plants. The biomass of these plants is low, but their productivity is high.

In common with many other aquatic biomes, detritus forms a significant energy source in coastal reefs. There are important plant communities too, providing food for herbivorous animals, particularly molluscs and fish. The plants are mainly algae, some of which are calcareous and consequently eaten only by specialized animals. Within the corals are Zooxanthellae (algae which live symbiotically within the coral polyps) and filamentous algae enmeshed in the coral skeletons. These are consumed by animals such as parrot-fish, which crunch coral. Planktonic algae are abundant in the water, and being easily digested like most of the detritus, help to support an unusually high biomass of animals (Table 3.2).

(c) Woodlands and forests have a large plant biomass (Table 3.2), but their *productivity* is comparatively small. This relationship can be expressed as the **'production : biomass ratio'** (P : B), which is much smaller for woody plants than for grasses and herbs (Table 3.3).

(d) Animals with high *metabolic rates* use comparatively large proportions of their energy for basic activities, thus leaving comparatively little for growth and reproduction, which together constitute their net

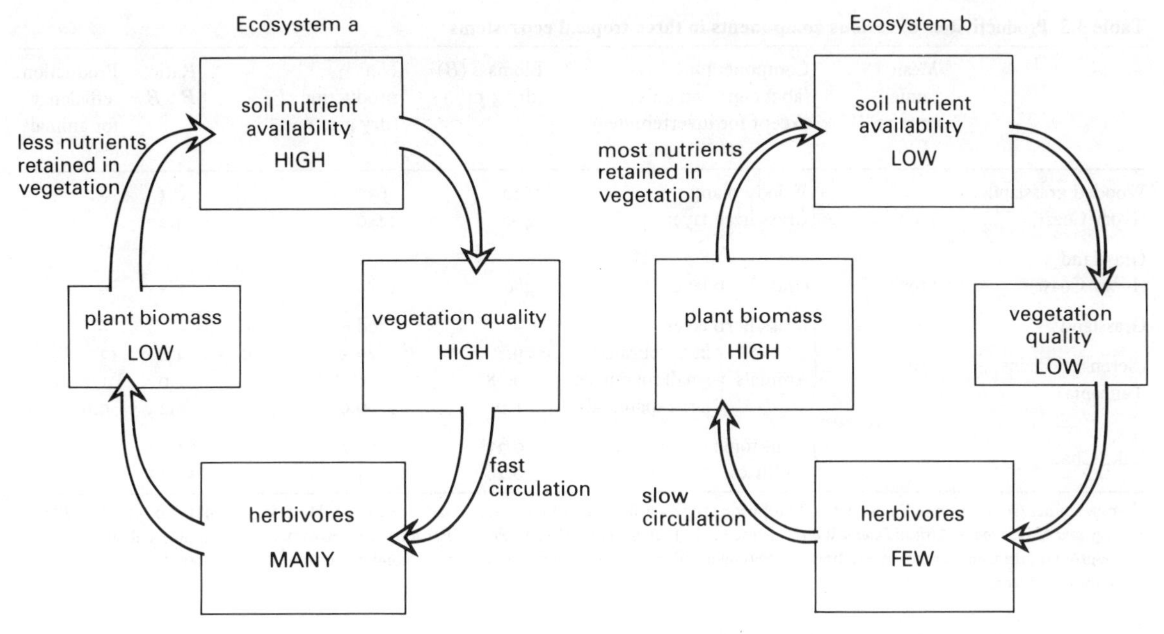

Figure 3.7 Models, based on a proposal of Bell (1982) to compare the effects of nutrient availability on non-forest ecosystems. The two contrasting ecosystems both have similar, moderate rainfalls (e.g. 500–600 mm yr^{-1}). Ecosystem (a) is on a soil rich in nutrients, whilst (b) is on an impoverished soil. Under these conditions, (a) will be a grassland, with palatable grasses and scattered woody plants; (b) will be a woodland with tall, unpalatable grasses. Most of the mammalian herbivores in (a) will be of medium size, such as antelopes and zebras. In (b) there will be less herbivores and most will be the large species – buffalo, rhinoceros and elephants. Notice in Table 3.2 that the animal biomass in grasslands is considerably higher than in woodlands. These models are of course generalized; much variation exists. However, the actual examples in Figure 3.20 (Serengeti) and Figure 3.19 (Lamto) correspond quite closely to models (a) and (b) respectively, except that their rainfalls are dissimilar (pages 58 and 59).

Table 3.2 Biomass estimates for plants and animals in various biomes. Although there is considerable variation within each biome, the mean figures given in the table can be taken as fairly representative. The biomass of animals, expressed as a percentage of the total, is shown in the last column

Biome	Biomass per unit area (dry g m^{-2})		
	Plants	Animals	% Animals
Rain forest	45 000	19	0.04
Seasonal forest	35 000	19	0.03
Woodland, bushland	6 000	5	0.08
Wooded grassland	4 000	15	0.38
Semi-desert	700	0.5	0.07
Extreme desert	20	0.01	0.05
Cultivated land	1 000	0.5	0.05
Swamp, marsh	15 000	10	0.07
Lake, stream	20	5	20
Open ocean	3	2	40
Coastal reefs	2 000	20	1.0

Adapted from Whittaker, 1975 – reference in chapter 1

production. The ratio of net production to consumption is known as the **production efficiency**, and it varies inversely with the metabolic rate. The effect is most obvious in comparisons of exotherms with endotherms.

Endotherms (a term much preferable to homoiotherms) maintain approximately constant body temperatures, usually between 37 and 40 °C, depending upon the species. These animals – the birds and mammals – commonly use between 60 and 90 per cent of their energy resources to maintain their body temperature, which is almost always above that of their surroundings. Understandably, the greater the temperature difference between an animal and its surroundings, the more energy has to be used in maintaining a relatively constant body temperature. There are examples from our own species: whereas a man in the tropics can manage on two meals a day, eskimos in Alaska need five!

In contrast, the body temperatures of **exotherms** (or poikilotherms) follow the ambient temperature quite closely. Hence exotherms use little or no energy in keeping themselves warm, and relatively more of their energy can be used for secondary production. The P : B ratio is higher for exotherms, especially for

Table 3.3 Productivities of various components in three tropical ecosystems

	Mean rainfall (mm yr^{-1})	Component[d] (above-ground only, except for invertebrates)		Biomass (B) (dry g m^{-2})	Net production (P) (dry g m^{-2} yr^{-1})	Ratio $P:B$	Production efficiency for animals (%)[e]
Wooded grassland (Ivory Coast)	1300	Woody plants		5820	630	0.1	–
		Grass/herb layer		460	1450	3.2	
Grassland (Ivory Coast)[a]	1300	Grass/herb layer		480	1180	2.5	–
Grassland (Serengeti Plains, Tanzania)[b]	500	Grass/herb layer		–	300	–	–
		Animals	invertebrates[f]	9.8	29.5	3.0	13
			small mammals	0.08	0.16	2.0	1.2
			large mammals	1.0	0.6	0.2	0.6
Lake Chad[c]	–	Planktonic crustaceans		0.33[g]	21.2[g]	64	–
		Molluscs		0.26	14.8	57	–

[a]Lamto studies (see caption to Figure 3.19). [b]Phillipson, 1973 – reference in Figure 3.20 [c]Davies, B. R., & Hart, R. C., 1981, pp. 51–68 in *The Ecology and Utilization of African Inland Waters*, Symoens J. J. *et al.* (eds), UNEP, Posis. [d]For terrestrial ecosystems, this applies only to above-ground organisms, except invertebrates at Serengeti. [e]Ratio of (Secondary production : Secondary consumption) × 100. [f]Includes decomposers. [g]Per m^3.

aquatic species, and they have much higher production efficiencies (Table 3.3).

Herbivores

The variety of herbivores reflects the variety of plant structures. Some of the more important groupings appear in Box 3.1. There are a few plant products – e.g. nectar and fleshy fruits – which have evolved specifically to attract animals to feed on them. But these are exceptions: most plant tissues are protected from herbivores. They may be hard to reach, or to digest, or they are poisonous or distasteful (pages 84 and 85). Leaves and shoots of most perennial plants are tender when they are young, although they are sometimes more toxic at this time. The growing season for most plants is synchronized with the rains; a sudden flush of young shoots and leaves appears simultaneously. Much of this material is palatable to herbivores, but they can only eat a small proportion of it in the short time before it matures. There is an interesting exception to this, where the herbivores migrate, following the rains. By doing so, the wildebeest consume much higher proportions of the NPP (pages 170 and 172). The same strategy has been adopted by pastoralists and their livestock in many semi-arid areas of Africa.

In arid ecosystems, with their short growing seasons, the proportion of hard tissues in plants is very high. In Israeli deserts, only 2–3 per cent of the NPP of leaf and shoot is taken by herbivores; but up to 80 per cent of the seed and fruit production is consumed, mainly by ants, rodents and birds.

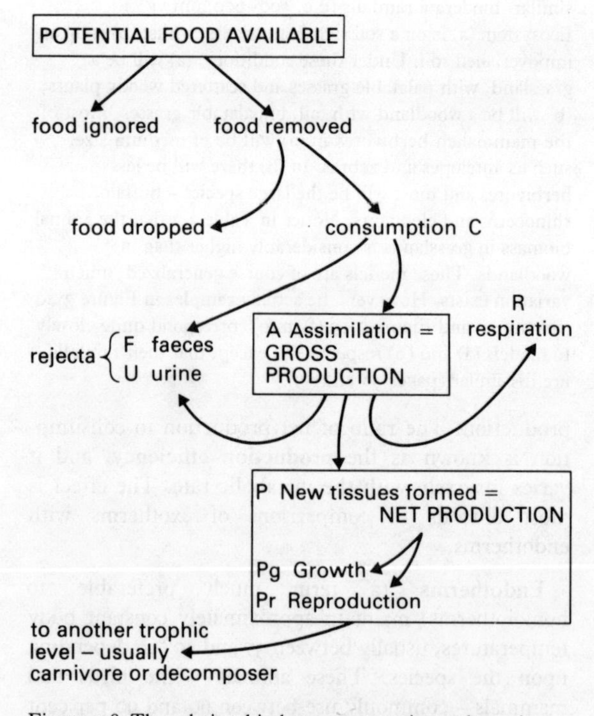

Figure 3.8 The relationship between a grazing or browsing herbivore and its food. C, P, F and U can be evaluated by keeping the animal in a metabolism chamber, R is usually obtained by difference; A = P + R. Overall, the food consumed is equal to P + R + F + U. (Modified from Petrusewicz & Macfadyen, 1970.)

Herbivores which eat leaves and young stems can be divided for convenience into two groups: **grazers** which eat grasses and herbs, and **browsers** which feed from shrubs or trees (Figure 3.9). The relationship between a grazer or a browser and its food is shown diagrammatically in Figure 3.8. Animals vary considerably in the proportion of their diet that is actually assimilated. The **assimilation (A) / consumption (C) efficiency** (or assimilation efficiency for short) is defined as

$$\left(\frac{\text{assimilation}}{\text{consumption}} \times 100 \right) \%$$

and it measures how much of the intake actually enters the animal's metabolic processes.

Assimilation efficiencies depend upon the animal's digestive system as well as on the type of food. The digestibility of plant materials is determined mainly by their content of structural carbohydrates, which ranges

a

b

c

Figure 3.9 Amongst large mammalian herbivores, a majority are grazers, such as the zebra (a); others, for example the gerenuk (b) are browsers; whilst some, including the eland (c) are mixed feeders. Eland graze when suitable grass is available, but in dry conditions they turn increasingly to browse.
(Photos (a) and (c) D. E. Pomeroy, (b) F. Hartmann.)

from none in nectar to nearly 100 per cent in wood.

Amongst mammals, the ruminants have higher assimilation efficiencies than non-ruminants. For cows on high-quality pastures, values of 80 per cent or even higher may be achieved, although for most ruminants a figure of 50 to 60 per cent is more typical. On the other hand, elephants assimilate only 30 to 40 per cent of what they eat. An average adult elephant weighing about 3000 kg requires some 40 kg (dry weight) of food per day, or nearly 15 tonnes a year. From this it will produce about 10 tonnes of faeces, which in turn supports a large population of dung beetles (Box 5.3). In addition, elephants frequently pull off much more vegetation than they actually eat, dropping the rest on the ground, where it dies and goes to the decomposers.

Invertebrates feeding on tissues of higher plants tend to have rather low assimilation efficiencies; 30 per cent is often quoted as a general figure, but it can be as low as 10 per cent. This is largely because of their inability to digest cellulose, so that, unless a cell wall is broken open during ingestion, its contents are inaccessible to their enzymes.

The greater efficiencies of mammalian herbivores depend mainly on their specialized grinding teeth, the molars and premolars. However, some herbivores have developed symbiotic relationships with organisms that can digest cellulose, and in some instances lignin. In higher termites, the relationship is with specific fungi (pages 50 and 111), but in most cases the symbionts are microorganisms, particularly bacteria, inhabiting the herbivores' guts. Amongst mammals, the ruminant's digestive system makes particularly effective use of these microorganisms.

Carnivores

The relationship between a carnivore and its food is essentially the same as that of a herbivore (Figure 3.9). However, a significant difference for many carnivores is the difficulty experienced in catching their food (i.e. prey), and many potential prey escape (pages 125–126 and 193–195). But once caught, the food, which is mainly protein, is much more digestible than the structural carbohydrates of higher plants. Hence assimilation efficiencies in carnivores are commonly between 60 and 90 per cent.

All herbivores belong to the same trophic level, the second. Carnivores can belong to the third, fourth or even higher levels, depending upon the length of the food chain (see the example on this page).

At each stage in the process there is a loss of energy. Typically in terrestrial ecosystems, herbivores utilize only 5–10 per cent of the NPP above ground. At succeeding levels, the proportionate offtake is higher,

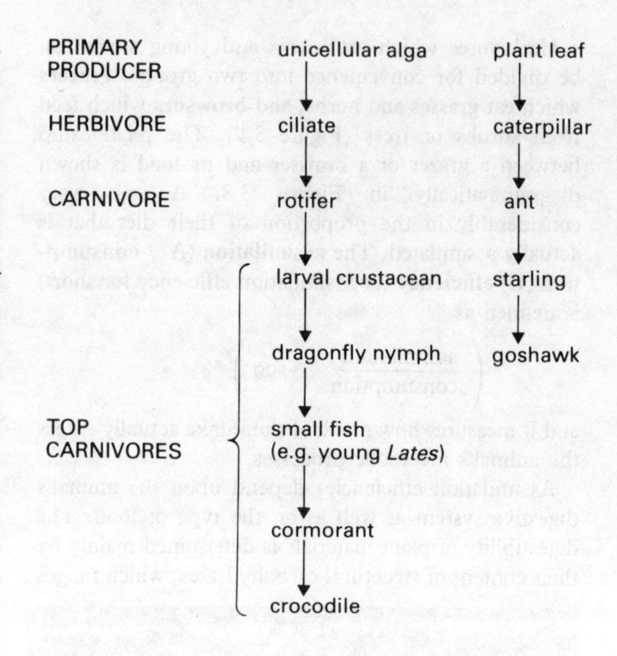

carnivores often taking 10–15 per cent of the net production by herbivores. Nevertheless, for a terrestrial ecosystem with an NPP of 1000 g m^{-2} yr^{-1}, the net production of herbivores will probably be about 1.0 g m^{-2} yr^{-1}, and of the first level of carnivores, 0.1 g m^{-2} yr^{-1}, at most. Clearly there is very little left for higher levels of carnivores. This is especially so if they are endotherms, using most of their energy maintaining their body temperatures. For instance, the goshawk included in the example above is considered to be a relatively common bird of prey in woodlands and wooded grasslands throughout tropical Africa. Yet its biomass in such places is no more than about 0.00002 dry g m^{-2}.

The term 'carnivore', which contrasts clearly with 'herbivore', is mainly used in the context of ecosystem processes. At other times, the word 'predator' is more widely used, contrasting with 'prey' (page 125). Indeed, it is becoming common to refer to herbivores as predators too, and the plants they eat as their prey. (We shall make it clear if we are using predator and prey in this sense.)

When considering trophic levels, parasites and carnivores are also equivalent, although clearly their significance in other respects is very different.

Decomposition

It is a truism that all organisms eventually die, and it is also true that in most cases their fate is to be consumed by other organisms. But even whilst alive they may lose parts – for example, trees shed leaves, and some animals moult their skins periodically. The term **detritus** is used for the dead organic material which falls to the ground – or sinks to the bottom in aquatic ecosystems. If the detritus accumulates on the ground it is sometimes called litter. The organisms obtaining their food from detritus are referred to as **decomposers** (Figure 3.10; cf. also Figures 3.2, 3.3). This term embraces animals and plants that consume dead material (detritivores, scavengers and dung feeders) plus those microorganisms which mineralize detritus and thereby release simple nutrients into the soil: these last are called **primary decomposers**.

Figure 3.10 Some of the more important pathways of materials and nutrients involved in decomposition processes in a terrestrial ecosystem. There are, of course, differences between ecosystems, but the general patterns are similar, and apply also to most aquatic ecosystems. In moist conditions, much of the detritus is broken down by the primary decomposers, but some is consumed by termites and other detritivores. Termites have symbiotic organisms which assist them to breakdown the dead plant materials of the detritus. They do this so thoroughly that very little of their original food material passes on to the primary decomposers; thus the dashed line represents a small flow. (Thicker lines represent a greater flow.) The smaller soil animals, including microarthropods such as mites and springtails, feed on the easily digested fungi and bacteria, or on detritus invaded by them.

Detritus is composed mainly of dead plant materials. Since herbivores in most ecosystems consume only a small fraction of the net primary production, the input of detritus is often high. To most animals, dead plant material is less nutritious than when it was alive. However, a few specialists, mostly arthropods, have evolved ways of using it. Whatever these animals do not consume is attacked by the primary decomposers, bacteria and fungi. In their turn, they are eaten by various small animals, again mainly arthropods. There is thus a degree of recycling within the decomposer system (Figure 3.10).

The rate of disappearance of detritus varies enormously. In a few ecosystems the rate is negligible, and dead organic material accumulates. For instance, in the oceans, the remains of organisms sinking to the deeper parts decompose only partially, contributing organic matter to the sediments on the ocean floor. Much of the world's oil originated in this way. Organic matter also accumulates on land, in the acid bogs characteristic of cool, wet regions. As it builds up, the remains of undecomposed plants form peat, a precursor of coal.

In most ecosystems the detritus disappears at a rate comparable to its input, so it does not accumulate. The high temperatures of most tropical ecosystems promote rapid decomposition of detritus so long as it

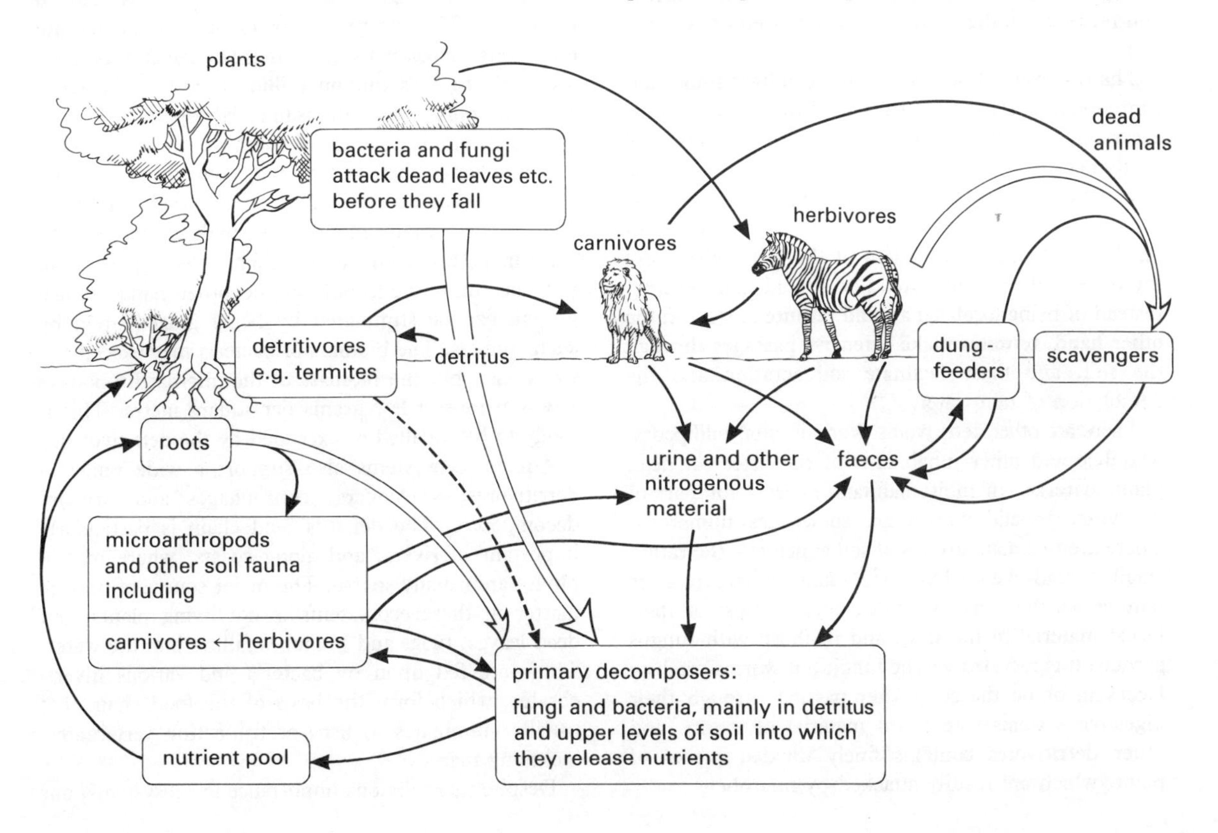

is moist. The type of material also affects the rate of disappearance. Dead animals (carcases), dung and the softer plant tissues can disappear within a few days or weeks, whereas twigs, branches and tree trunks can take several years, especially in the absence of termites or fire.

Amongst the many animals involved in decomposition, **detritivores** are usually significant. They eat dead plant material; in tropical terrestrial ecosystems, termites are often the most conspicuous. These insects sometimes remove as much as 90 per cent of the dead grass, bark, twigs and branches in various non-forest ecosystems, although 30–50 per cent is probably more typical. In moister ecosystems, such as in forests, the proportion is lower, usually only 20–30 per cent. Termites have symbiotic associations with other organisms which may be bacteria, fungi or protozoans, depending upon the group of termites (pages 111 and 117). The net effect, however, is always the same. The organic matter in their food is decomposed completely; and although some nutrients leach into the soil nothing else remains. The termites' overall assimilation (including the activities of the symbionts) can approach 100 per cent (Figure 3.10). But termites have many predators, some such as the aardvark and various ants (e.g. 'Stink ants': subfamily Ponerinae) being exclusive termite-feeders. Some of the energy and nutrients in the termites is thus passed on to other animals.

The removal of litter by termites can have important consequences for soil fertility. Most tropical soils, except in some forests, have rather little humus, and partly because of that, their nutrient status is low. The removal of litter by termites is one reason for the low humus content. In their absence, decomposition would be slower, the humus content of the soil higher, and the release of nutrients would be spread more evenly, instead of being localized around termite nests. On the other hand, termites make extensive passages through the soil, and these facilitate soil aeration and the percolation of rain-water.

There are other detritivores. Various ants, millipedes, woodlice and other inhabitants of the litter eat dead plant material. In moist habitats, at least for part of the year, annelid worms are sometimes numerous. There are two main groups of soil annelids – the rather small, thread-like enchytraeids – and the larger, more familiar earthworms. Termites deposit most of their faecal material in the nest, and in those with fungus gardens it is recycled via the fungi; but worms produce faeces in or on the soil. After passing through their digestive systems, the faecal material of worms and other detritivores contains finely divided pieces of plant, which are readily attacked by microbes.

Faeces of larger animals form the diet of the dung feeders, or **coprophages**. The ejected material, which probably amounts to about half of all the vegetation consumed by herbivores, is often referred to as dung when produced by mammals, or frass if it comes from arthropods. Dung-beetles and termites are important consumers of dung.

The bodies of dead animals, although their total quantity is relatively small, form the food of animals known as scavengers. Examples are some ants, marabou storks, vultures and hyaenas.

Anything not eaten by detritivores, coprophages or scavengers is attacked by the primary decomposers – unless it has been destroyed by fire. Herbivores and detritivores together commonly remove between 30 and 60 per cent of the NPP in terrestrial ecosystems. The amount remaining in unburnt areas is therefore between 40 and 70 per cent, and this is decomposed by microorganisms, especially by bacteria and fungi. There are sometimes large seasonal differences in their activity: unlike many detritivores, microorganisms are only active when the upper layers of the soil, and the litter, are moist, although non-surface microorganisms may be active when litter and soil surface are dry.

Detritus food chains can be surprisingly complex, and often involve a whole range of small and microscopic animals feeding on the microrganisms, and on each other. The numbers of soil organisms can be quite enormous. Beneath a square metre of soil surface there are likely to be a million million (i.e. 10^{12}) bacteria, although only a few thousand arthropods. The soil microarthropods, so called because most are only 1–2 mm long when full grown, comprise two major groups. These are the mites (Acari) and springtails (Collembola). Many members of both groups feed on soil fungi, whose effects as primary decomposers can thus be diminished, but on the other hand, fungal growth can be stimulated by being grazed upon by such animals. The biomass of bacteria may exceed 100 wet g m^{-2}, but the biomass of the microarthropods is low – at most a few grams per square metre. This is likely to be equalled or exceeded by the detritivores.

Aquatic ecosystems also support a wide range of detritivores, scavengers, coprophages and primary decomposers. The detritus food chain is particularly important in rivers, and along coasts, where macrophytes are usually sparse. The major source of organic matter in these ecosystems is not living plants, but dead leaves, twigs and branches falling into the water. These are fed upon by bacteria and various invertebrates, which form the basis of the food chain that usually culminates in fish, or fish-eating vertebrates, including man.

Despite their obvious importance in ecosystems, our

present knowledge of decomposition processes is still very scanty, and accurate quantitative data are particularly scarce. This is a topic requiring much more research. For although man is removing much of the above-ground natural vegetation to grow food crops, pasture grasses and trees, he is still heavily dependent upon the natural soil organisms for decomposition and hence soil fertility.

Nutrients in ecosystems

There are about ninety-two naturally occurring elements, and of these approximately fifteen occur almost universally in living organisms. Each of these elements takes part in a characteristic cycle within the biosphere, its atoms alternating between living tissues and the atmosphere, soil or water. The elements contributing most to living tissues are hydrogen (H), oxygen (O), carbon (C) and nitrogen (N). The approximate ratio of the atoms of these four is 200 : 100 : 100 : 1. Phosphorus (P), calcium (Ca), sodium (Na), potassium (K), magnesium (Mg) and sulphur (S) are also important, but their total quantity is less than that of nitrogen. A few other elements are involved, but all in minute quantities and they are rarely limiting.

Nutrient cycles within the biosphere are termed **biogeochemical cycles**, and they are of two main kinds. In **closed cycles**, the elements circulate mainly within each ecosystem, whereas **open cycles** involve considerable interchange with the earth's atmosphere.

Phosphorus provides an example of a closed cycle (Figure 3.11). It is scarce in most soils, and also in many aquatic ecosystems. Its shortage in soils is partly a result of the rapidity with which it is taken up by plant roots, thus preventing its accumulation in the soil. However, there is increasing evidence that its low levels in many aquatic as well as terrestrial ecosystems limit plant growth (Figure 3.12). Adequate phosphorus is vital to all living organisms, particularly because of its ability to form high-energy compounds such as ATP (adenosine triphosphate). The cycles of sulphur, magnesium, potassium and other minor elements are also closed, and resemble the cycle for phosphorus.

Figure 3.11 Most mineral nutrients, such as phosphorus, calcium, potassium and sulphur, take part in **closed cycles**, that is mainly within an ecosystem (here a wooded grassland). However, there are some gains ('imports') and losses ('exports'), involving other ecosystems, although in most cases the amounts concerned are relatively small. With ions such as those of P and K, which are highly soluble, a very large proportion of the total stock is within the living organisms. Nutrient ions are released rapidly from dead and decaying organisms and quickly absorbed by plants, leaving low concentrations in soil or water.
(Based on Colinvaux, 1973, reference in chapter 1.)

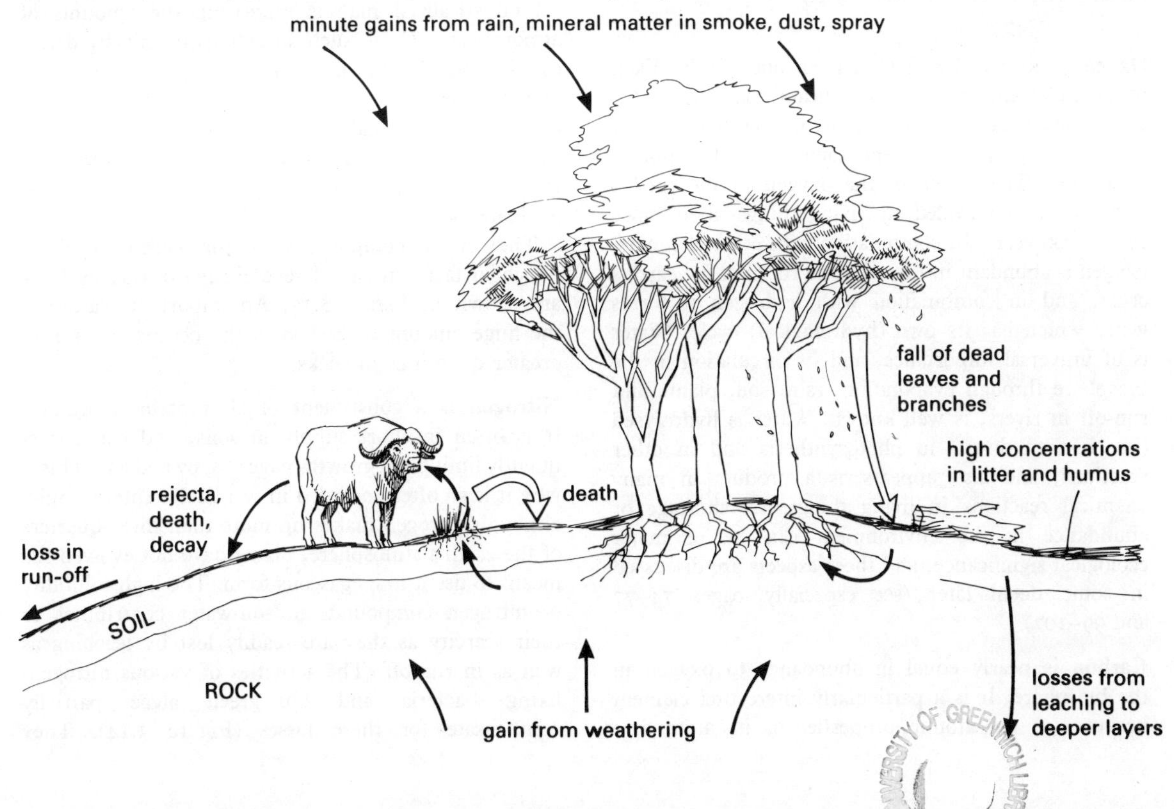

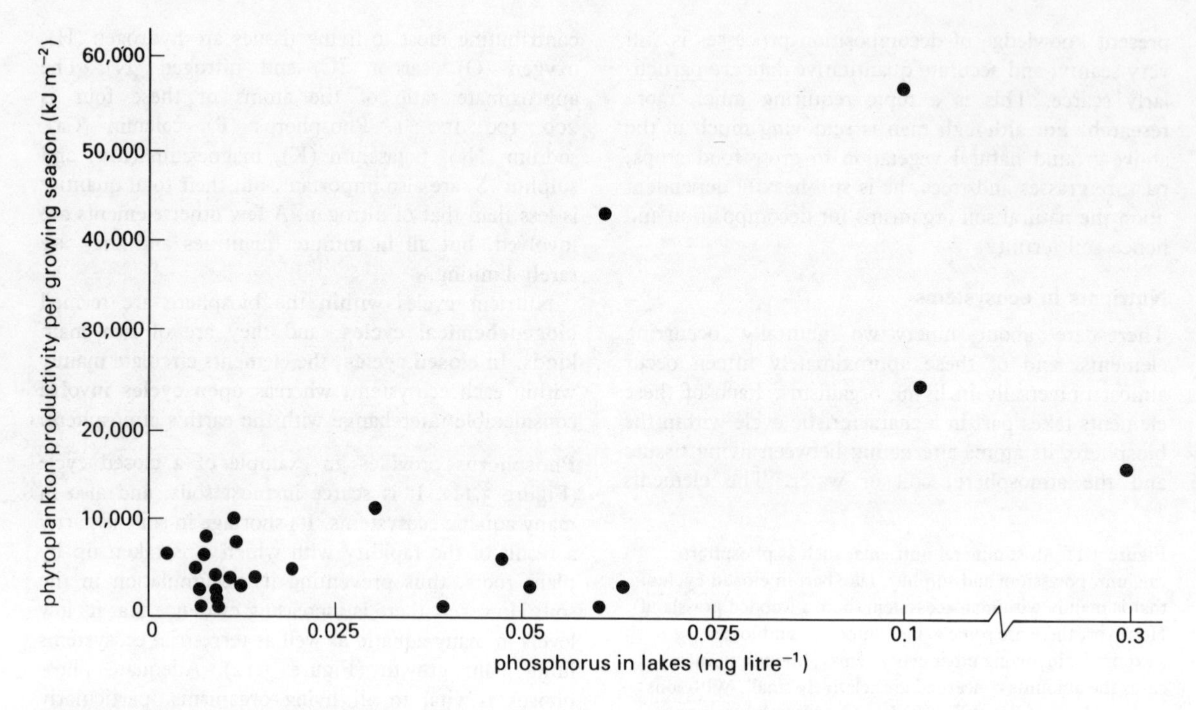

Figure 3.12 Gross productivity of the phytoplankton (microscopic plants) in a series of lakes throughout the world, related to the phosphorus content of the water. The higher productivities were all in lakes with more than 0.05 mg l^{-1} of phosphorus.
(Modified from Barnes & Mann, 1980.)

Oxygen, like C, H and N, has an open cycle. Each of these elements occurs in significant quantities in the atmosphere. Oxygen, as a gas or in solution, is used by all living things except some anaerobic micro-organisms. The whole of the oxygen in the earth's atmosphere is recycled by living things about once every 2000 years. In combination with other elements, oxygen is abundant in the soils and rocks of the earth's crust, and in combination with hydrogen it forms water which has its own (hydrological) cycle. Water is of universal importance, and its circulation in the biosphere through evaporation, rain, soil, plants and run-off in rivers, is well known. Water is hydrolysed (i.e. broken down) in photosynthesis and in other reactions, and also appears as a product in many chemical reactions in living things. Its shortage or abundance in the environment can be of major ecological significance, and these aspects are discussed in some detail later (see especially pages 74–77 and 99–102).

Carbon is nearly equal in abundance to oxygen in the biosphere. It is a particularly interesting element because (a) the atomic properties of its atom give carbon a central place in the structural molecules of living things, as in all organic compounds; (b) although fairly abundant in the earth's crust (especially in carbonates), carbon dioxide gas (CO_2) forms only about 0.033 per cent of the earth's atmosphere; and (c) industrialized man is increasing the amounts of atmospheric CO_2 to such an extent (mainly by deforestation and by burning fossil fuels) that significant climatic changes are feared within the next century – this is the so-called 'greenhouse effect' (page 220). At present, the whole of the atmospheric CO_2 is recycled roughly once in 300 years, but this too is changing.

The major components of the carbon cycle – which in fact consists of several interlocking cycles – are shown in Figure 3.13. An important feature is the huge amount of carbon in the oceans, with even greater quantities in rocks.

Nitrogen is a constituent of all protein molecules. It is often in short supply in soils, and can consequently limit plant growth (pages 18, 67 and 85). This is why it is so often included in fertilizers. Interestingly, although nitrogen makes up more than three-quarters of the earth's atmosphere, plants have not evolved the means to use it in its gaseous form. The high solubility of nitrogen compounds in soil-water contributes to their scarcity as they are readily lost by leaching as well as in run-off. The activities of various nitrogen-fixing bacteria and blue-green algae partially compensate for these losses (Figure 3.14). They

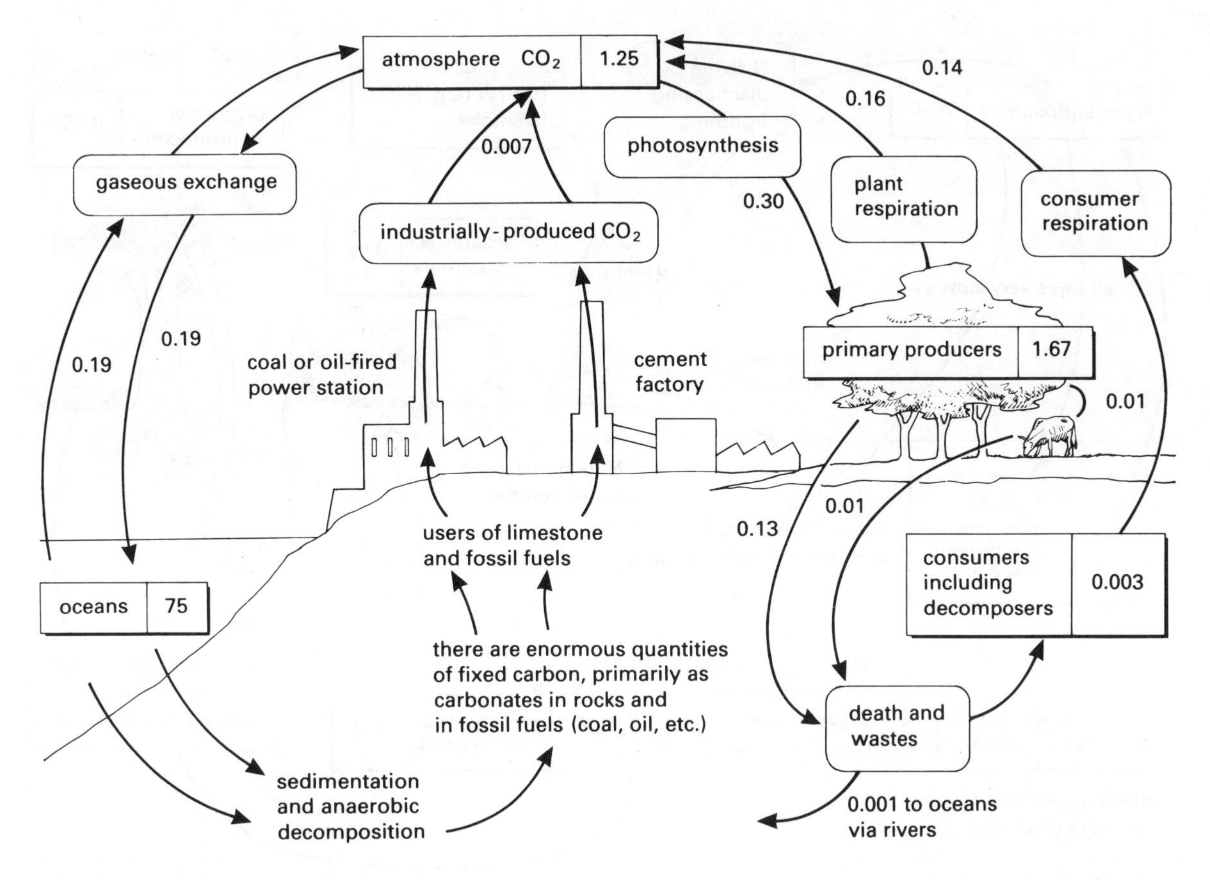

Figure 3.13 Representation of the circulation of carbon, which is an example of an **open cycle**, involving the whole biosphere. The boxes are **stocks** of C and show the quantities of the element in kg m^{-2} of the earth's surface. Arrows indicate **transfers** of carbon by the processes shown in boxes, measured in kg m^{-2} of the earth's surface per year. Until the industrial 'revolution', the amounts of CO_2 entering and leaving the earth's atmosphere were almost exactly equal. Since then, various industrial processes, including those shown, and also motor vehicles, bush fires, charcoal burning, etc. have provided an additional supply to the atmosphere, with no return. Meanwhile, destruction of forests is continually reducing the amount of CO_2 taken in by plants. At present, atmospheric CO_2 is variously estimated as increasing by between 11 and 28 Gt per year (1 Gt $\equiv 10^9$ tonnes or 10^{12} kg).
Based mainly on Colinvaux, 1973, and Whittaker, 1975, references in chapter 1.)

convert atmospheric nitrogen gas into ammonium ions, which can be taken up directly by plants from the soil solution. Some ammonium ions are converted to nitrates by bacteria, and can then be taken up again by plants. Symbiotic *Rhizobium* bacteria in the root nodules of legumes fix atmospheric nitrogen, which is used by the plants whose roots they inhabit.

These examples illustrate the importance of understanding the circulation of the various elements contained in plants and animals, both within ecosystems and through the biosphere as a whole. Scientists are beginning to realise that it is necessary to take a global view of such matters: cutting down of forests in Nigeria or Angola has repercussions for everybody, not just for Nigerians or Angolans. However, neither regional nor global aspects of nutrient cycles are fully understood yet, so that changes induced by man, whilst undoubtedly happening, cannot be properly assessed. As these effects are more likely to be detrimental than otherwise, research into them must have a high priority.

Productivity in African ecosystems

Overall pattern

There are large regional differences in net primary production of natural ecosystems in Africa and around its coasts (Figure 3.15). The contrast between the generally low oceanic values and the much higher productivities of most land areas is particularly noticeable. Some of the factors limiting net primary

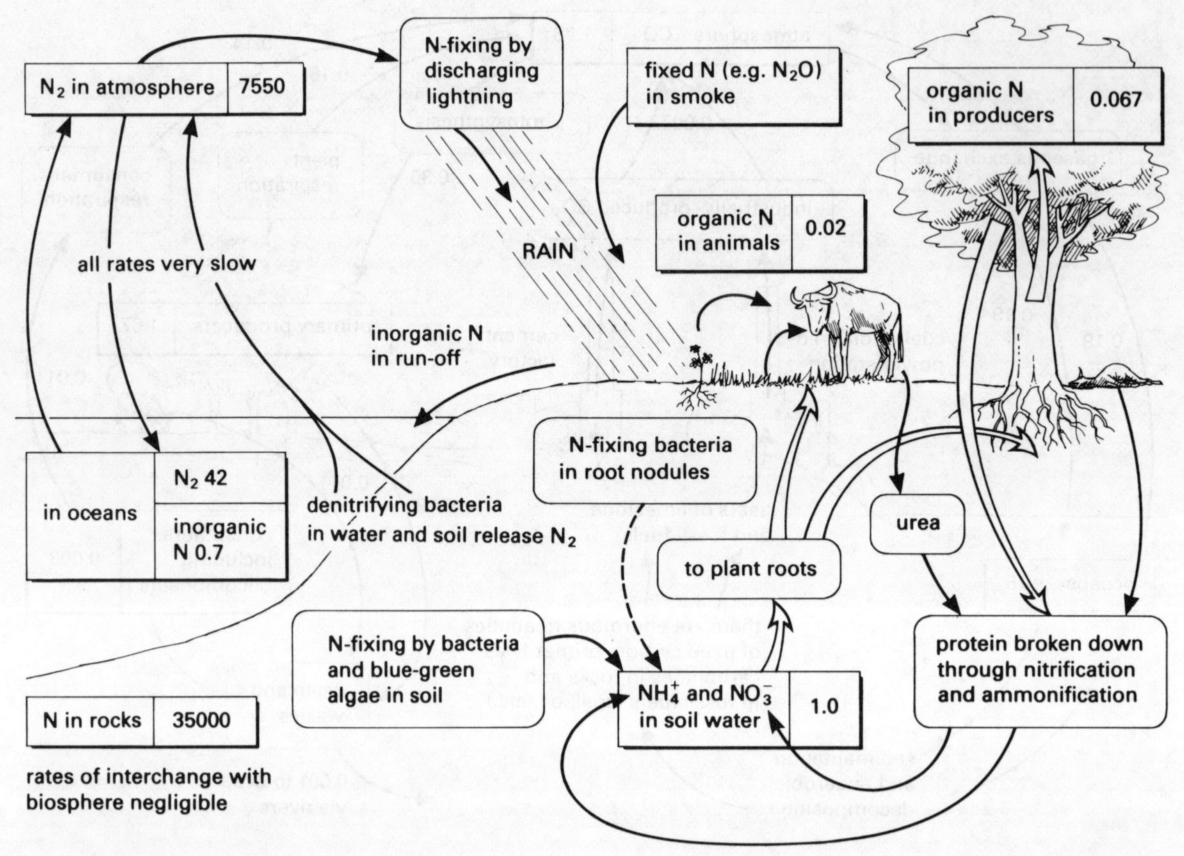

Figure 3.14 Representation of the nitrogen cycle. The general features are arranged to facilitate comparison with Figure 3.13, with figures again in kg m^{-2} of the earth's surface. However, there are important differences between the carbon and nitrogen cycles. Nitrogen is abundant but almost inert in the atmosphere, whilst the chemically active compounds of nitrogen (mainly the soluble ions, nitrate $-NO_3^-$ and ammonium $-NH_4^+$) are comparatively sparse. Nitrogen is lost from terrestrial ecosystems by run-off in water of these highly soluble ions, but this is approximately compensated for by the activities of nitrogen-fixing bacteria in the soil and in root nodules of plants; and also by blue–green algae in wetter soils. The rates of the various processes are not shown because they are extremely variable as well as difficult to determine, but the widths of the arrows give an indication of relative rates. Industrial processes are not shown, but smoke often contains nitrogen (e.g. as N_2O), and processes of industrial fixation are expanding rapidly (especially the production of fertilizers from atmospheric nitrogen).

Globally, industrial fixation is approaching the rate achieved by bacteria.

(Based on Colinvaux, 1973, Whittaker, 1975, references in chapter 1.)

production (NPP) in the oceans, such as temperature, are similar to those on land, but others are quite different. Thus, two of the three areas of highest marine productivity, off the coasts of Somalia and Namibia, are adjacent to land areas of low productivity. Elsewhere, areas of high terrestrial productivity are next to unproductive seas, as around the Gulf of Guinea, and along much of the eastern seaboard of Africa. Several factors explain these patterns, including the difference between the media – water and air.

Low values of NPP in the oceans are mainly caused by shortages of nutrients within the euphotic zone. With increasing depth, nutrients are more plentiful, but there is insufficient light for photosynthesis. The circulation of water within the oceans results in localized **upwelling zones** where the cold, nutrient-rich deeper waters rise to the surface. Where this happens, the productivity is much higher. Globally, the most extensive upwelling zones are between the latitudes of 40° and 60°, both north and south of the equator, and these support most of the world's major fisheries. The two areas of higher productivity off the west coast of Africa are also upwelling zones, though comparatively small ones (Figure 3.15).

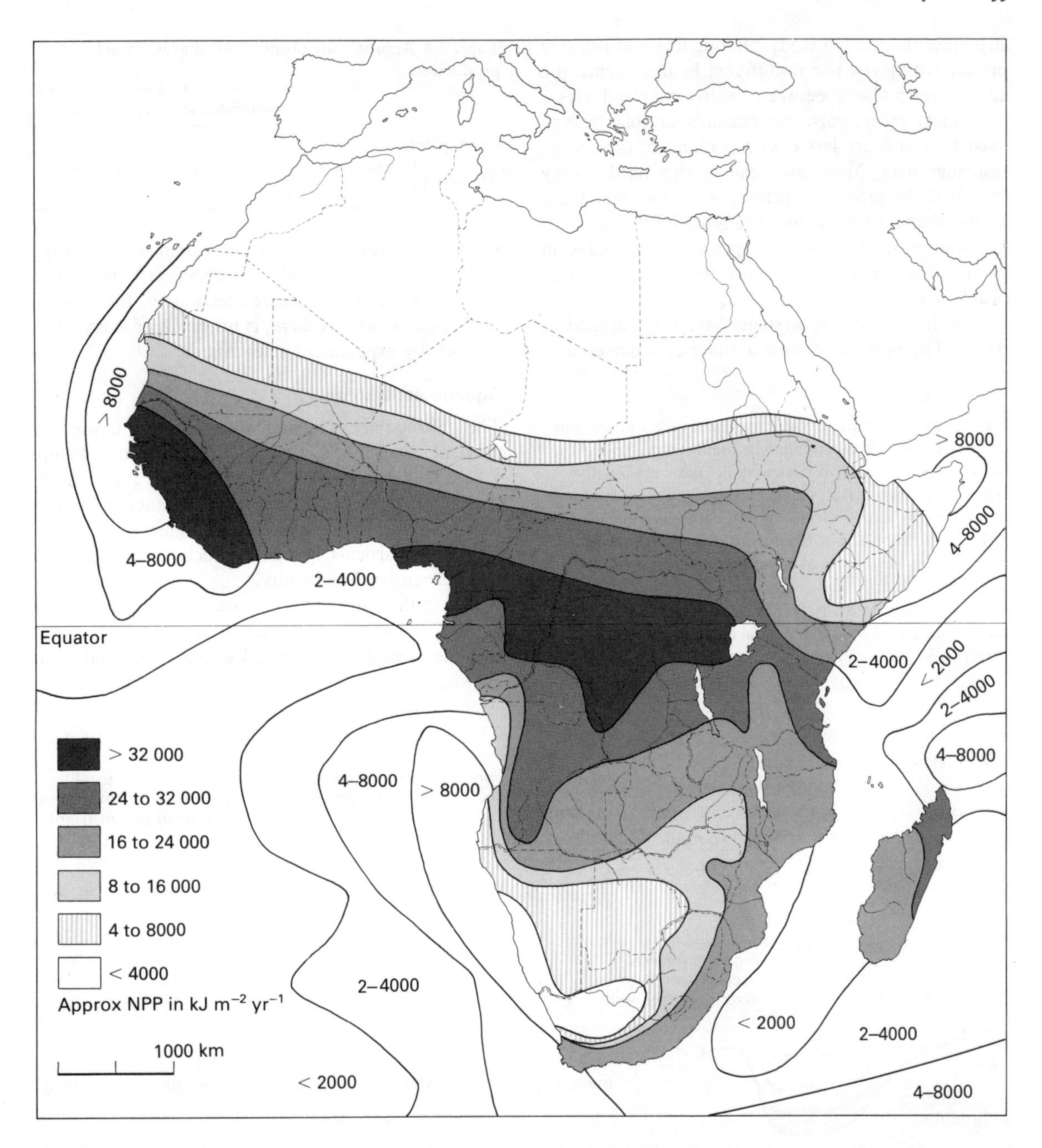

Figure 3.15 Estimated annual NPP in tropical Africa and the surrounding oceans. Production is shown as energy fixed in kJ m^{-2} yr^{-1}. The corresponding quantities of dry matter in g m^{-2} yr^{-1} can be obtained by multiplying the energy values by 0.056. Considerable variability occurs locally, especially in hilly regions, and also from year to year.
(Modified from Kormondy, E. J., 1976, *Concepts of Ecology*, Prentice-Hall: New Jersey.)

Worldwide, the annual production of new plant material in the oceans is less than half as much as on the land, despite the fact that 70 per cent of the world's surface is sea. The difference in plant biomass is even greater – more than 99.5 per cent of the world's total plant biomass is on land (Table 3.2). The production to biomass (P : B) ratios for aquatic ecosystems are often much higher than for terrestrial ones, because, unlike water plants, many land plants are multicellular and contain a high proportion of structural tissues. On land, differing levels of primary productivity given rise to different biomes (hence the general resemblance between Figures 2.11 and 3.15).

In most ecosystems, the amounts of solar energy fixed by plants are less than 2 per cent of the energy reaching them. More than half of this fixed energy (which is the gross production) is used metabolically by the plant, the remainder being the NPP. However, the proportion of gross to net production is higher in plants than in animals, especially of endotherms (Table 3.4).

Very few African ecosystems have been studied in detail. The making of such a study is a major task, since it involves productivity measurements of all the major components, and in some cases individual species have to be considered separately. Each of the three studies outlined here, is the result of many years of work by experienced scientists.

Table 3.4 Approximate proportions of gross to net production

		Gross production : Net production
Green plants		2.7 : 1
Animals	Exotherms	4 : 1
	Endotherms	50 : 1

Aquatic ecosystems

The productivity of inland waters is often limited by nutrient availability, especially phosphorus (Figure 3.12). **Oligotrophic** lakes – those with low concentrations of nutrients – have lower productivities than nutrient-rich, or **eutrophic** lakes (Figure 3.16). Saline lakes have particularly high productivities, at least up to a certain level of salinity.

One of the few aquatic ecosystems in tropical Africa whose productivity has been thoroughly investigated is Lake George in western Uganda. The study was

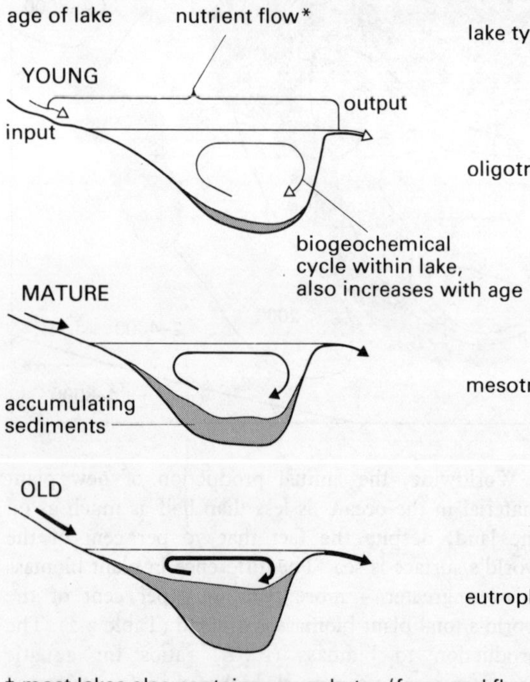

age of lake
nutrient flow*
YOUNG
input
output
biogeochemical cycle within lake, also increases with age
MATURE
accumulating sediments
OLD

lake type	annual NPP (dry g m^{-2} yr^{-1})	phytoplankton biomass† (dry g m^{-2})	depth of light penetration (m)	total inorganic nutrients (ppm = mg l^{-1})
	approximate values for:			
oligotrophic	15–50	<0.2	20–120	2–20
mesotrophic	50–150	0.2–0.6	5–40	10–200
eutrophic	150–500	0.6–10	3–20	100–500

† most lakes also contain macrophytes (ferns and flowering plants)

* increasing input as drainage basin weathers. As the input (and output) increase the lake ecosystem changes from (relatively) closed to open.

Figure 3.16 The nutrient status of a lake is a key factor (but not the only one) in determining its productivity. Young lakes, or those on ancient metamorphic rocks, are usually oligotrophic. In contrast, old lakes, especially those fed by rivers draining across volcanic or nutrient-rich sedimentary rocks, are eutrophic. The latter usually have a rich flora of macrophytes. Some sorts of pollution accelerate the process of eutrophication.
(From Boughey, A. S., 1973, *Ecology of Populations*, Macmillan: New York, and Whittaker, 1975, reference in chapter 1.)

made by an international team over a 4-year period around 1970. Lake George is comparatively small, having a surface area of about 250 km^2, and is unusually shallow, its mean depth being about 2.2 m. It lies in the western Rift Valley and is fed by rivers from the rift wall, which to the west includes the Rwenzori Mountains. The rivers entering the lake flow across areas of recent volcanic origin which are rich in nutrients. Consequently the lake is strongly eutrophic. Because it is shallow and open, the winds blowing over the lake surface are sufficient to ensure thorough mixing of the water. These factors, together with the lake's warmth, lead to a high biomass of algae. The algae are so abundant that those lower down receive insufficient light for photosynthesis, because of shading by the ones above them. The euphotic zone is only 50 cm deep, and were it not for currents within the lake turning over the water, the algae would be confined to that zone.

Most of the algae belong to the blue–green groups, as is typical of eutrophic lakes; in oligotrophic lakes the phytoplankton consists mainly of green algae. The predominant genus of algae in Lake George is *Microcystis*, which forms the principal food of various planktonic animals, especially crustaceans, and some fish, mainly *Tilapia nilotica* and *Haplochromis nigripinnis* (Figure 3.17). This herbivorous food chain supports various species of predatory fish, e.g. the lungfish *Protopterus* and catfish *Clarias*. Local fishermen catch a high proportion of the total fish production.

The algae are easily caught by the filter-feeding herbivores, which literally live in a soup of food. This accounts for the high proportion of NPP going to herbivores, and correspondingly less to the detritivores (which mostly live in the mud at the bottom). The overall yield to man from Lake George amounts to about 100 g fresh weight (about 25 g dry weight) of fish per year for every square metre of the lake surface. Yields from other African lakes are much lower, mostly falling within a range of 0.4 to 50 g fresh weight per year.

Figure 3.17 Energy-flow in the aquatic ecosystem. The data are from Lake George, which lies on the equator as it crosses the western Rift Valley in Uganda. Each box represents a trophic level, except that carnivores and top carnivores are combined. 'Decomposers' refers to all of the decomposer groups except detritivores (cf. Figure 3.10), of which the primary decomposers are the most important. Figures within the boxes show biomass B in dry g m^{-2}, and production P in kJ and dry g m^{-2} yr^{-1}. Energy-flow from one trophic level to the next is indicated by arrows; the figures are again kJ m^{-2} yr^{-1}. (Absence of figures from some boxes and arrows indicate that no estimate was made.) The diagram is simplified in various ways; e.g. energy losses through respiration are not shown, nor are the organisms which feed on detritivores.
(Based on data from Whittaker, 1975, reference in chapter 1.)

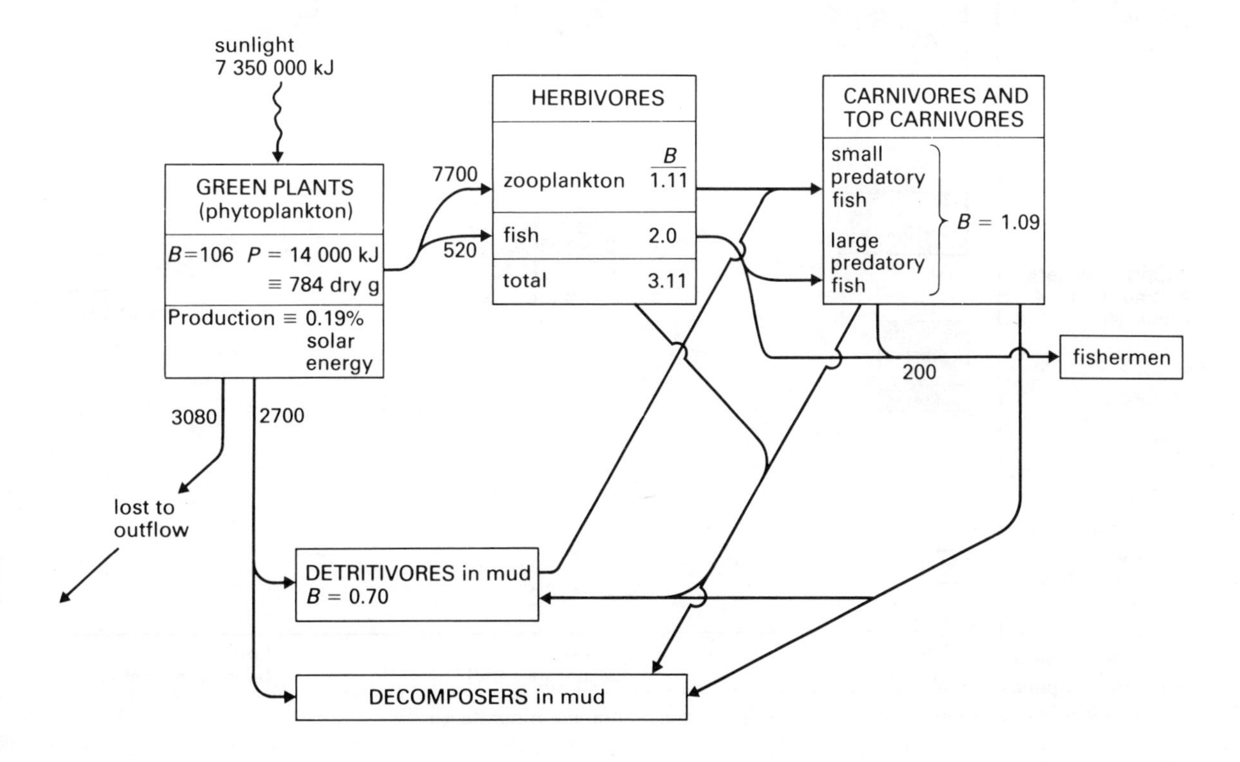

Terrestrial ecosystems

Terrestrial ecosystems have received rather more attention than aquatic ones. Estimates of NPP have been made for all of the major biomes of the world, and some of these are represented in Figure 3.18. They reflect the dependence of plant growth on temperature, water and nutrients (pages 39–42, 74–79).

A few African ecosystems have been investigated in some detail. Two of the most complete studies are considered here. Both are wooded savannas, but they differ markedly from each other.

At Lamto, in the Ivory Coast (Figure 3.19) more than 70 per cent of the herb layer consists of two species of *Hyparrhenia* grass, which are 2 to 2.5 m tall when flowering. There are also scattered shrubs and trees, mainly belonging to the genera *Annona*, *Bridelia* and *Crossopteryx*. A mean annual rainfall of about 1300 mm results in a high value for NPP. However, much of the vegetation is extremely tough. During the study, the estimated offtake by all herbivores was a mere 2 per cent, a very low figure. A further 11 per cent was lost by fire, the remaining 87 per cent going to the detritivores and other decomposers, of which the former took about 20 per cent. There was a considerable standing crop of humus, amounting to about 5000 g m^{-2}. In terms of biomass, the main group of carnivores comprised arthropods, especially spiders; vertebrates contributed only 20 per cent or so. However, the carnivores contributed only a very small fraction of the biomass and production of the ecosystem as a whole.

Amongst the detritivores, termites were estimated as consuming about 120 g m^{-2} of litter each year, whilst earthworms took another 100 g m^{-2}. The worms also moved incredible amounts of soil – calculated to be 100 kg m^{-2} yr^{-1}, equivalent to a layer about 5 cm thick over the complete surface! Of this, 2.5 kg m^{-2} yr^{-1} was actually brought to the surface as worm casts, together with piles of soil from the foraging passages of ants and termites.

Figure 3.18 Above-ground NPP of various terrestrial biomes. The figures (which are approximate modes) are in dry g m^{-2} yr^{-1}; but there is considerable variation. Thus although the value of 2200 is given for tropical rainforests, observed values actually varied from approximately 1000 to 6000 dry g m^{-2} yr^{-1}, and there are similar ranges for other biomes. Nevertheless, the combined effects of nutrients, water and temperature on NPP are very clear. (Based on data from Whittaker, 1975 reference in chapter 1.)

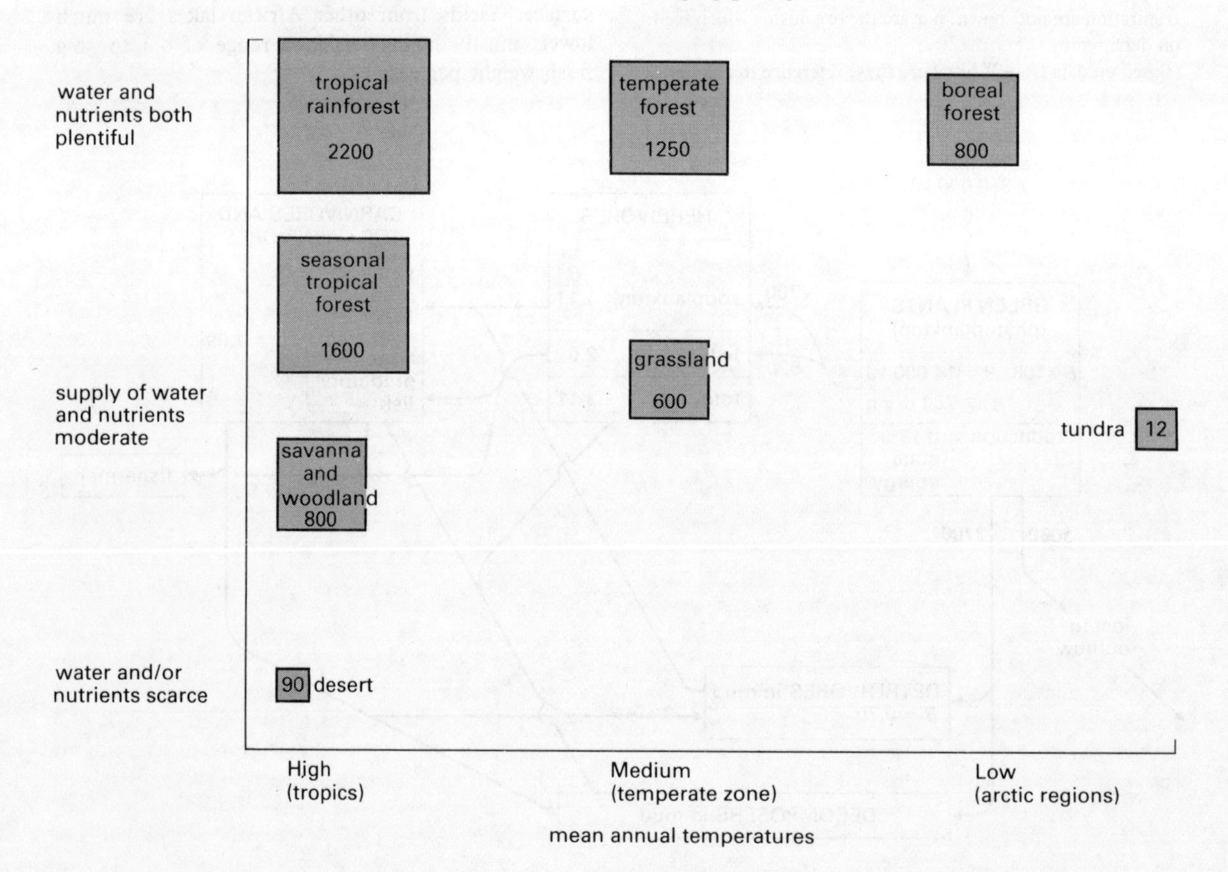

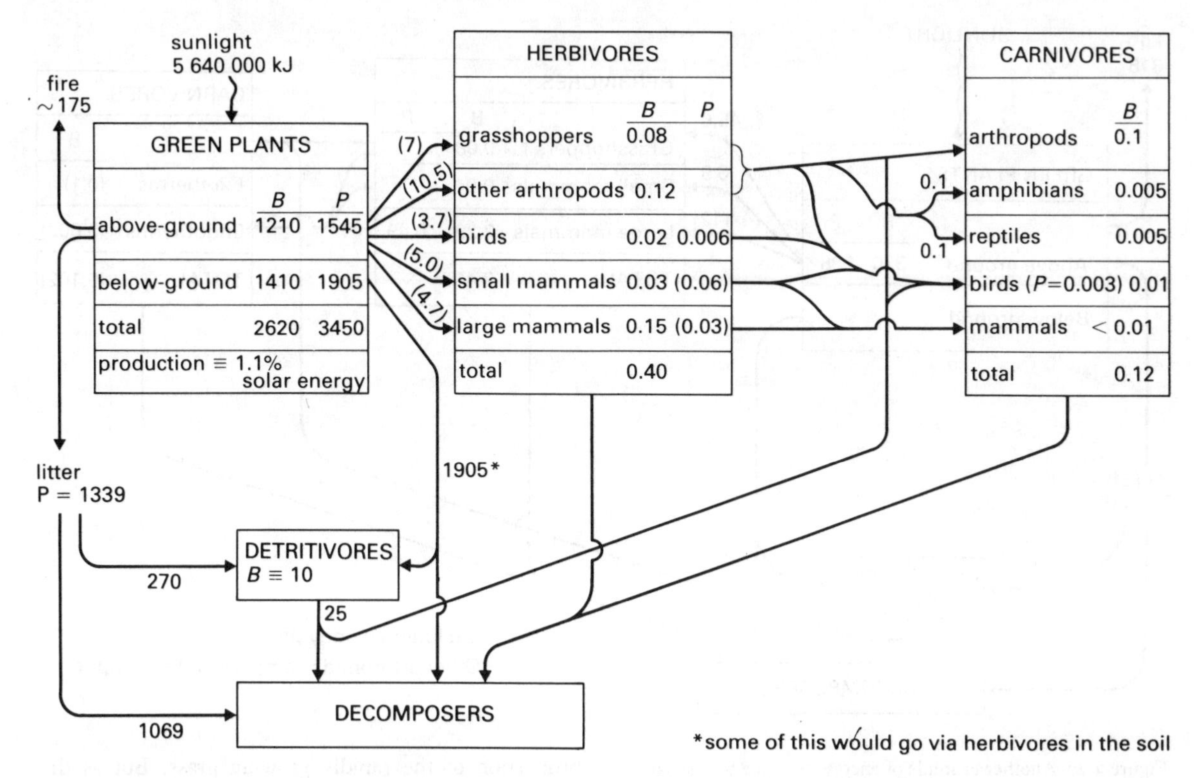

Figure 3.19 Energy-flow in a terrestrial ecosystem at Lamto, in the moist savanna of the Ivory Coast. Data are given for biomass B in dry g m^{-2}; and for net production P in dry g m^{-2} yr^{-1} (cf. Figure 3.17). Figures alongside arrows show partitioning of the production in some cases. Those in brackets are estimates based on the assumptions of Phillipson, 1973 reference in figure 3.20. (Absence of a figure indicates that no value was obtained.) A few minor pathways are omitted. As with Figure 3.17, the term 'decomposers' includes all groups except detritivores. The estimates of total primary production or a proportion of solar energy received assumes that one group of plant tissue is equivalent to 18 kJ. (Based upon data in the 'Bulletin de liaison des chercheurs de Lamto', parts I to V publ. by ORSTOM, 1974, Paris.)

The Lamto study also revealed that below-ground primary production exceeded that above ground. However, it was difficult to find out what happened to it; some was eaten by herbivores such as nematodes and various arthropods, and by mammals such as voles and rodents. But in all probability, most went to detritivores and other decomposers.

On the Serengeti Plains of northern Tanzania, the mean annual rainfall is only about 470 mm, and consequently the NPP above ground is much less than that at Lamto. In the tall grasslands of Serengeti, where the grass is about a metre high, the main grass species are *Themeda triandra*, *Sporobolus pyramidalis* and *Pennisetum mezianum*. The first two of these are relatively palatable, especially when young. There are scattered trees, mostly species of *Acacia* and *Commiphora*. During the study period, an average of about 62 per cent of the land was burnt each year, and it was estimated that fire consumed about 53 per cent of the NPP above ground (Figure 3.20). Since the herbivores took a further 27 per cent only about 19 per cent remained for the detritivores (mainly termites) and decomposers.

The major difference between Lamto and Serengeti was in their populations of large mammals. Although the P : B ratio for plants above ground in Serengeti (1.62) was not much higher than in Lamto (1.28), the Serengeti Plains supported more than ten times the biomass of mammals. In part, this was because the Serengeti mammals were protected (it is a National Park), but the palatability of the forage must have been a factor too (cf. Figure 3.7). There were formerly quite large populations of elephants, hippopotamuses and other ungulates at Lamto, but although the area is now a reserve, there are very few large mammals since they are vigorously hunted in the surrounding areas. Nevertheless, the Lamto study is of special interest because of its relative completeness.

The total biomass of animals at Lamto (most of which were detritivores) came to about 10.5 g m^{-2} which was only 0.87 per cent of the above-ground plant biomass. In Serengeti, the total animal biomass

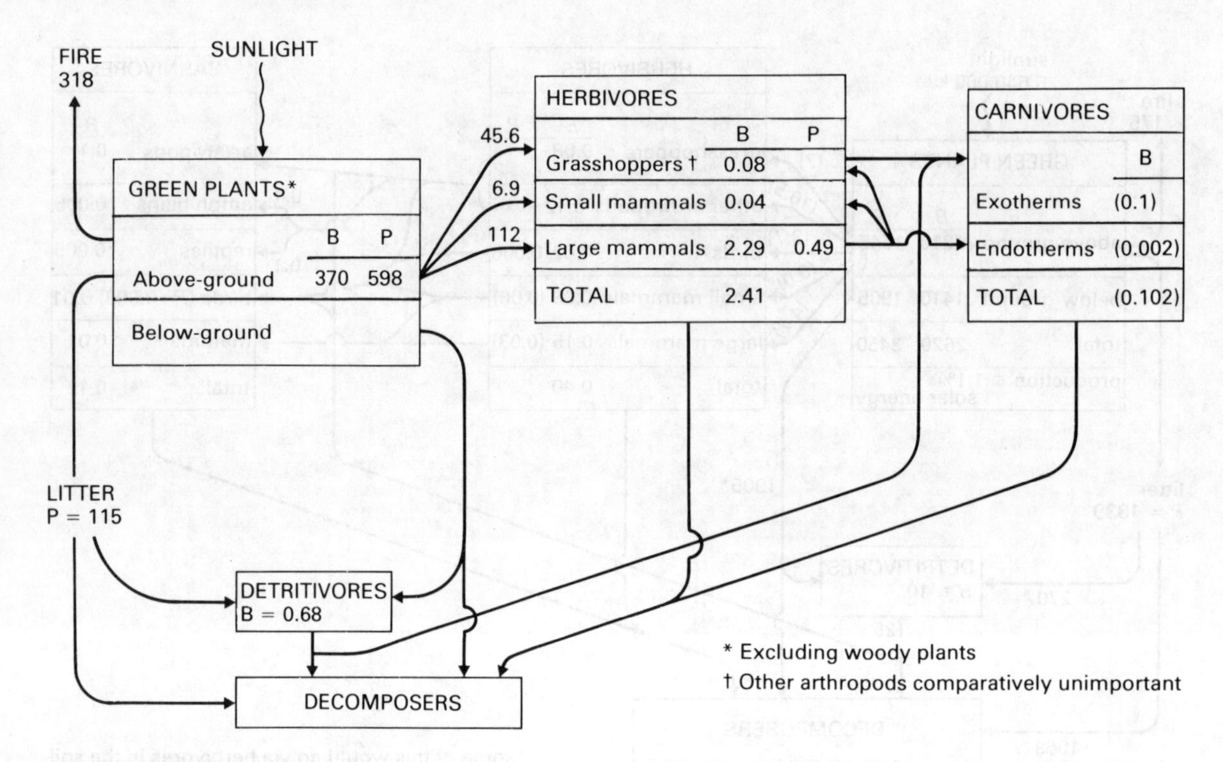

Figure 3.20 Another example of energy-flow in a terrestrial ecosystem, here the tall grasslands of Serengeti National Park, northern Tanzania. Biomass B is in dry g m^{-2}; and production P in dry g m^{-2} yr^{-1}. Data were less complete than those used for Figure 3.19; for example, the only arthropod herbivores included were grasshoppers (the most important group), and although there were some woody plants, there are no data for them, or for below-ground production. There was also considerable variation over a period of years. Comparison with Figure 3.19 nevertheless shows some interesting differences, notably in the biomass of mammalian herbivores.
(Figures in brackets based upon assumptions of Phillipson, J., 1973, pp. 217–35 in J. G. W. Jones (ed.) *The Biological Efficiency of Protein Production*, Cambridge University Press, all other figures based on Sinclair, A. R. E., 1975. *Journal of Animal Ecology*, **44**, 497–520.)

was about 3.2 g m^{-2}, which at 0.86 per cent of the above-ground plant biomass was an almost identical proportion to that at Lamto. The large biomass of detritivores at Lamto seems to have compensated for the lack of mammalian herbivores there.

The Serengeti is also unusual in that its mammal population is exceptionally large. A possible explanation is that most of the large mammal biomass in the grasslands (about 80 per cent of it) consists of two migratory species, the wildebeest and the common zebra. These two arrive in such large numbers early in the wet season that they are able to consume a high proportion of the rapidly growing grass. But as the grass dries out they move to other parts of the park where the rainy season comes later (pages 170, 172). Other species, such as elephants, topi and impala are mainly sedentary. They remain in the long grasslands throughout the year, but are relatively few in numbers. It is unlikely that the area could support such a high biomass of large mammalian herbivores if none of them migrated, for although there would be an excess of food in the wet season, severe shortages would follow as the dry season progressed.

Despite the big differences in their mammalian populations, the Serengeti and Lamto ecosystems have much in common. It is thus possible to use data from these to construct a model of a generalized grassland ecosystem, and this has been done in Figure 3.21 for a hypothetical area with a mean annual rainfall of 600 mm. The expected above-ground NPP corresponding to this is about 1000 dry g m^{-2} yr^{-1} (Figure 3.5). Construction of the model necessitates making a number of assumptions, but nevertheless the figures shown in the model represent the approximate relative values to be expected. They would vary proportionately to the NPP, and they might not apply so well to forested and thickly wooded ecosystems.

A model such as this can be used to predict likely values in an ecosystem for which only some of the components have been studied. For areas of higher or lower rainfall, appropriate values of expected NPP can

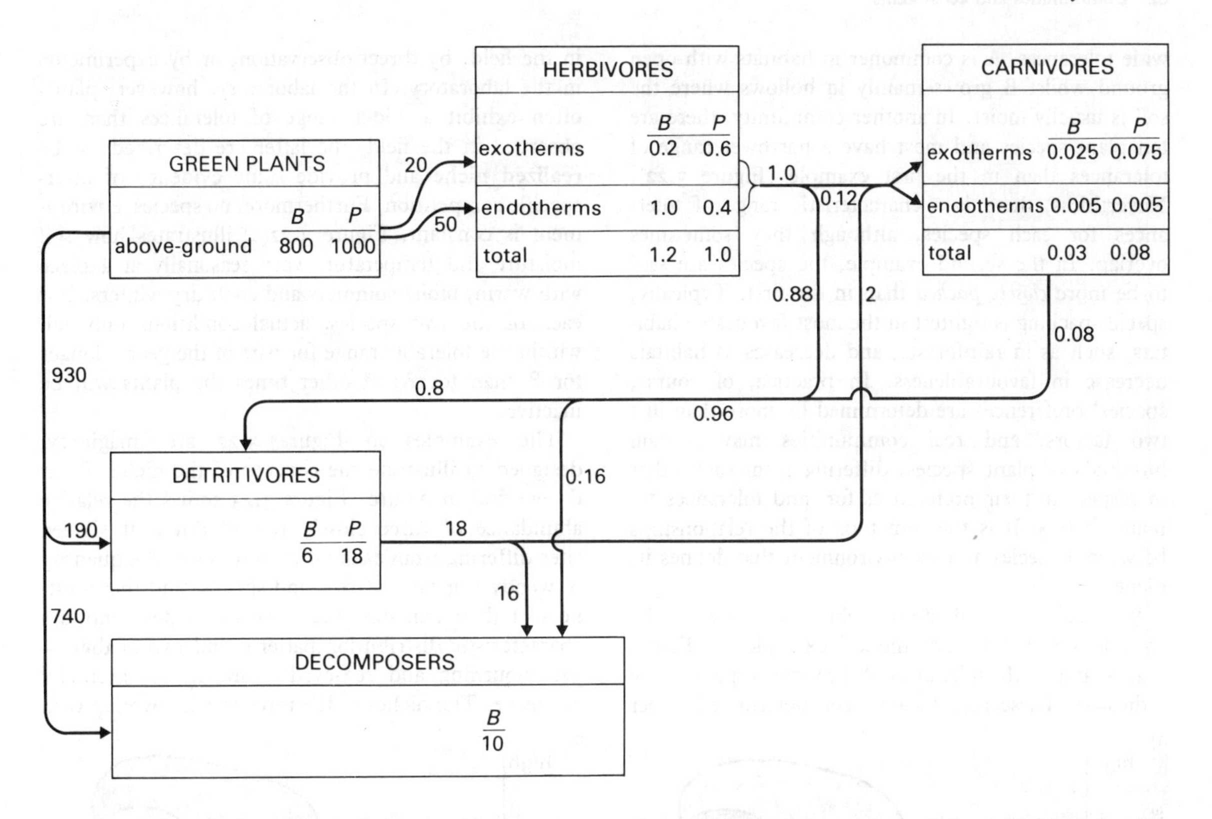

Figure 3.21 A model for a hypothetical terrestrial ecosystem in tropical Africa, in an unburnt natural or man-modified savanna with some woody vegetation, and receiving about 600 mm yr^{-1} of rain. Biomass B is in dry g m^{-2}; and production P in dry g m^{-2} yr^{-1}. The ratio of $B : P$ in green plants would decrease (and the biomass would increase) with greater amounts of woody vegetation. Below-ground production is omitted, because of lack of data. It is assumed that there is no burning (if there was, then correspondingly less of the NPP would go to the decomposers and detritivores). Similarly, as the proportion of unpalatable vegetation increases, so the fraction taken by herbivores decreases, whilst that going to detritivores and decomposers increases.

be predicted from Figure 3.5, at least within the range of about 200 to 1500 mm of rain per year.

Communities

Ecosystems sometimes contain very large numbers of species and this adds to the difficulties of studying them as whole systems. But the organisms of a particular group, such as flowering plants or beetles, are more manageable, and these **communities** are often taken as units for study (cf. page 35–37). Thus every species can be considered as belonging to a community, as well as to the whole ecosystem. We will first consider the position of the individual species, and will then examine some features of communities.

Species in communities

Communities frequently comprise many species, and a question that has long intrigued ecologists is 'how do they all coexist? – are their relationships harmonious or are they all competing with each other?' The results of many studies, covering both natural and semi-natural ecosystems, suggest that there is competition between species in nature, but perhaps less often than one might expect. We shall return to this subject later (pages 114–115, 118–125 and 195–197); in this section we examine some ideas on the broader question of how the various members of a community 'fit in'. To put it another way, how are the available resources shared between them?

We introduced the term **niche** on page 37, saying that it described an organism's role in the ecosystem (and thus also in the community). If we could show that different species occupied different niches, that would give us some understanding of how communities are made up, and at the same time see how competition within the community is minimized.

Imagine a simple plant community, with only two species. One, A, prefers warm, dry places whilst the other, B, thrives in cool, moist places. Both have fairly wide tolerances, A is commoner in habitats with open ground whilst B grows mainly in hollows where the soil is usually moist. In another community, there are ten plant species and most have a narrower range of tolerances than in the first example (Figure 3.22). Nevertheless, there is a characteristic range of tolerances for each species, although they sometimes overlap. In the second example, the species are said to be more *closely packed* than in the first. Typically, species-packing is tightest in the most favourable habitats, such as in rainforests, and decreases as habitats decrease in favourableness. In practice, of course, species' preferences are determined by more than just two factors, and real communities may contain hundreds of plant species, differing from each other in respect to their preferences for, and tolerances to, many factors. It is the sum total of the relationships between a species and its environment that defines its niche.

We can best illustrate what this means by considering another hypothetical example. In Figure 3.22(a) and (b) the tolerances of the various species were indicated. These might have been determined either in the field, by direct observation, or by experiments in the laboratory. In the laboratory, however, plants often exhibit a wider range of tolerances than are observed in the field; the latter are described as the **realized niche** and provide some evidence of interspecific competition. Furthermore, no species' environment is constant. Figure 3.22(c) illustrates how soil moisture and temperature vary seasonally at a place with warm, moist summers and cool, dry winters. For each of the two species, actual conditions only fall within the tolerable range for part of the year – longer for B than for A; at other times the plants will be inactive.

The examples in Figure 3.22 are imaginary, designed to illustrate the concept of the niche. What do we find in nature? Figure 3.23 shows the relative abundance of three closely related doves at sixteen sites differing from each other in terms of the quantity of woody vegetation (trees and shrubs) and the moistness of their climates. Each species of dove shows a characteristic distribution pattern, and two of them – the mourning and red-eyed – are almost mutually exclusive. The niches of the three species overlap with

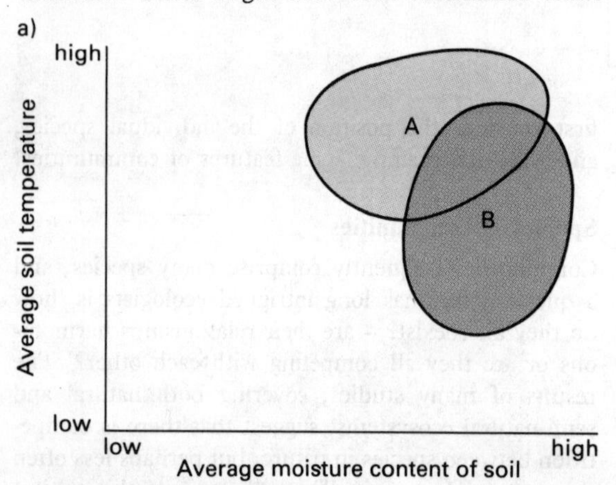

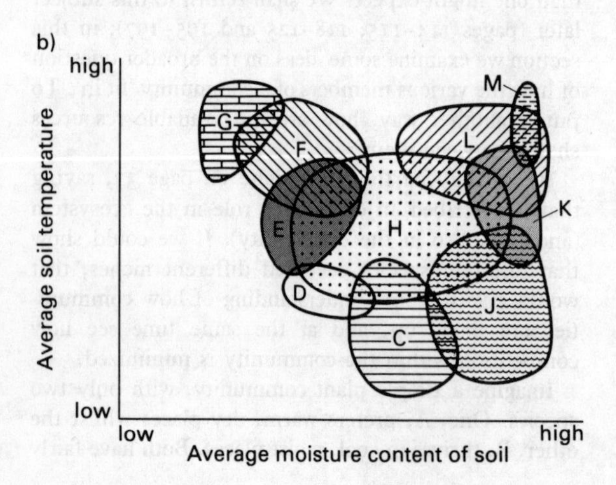

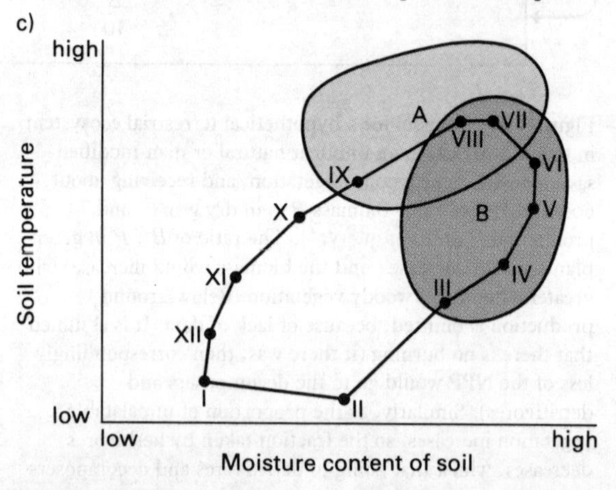

Figure 3.22 The range of tolerances of plants to various levels of soil moisture and temperature in two hypothetical communities. The elipses encompass those combinations of moisture and temperature which allow the various plant species to grow actively.

Community (a) has only two species, both with quite wide tolerances, whereas (b) contains ten species, more *closely packed* with respect to their tolerances. Consequently, despite overlaps, each species has a unique combination of preferences. Some of them, e.g. D and M, only exist in places experiencing a narrow range of average values for temperature and moisture. Others, notably H, have a wider range of tolerances, in fact similar to plant B above. Figure (c) shows the same two-species community as (a) but here the mean monthly values of temperature and soil moisture are shown for a particular place. (I, II, III. . . . indicate January, February, March . . .)

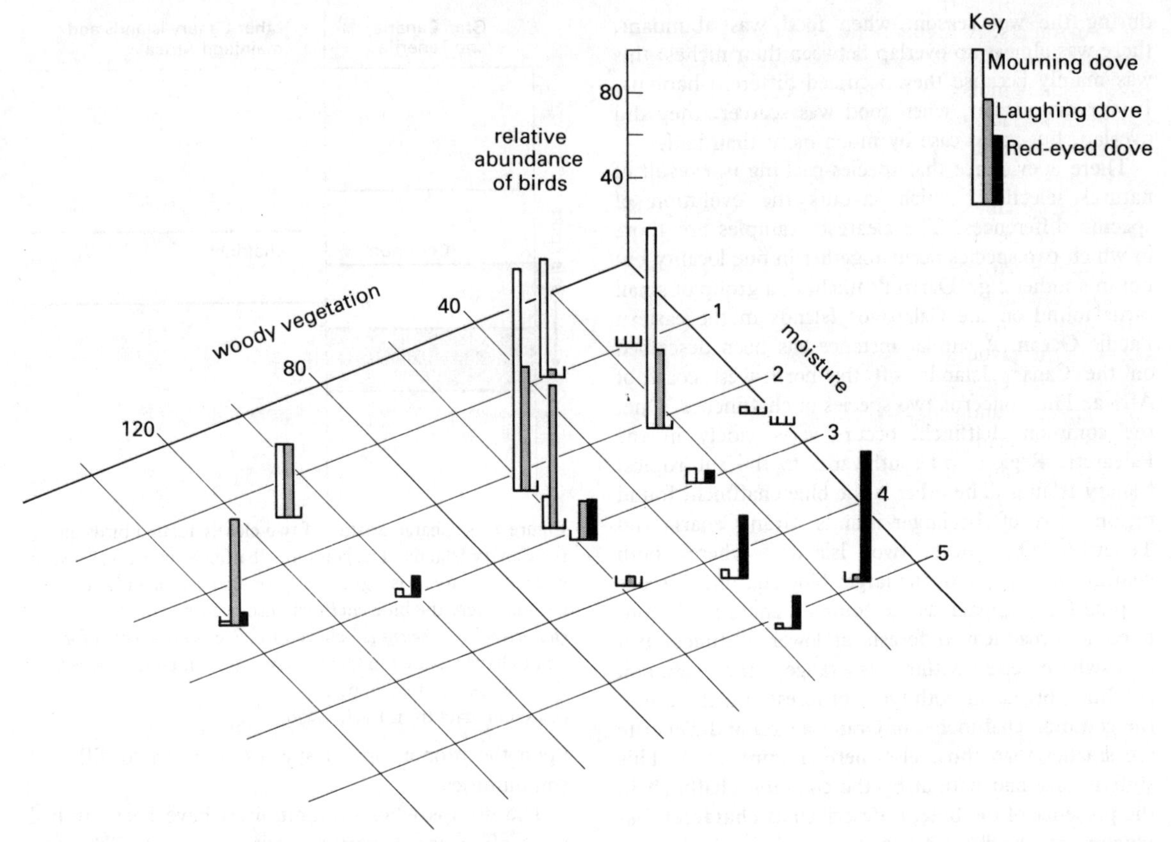

Figure 3.23 The relative abundance of three species of doves at places varying in their climate and vegetation. All three species are widespread in tropical Africa, both west and east; the data here come from Kenya. The index of woody vegetation measures the sum of the percentage cover in each of several layers, whilst the moisture index (Bailey's) combines rainfall with temperature.
(D. E. Pomeroy and B. Tengecho, unpublished.)

respect to the two environmental variables represented in Figure 3.23, but as we have already mentioned – and shall discuss in detail in chapter 5 – there are many environmental variables.

To represent just two environmental variables, together with the species' abundance, requires a three-dimensional diagram such as in Figure 3.23, but theoretically a niche can have many dimensions, even though we cannot plot them on a piece of paper.

Another study, involving four species of browsing mammals characteristic of the drier regions of eastern Africa, considered three niche dimensions. Each species was compared to each of the others with respect to their food, the height at which they fed, and their precise habitat. These three niche dimensions were then combined into a single measure of niche overlap, which was calculated for both wet and dry seasons. The result (Figure 3.24) shows that

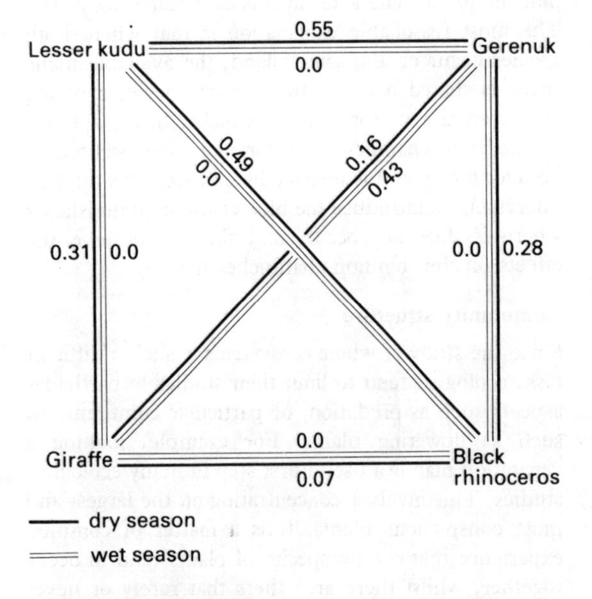

Figure 3.24 Measures of niche overlap between four species of browsers in wet and dry seasons. Zero represents no overlap: the maximum score is 1.0.
(Based on data in Leuthold, W., 1978 *Oecologia*, 35, 241–252.)

during the wet season, when food was abundant, there was almost no overlap between their niches; this was mainly because they occupied different habitats. In the dry season, when food was scarcer, they did overlap, but in no case by much more than half.

There is evidence that species-packing is a result of natural selection, which favours the evolution of specific differences. The clearest examples are those in which two species occur together in one locality but not in another e.g. 'Darwin's finches', a group of small birds found on the Galapagos Islands in the eastern Pacific Ocean. A similar instance has been described on the Canary Islands, off the north-west coast of Africa. This concerns two species of chaffinches. One, the common chaffinch, occurs very widely in the Palearctic Region and southwards to the subtropical Canary Islands. The other is the blue chaffinch, found on only two of the larger islands, Gran Canaria and Tenerife. On these two islands, where both chaffinches are found, the larger blue chaffinch breeds in pine forests, whereas the common chaffinch is confined to broad-leaved forests at lower altitudes. But everywhere else within its range, the common chaffinch breeds in both types of forest. Furthermore, the common chaffinches of Gran Canaria and Tenerife are smaller than those elsewhere (Figure 3.25). This shift in size and habitat by the common chaffinch in the presence of the blue, is described as **character displacement**, the character in this case being body size. The most reasonable explanation is that where both species occur on the same island, the available niche space is shared between them, each species evolving characteristics appropriate to its niche. In the absence of the blue chaffinch, the common one occupies a broader niche, and is intermediate in size. It would be interesting to introduce the blue chaffinch onto islands where it does not occur, and then to observe the effects on the common chaffinches there.

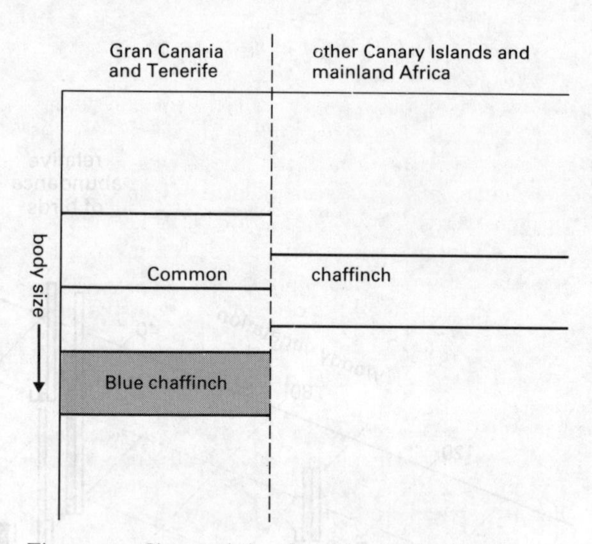

Figure 3.25 Characteristics of two closely related birds on the Canary Islands. The common chaffinch, which has a very wide distribution, is significantly smaller on the only two islands where the blue chaffinch also occurs. This *character-displacement* is believed to have been selected for and to have resulted in less interspecific competition where the two species are sympatric.
(Based on data from Lack, 1976.)

Community structure

Since the study of whole ecosystems is such a difficult task, ecologists tend to limit their studies to particular aspects such as predation, or particular communities, such as flowering plants. For example, making a vegetation map is a useful first step in many ecological studies. This involves concentrating on the largest and most conspicuous plants. It is a matter of common experience that certain species of plants tend to occur together, whilst there are others that rarely or never do so. The plant ecologist can build up patterns of plant communities in this way, based on **associations** between species. There are quite sophisticated numerical analyses available for distinguishing and delimiting communities. These are mainly based upon the proportions of species common to different communities.

The groups whose communities have been studied most often are flowering plants and birds. We shall mainly use these groups to illustrate some interesting generalisations about communities.

Commonness and rarity

In almost all plant communities, whether natural or altered by man, one finds that there is a small number of common species and a much larger number of uncommon ones. (Botanists tend to use the term 'dominant' for the commonest or most conspicuous species; it should be noted that dominance has another meaning when applied to animals – see pages 137–138.) A similar situation exists with animals, and Figure 3.26 shows an example of the inverse relationship between numbers of species and their relative abundance for two bird populations. Notice, in the example illustrated in the figure, how changes in the plant community, as land was cleared for cultivation, also led to changes in the bird community too. Data of this sort have given rise to numerous mathematical models which are of special interest in the field of evolutionary ecology.

Species richness and species diversity

The number of species of plants (or beetles, or frogs or almost any other group) in a square kilometre of forest, is many more than in a similar area of desert. A more concise statement would be that the species richness of forests is higher than that of deserts, and this is taken into consideration in the concept of **species diversity**, which embraces both number of individuals per species and number of species. For instance, a community of 100 individuals belonging to only two species is considered to be less diverse than another, also of 100 individuals, but comprising twenty species. A third community, with only fifty individuals of twenty different species, is more diverse still. There are several ways of quantifying this relationship; two widely used ones are shown in Box 3.4.

Figure 3.26 Commonness and rarity of birds in two adjacent communities near Kibwezi, Kenya. (a) is an area of natural bushed woodland, whilst in (b) all shrubs and most trees have been removed to allow for cultivation. In each habitat, birds were counted fifteen times along a fixed transect, 800 m long and 20 m wide. The frequency diagram shows the number of species represented by one, two, three, etc. individuals. Summing all the data for each area we have:

(a) uncleared – 74 species, 446 individuals
(b) cultivated – 50 species, 516 individuals.

More species were recorded from transect (a) but many were rare – half of them were represented by only one or two individuals. There were fewer rare species in transect (b), and more common ones: about one-third of the species had fifteen or more individuals, as compared to one-eighth for transect (a).

We can use the data in Figure 3.26 to see how habitat structure affects diversity. The diversity of birds in area (a) (uncleared bush) is higher than that of area (b) (cleared and cultivated land): the values of α reflect this clearly (Box 3.4). When the bush is cleared, the herb and shrub layers are completely removed, together with most of the trees. A new herb layer is formed by the farmers' crops, but it contains far fewer plant species than the natural plants it replaces, and its biomass is also lower. As a result, some species of birds disappear, and others become rarer, but a few increase. In the cleared land (b), there are fewer rare species and more common ones. Ten species are represented by twenty or more individuals, as compared to only four such species in area (a). The common species in area (b) are the ones attracted by the new food source – the farmers' crops; hence, some of them are pests.

The tropical regions of the world are, in general, much richer in species than the temperate and arctic regions, and their species diversity is correspondingly higher. There are, for example, many more species of

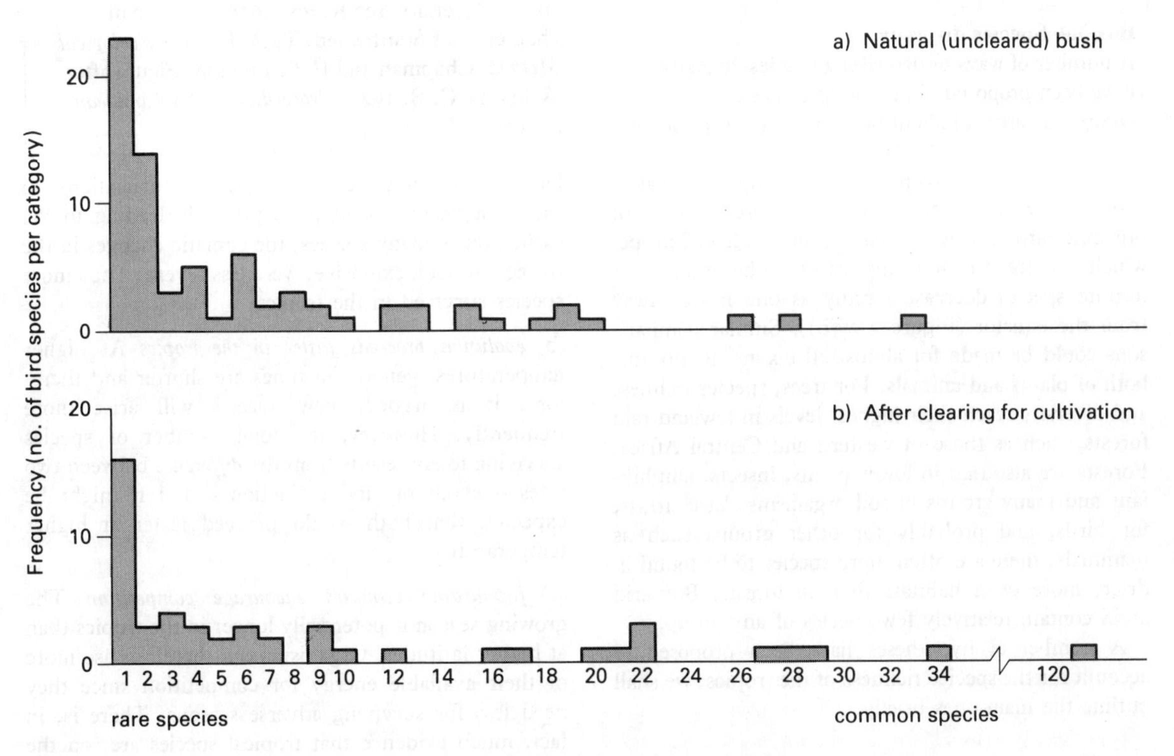

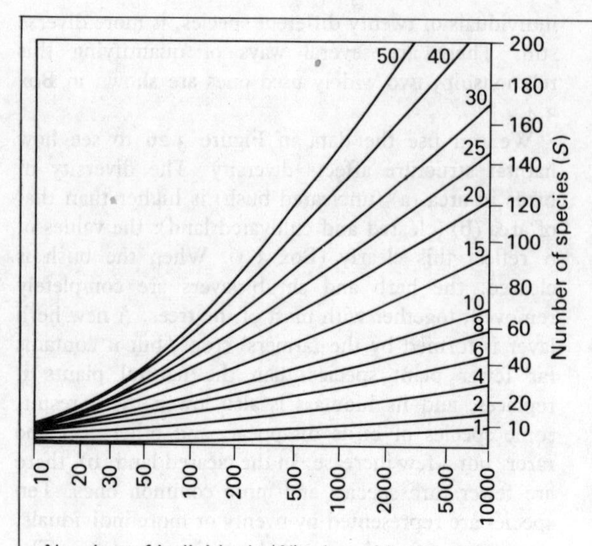

Chart for finding the Index of Diversity (α), which is shown as a series of curves. For example, taking the data in Figure 3.26, we have:

Area (a) $S = 74$ $N = 446$ $\alpha =$ approx. 25
Area (b) $S = 50$ $N = 516$ $\alpha =$ approx. 14

(Modified from Lewis T. and Taylor L. R. 1967, *Introduction to Experimental Ecology* Academic Press: London.)

Box 3.4 Species diversity
A number of ways of describing species diversity have been proposed. The simplest, when comparing areas of about the same size, is simply to take the total number of species (S) in each area; this is 'species richness'.

Two methods of evaluating diversity, which are almost independent of sample size, are the 'Index of Diversity' (α), and the 'Information Index' (*H*). (The second is also called the 'Shannon-Wiener Index'.) The Index of Diversity is more difficult to calculate, but approximate values can be read from the graph. The Information Index has been criticized by statisticians because it has no theoretical basis, but it is nevertheless widely used; it can be calculated from the formulae:

$$H = -\Sigma\ (p_i \log_2 p_i) \quad \text{(using logs to base 2)}$$

$$\text{or } H' = -\Sigma\ (p_i \log_{10} p_i) \quad \text{(using common logs)}$$

The term p_i means the proportion of the total population made up by any particular species. Suppose, for example, that we wish to calculate H' for a community containing ten individuals of three species, with six individuals of species A, three of species B, and one of species C. The respective values of p_i would be 0.6, 0.3 and 0.1. So we should have three values of ($p_i \log_{10} p_i$), namely (0.6×-0.222), (0.3×-0.523) and (0.1×-1.000). The total (Σ) of these terms is -0.3901; hence $H' = -0.3901$. For a more detailed discussion of species diversity, see Krebs, 1978 reference in chapter 1, or Southwood, T. R. E. 1978 *Ecological Methods* Chapman and Hall: London. Figure after Williams, C. B. 1947, *Proceedings of the Linnaean Society*, **158**, 104–8.

freshwater fish in tropical than temperate waters (Figure 3.27(a)), and there are more species of breeding birds in Nigeria than in the whole of Europe, which is more than ten times as big. The numbers of termite species decrease rapidly as one moves away from the equator (Figure 3.27(b)). Similar comparisons could be made for almost all taxonomic groups, both of plants and animals. For trees, species richness and diversity reach their highest levels in lowland rain forests, such as those of western and Central Africa. Forests are also rich in lower plants, insects, amphibians and many groups of soil organisms. In contrast, for birds, and probably for other groups such as mammals, there are often more species to be found in drier, more open habitats than in forests. But arid areas contain relatively few species of any group.

A number of hypotheses have been proposed to account for the species richness of the tropics; we shall outline the main ones briefly.

(a) *climatic stability* Whereas the temperate zones have been subjected to devastating glaciations in the geologically recent past, probably leading to the extinction of many species, the climatic changes in the tropics, though extensive, were less severe. Thus more species survived in the tropics.

(b) *evolution proceeds faster in the tropics* At higher temperatures, generation times are shorter and therefore, it is argued, new species will arise more frequently. However, the total number of species surviving today results from the *difference* between two rates – evolution and extinction – and it might be expected that both would proceed faster at higher temperatures.

(c) *favourable climates encourage competition* The growing season is potentially longer in the tropics than at higher latitudes. Organisms can therefore use more of their available energy for competition since they need less for surviving adverse seasons. There is, in fact, much evidence that tropical species are, on the whole, more specialized than temperate ones, i.e. the

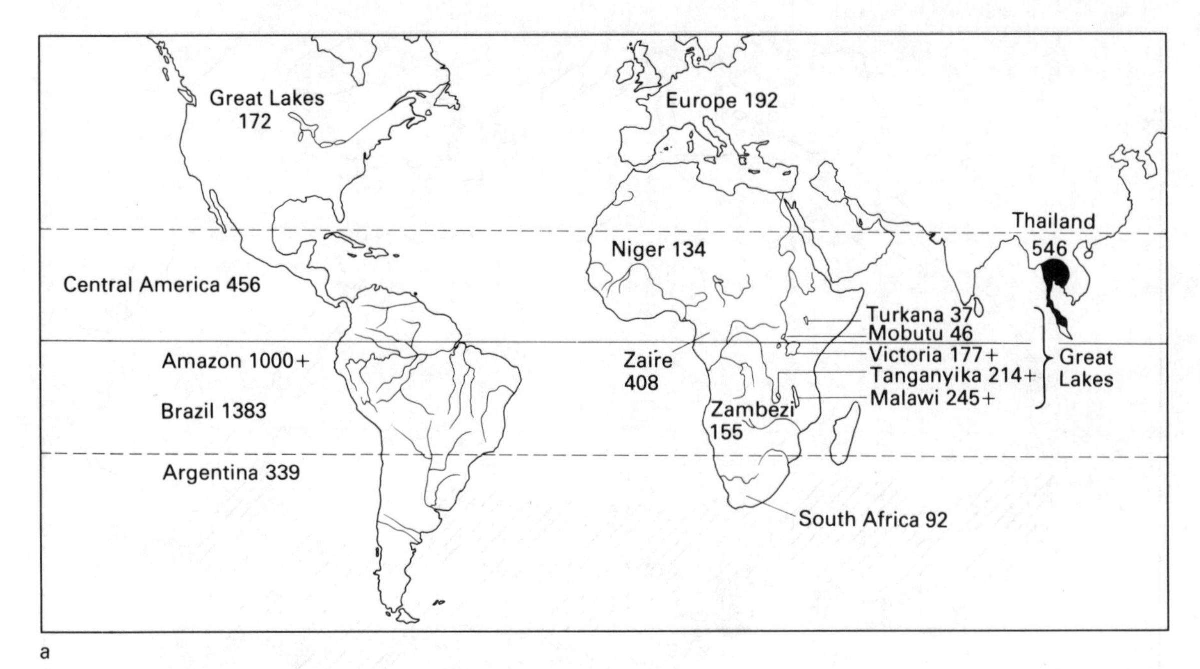

a

Figure 3.27 (a) Numbers of fish species for inland waters in various parts of the world. Although the regions considered vary greatly in extent, the pattern is quite clear: tropical waters (between the dashed lines) have far more species than non-tropical ones. A plus (+) indicates that more species are likely to be found; in the Amazon basin the final total will probably be double. (from Lowe-McConnell, R. H., 1969, *Biological Journal of the Linnean Society*, **1**, 51–75).

(b) The numbers of species of termites in a series of localities varying in distance from the equator. (Redrawn from Lepage, M., 1983. *Acta Oecologia/Oecologia Generalis*, **4**, 65–87.)

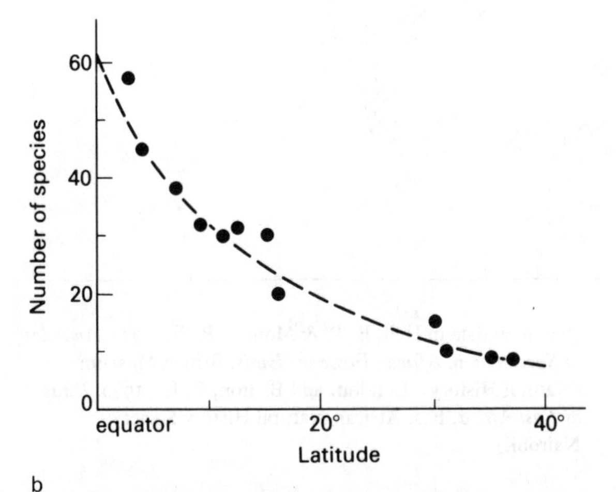

b

species packing is closer in comparable tropical habitats (cf. Figure 3.22). This situation could be explained in terms of competition. For instance, in tropical Africa there are many species of birds which are specialist feeders on nectar (e.g. sunbirds), fruit (e.g. parrots) and on mammalian carcasses (e.g. vultures). In Europe, birds with these specializations are found only in the most southerly parts; they are absent over most of the continent. Nectar, fruit and carcases occur there, and are eaten, but only as part of a more catholic (i.e. general) diet. Similarly, a variety of specialized mammals, characteristic of tropical Africa, have no counterparts at higher latitudes. Examples are pangolins, fruit bats, hyaenas, elephants and giraffes.

Many of these tropical specialists belong to what are known as **superspecies**: groups of two or more closely related species which often have non-overlapping distributions: that is, they are **allopatric** (Figure 3.28). This form of separation, with one species replacing another across the continent, is likely to have come about through comparatively recent evolution. When the distributions of two species overlap, they are said to be **sympatric**. When closely related species, such as the members of a superspecies, are sympatric, they are usually found to differ ecologically, for example by occupying different habitats. A likely explanation is that because they are very similar to each other, they would compete if they were sympatric, and one would then eliminate the other.

(d) *environmental variability* It has been suggested that there is a greater variety of habitats in the tropics than at higher latitudes. This results in a wide range of values for temperature, light intensity and moisture; and the availability of nitrogen, phosphorus and other

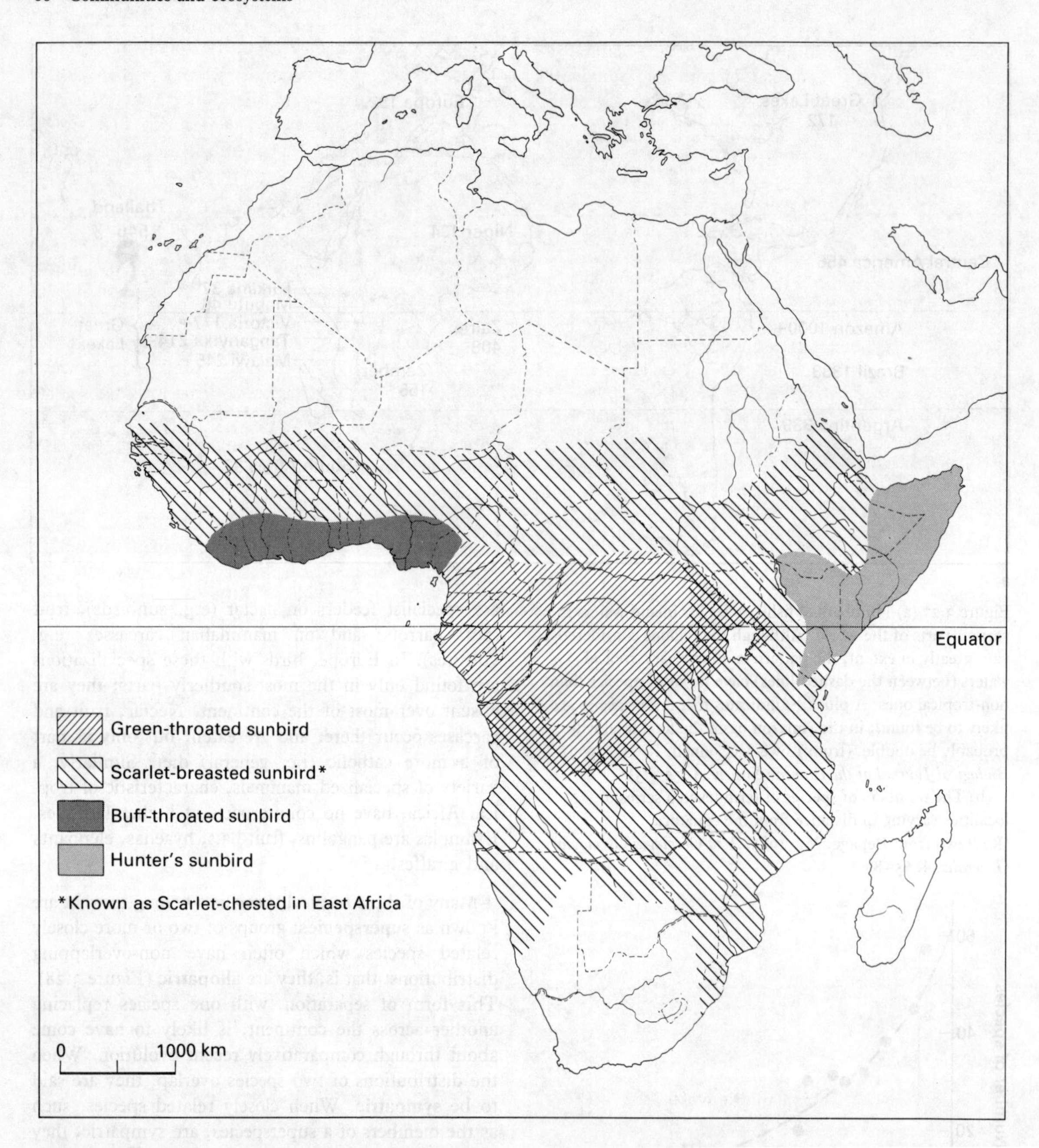

Figure 3.28 Four species of sunbirds, three of which belong to a superspecies, the exception being the scarlet-breasted sunbird. The members of the superspecies have distributions which meet but do not overlap – they are said to be *allopatric*. The scarlet-breasted sunbird, whilst closely related to the other three, has a distribution overlapping that of the green-throated in a broad zone of central Africa. Where these two occur together they are said to be *sympatric*. However, the scarlet-breasted is allopatric over most of its range with the green-throated, and completely so with the other two species.

(Based on data in Hall, B. P. & Moreau, R. E., 1970. *An Atlas of Speciation in African Passerine Birds*, British Museum (Natural History), London, and Britton, P. L., 1980. *Birds of East Africa*, East African Natural History Society: Nairobi.)

plant nutrients, all differ too, from one habitat to another. Different combinations of environmental conditions favour different plant activities – growth, seed production, germination and so on. For each species of plant there is an optimal combination of environmental variables for each activity (cf. Figure 3.23). Only when conditions are optimal will a particular species grow and reproduce more efficiently than others in its vicinity. The resulting diversity of plants could in turn help to explain the large number of animal species (cf. Figure 3.26).

What can we conclude? Over the years, each of these hypotheses has had its supporters. One difficulty in debating their relative merits is that each hypothesis is such a broad generalization as to be incapable of rigorous testing. (Some, however, can be shown to be true in particular aspects, e.g. the development times of insects certainly decrease with rising temperatures.) In all probability, a general theory attempting to relate species richness to latitude will need to incorporate all four hypotheses, although some may apply more widely than others.

Species–area curves

Since every community has a characteristic set of species, it follows that the species compositions of adjacent communities will differ, to some extent at least. We should therefore expect an area of (say) 100 km^2 to include more species than an area of

Figure 3.29 A species–area curve. The number of species of any particular taxonomic group is greater in larger than smaller areas, as might be expected. The graph shows the number of species of breeding birds in areas of various sizes. Note that both scales are logarithmic, giving rise to an apparently straight-line relationship. However, within the Afrotropical region (solid circles and fitted line) there are rather fewer species than in other tropical regions of the world, such as Ecuador, a South American country half the size of Nigeria, which has more species of breeding birds than the whole of the Afrotropical Region. The reasons for this remarkable difference are unknown, but similar differences are found in other communities, e.g. trees. (Based on data in Britton, P. L., 1980 – see reference in Figure 3.28; Moreau, R. E., 1966. *The Bird Faunas of Africa and its Islands*, Academic Press: London; van Tyne, J. & Berger, A. J., 1976. *Fundamentals of Ornithology*, Wiley: New York, and Welty, C., 1981. Birds, pp. 14–22 *Scientific American* book: San Francisco.)

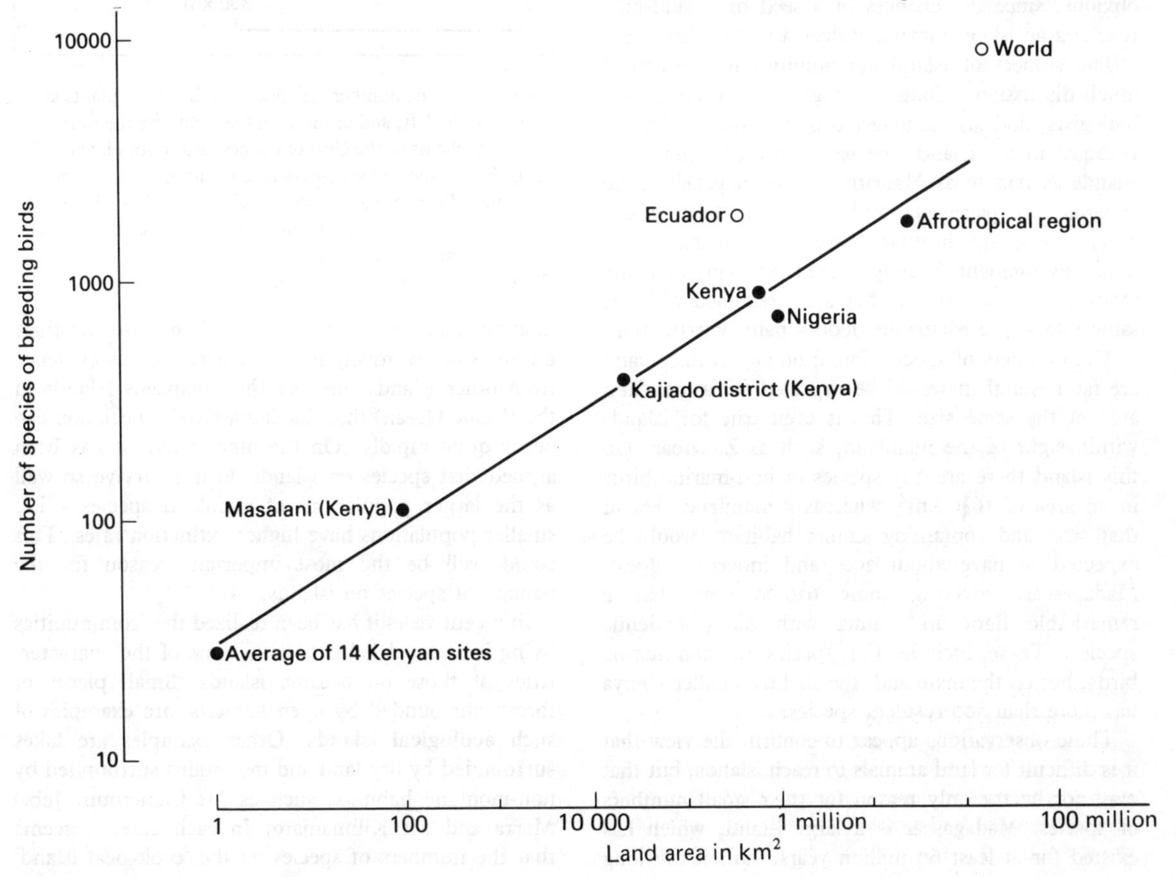

1 km^2, simply because the larger area is likely to embrace more communities. This indeed does happen, and in a characteristic way. Figure 3.29 shows an example of a **species-area curve** for birds. Similar results can be obtained for other groups, such as flowering plants, or butterflies.

Island communities

For a long time, naturalists have realized that oceanic islands have fewer species living on them than comparable areas of the mainland. Birds and flowering plants again provide good illustrations of this. Figure 3.30 reveals two trends in the numbers of plant and bird species on the islands in the Gulf of Guinea. Firstly, they decrease with distance from the mainland, and secondly there are more species on larger than on smaller islands. Thus the large, near island of Bioko (formerly called Fernando Póo) has seven times as many plants, and over twenty times more land-birds, than the small, remote island of Annobon. The phenomenon of fewer species on smaller islands is another instance of the species-area curves described previously. The decrease in numbers with remoteness is another matter. However, one explanation is fairly obvious, since the chances of a seed or a land-bird reaching an island inevitably decrease with distance.

The subject of island communities has provoked much discussion amongst ecologists and evolutionary biologists, and also amongst conservationists. This is because many island species, especially those on islands as remote as Mauritius or the Seychelles, are endemic (cf pages 93 and 94). But these island species seem much more susceptible to changes in their environment (mainly caused by man) than are mainland species. Hence they are threatened with the same fate as the Mauritius dodo – namely extinction.

The numbers of species found on an oceanic island are far fewer than would be expected for a mainland area of the same size. This is even true for islands within sight of the mainland, such as Zanzibar. On this island there are 105 species of non-marine birds in an area of 1640 km^2, whereas a mainland area of that size, and containing similar habitats, would be expected to have about 200 (and indeed it does). Madagascar, covering some 620 000 km^2, has a remarkable flora and fauna with many endemic species. These include 184 species of non-marine birds, but on the mainland, the slightly smaller Kenya has more than 800 resident species.

These observations appear to confirm the view that it is difficult for land animals to reach islands, but that may not be the only reason for their small numbers of species. Madagascar is a large island, which has existed for at least 60 million years, sufficiently long

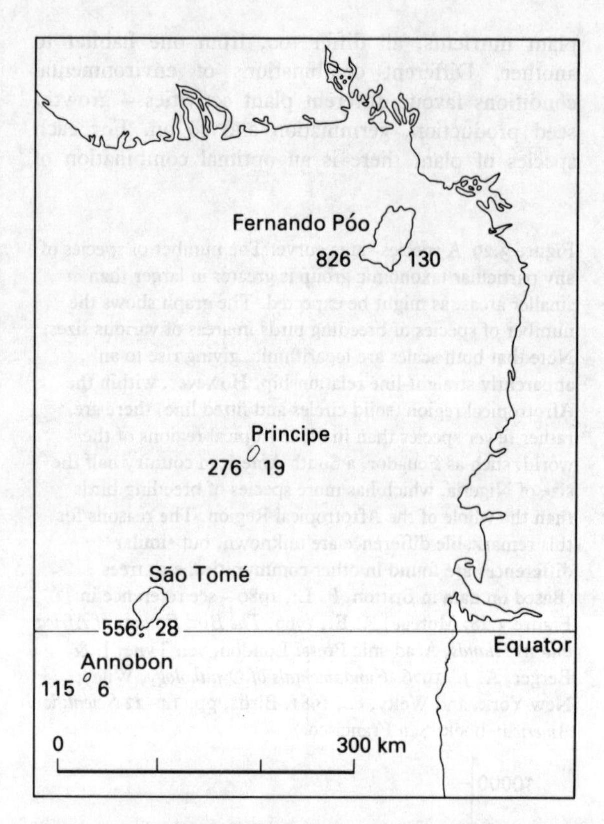

Figure 3.30 The numbers of species of flowering plants are shown on the left, and of birds on the right, for the four off-shore islands of the Gulf of Guinea. For both plants and birds the numbers of species decrease with distance from the mainland. There are also fewer species on smaller islands; the main reason for this is probably that smaller islands support fewer habitats than larger islands.
(Based on data from Lack, 1976.)

to allow many species to invade and colonize, yet their numbers are relatively few. Further, there is evidence from other islands (notably the Galapagos Islands in the Pacific Ocean) that, having arrived, speciation can occur quite rapidly. On the other hand, it has been argued that species on islands do not survive so well as the larger populations of mainland species – i.e. smaller populations have higher extinction rates. This could well be the most important reason for the paucity of species on islands.

In recent years it has been realized that communities living in isolated habitats have some of the characteristics of those on oceanic islands. Small pieces of forest, surrounded by open habitats, are examples of such **ecological islands**. Other examples are lakes surrounded by dry land and mountains surrounded by non-montane habitats, such as Mt Cameroun, Jebel Marra and Mt Kilimanjaro. In each case, it seems that the numbers of species on the 'ecological island' are related to its size, and distance from the next nearest piece of the same habitat. Obviously the ability of a species to disperse will affect its chances of colonizing island habitats. Amongst flowering plants, wind-dispersed seeds may occasionally be carried great distances, whilst insects and birds can fly from island to island. However, the fact that they *can* does not necessarily mean that they *do*; the behaviour of forest birds is such that they only fly within the forest, or across a quite short gap between one forest and the next. So for example, several species of birds found in the Usambara Mountains of northern Tanzania are absent from Mt Kilimanjaro, less than 200 km away, and from other mountains to the west and south. In this case the explanation is thought to be historical. During the great ice age, most African forests disappeared, because the glacial periods were generally dry. The last glacial ended some 10 000 years ago, since when the forests have spread. In Tanzania, the Usambaras remained forested during the last ice age, but Kilimanjaro did not. During the past 10 000 years, the forest has returned to Kilimanjaro, but this period has apparently not been long enough for some of the forest birds to recolonize it.

Studies of the kind described in this section have a practical importance in the field of conservation. For instance, the continual reduction in the size of indigenous forests, caused by their destruction, will presumably end when there are only a few scattered 'islands' of forest remaining. Since species survival on islands, and particularly remote islands, seems difficult, the extinction of many forest species seems almost inevitable (pages 212–213).

Suggested reading

Most of the topics discussed in this chapter can be found in the general ecological textbooks listed at the end of chapter 1 (p. 4), although few of their examples relate to Africa, for which Ewer & Hall, 1978, is useful.

Aquatic ecosystems are particularly well covered by Barnes & Mann, 1980; Beadle, 1981, and Symoens *et al.*, 1981, provide extensive data for African waters. Decomposition and related topics are described in some detail by Richards, 1974, and Swift *et al.*, 1979, they include some tropical situations.

Amongst studies on African ecosystems, two UNESCO publications, UNESCO, 1978, 1979, give wide-ranging reviews of forests and savannas respectively, the latter containing good accounts of both Lamto and Serengeti studies. The series 'Ecosystems of the World' contains several volumes of relevance to Africa; that on tropical savannas, edited by Bourlière, 1983, having chapters of a regional nature and others on the various plant and animal communities. The many years of work in Serengeti are nicely summarized by Sinclair & Norton-Griffiths, 1979.

Phillipson, 1966, provides a good introduction to ecological energetics: Brafield and Llewellyn, 1982, deal more broadly with animal energetics. More detailed accounts, especially on methodology, are found in the IBP (International Biological Programme) series of Handbooks; for example Milner & Hughes, 1968, on grasslands; Petrusewicz & Macfadyen, 1970, on terrestrial animals; and Grodzinski *et al.*, 1975, on ecological bioenergetics. The latter includes a table (pp. 277–78) giving calorific values for animals; data for plants and plant tissues appear in Leith & Whittaker, 1975. A useful source of information on nutrient cycling is the Scientific American book, 'The Biosphere', 1970.

Useful sections on the theory of niches are to be found in May, 1981 (chapter 8 by Pianka) and Putman & Wratten, 1984.

Important aspects of communities are discussed theoretically by Cohen, 1978, whilst islands and some related topics are well covered by Gorman, 1979, and Lack, 1976.

Essays and problems

1 What factors limit secondary production by herbivores?

2 Outline the hydrological cycle and discuss its significance to tropical Africa. Take account of both wet and dry regions.

3 Using Figure 3.21 as a guide, construct a possible energy-flow model in which a) the annual rainfall is 900 mm, and b) 30 per cent of the net primary production above ground is lost in fires. What would be likely to happen if you removed half of the mammalian herbivores, and all mammalian carnivores? (Use Figure 3.5 to estimate NPP.)

4 Give a detailed account of ways in which water may limit primary production.

5 Construct a food-web diagram for a natural or semi-natural ecosystem of which you have personal knowledge. Restrict yourself to between ten and fifteen of the commonest species of plants and animals. Which important aspects of your ecosystem are revealed by such a diagram; and which are not?

6 Discuss the usefulness of species–area curves in relation to birds, mammals and trees. For birds, refer to pages 64–69, for mammals to Western, D. & Ssemakula, J. 1981. *African Journal of Ecology*, **19**, 7–19, and for trees to Hall, J. B., 1981. *African Journal of Ecology*, **19**, 55–72.

7 Draw histograms to show the abundance of the three species of doves included in Figure 3.23 in relation to amounts of (a) woody vegetation and (b) moisture. Interpret the results in terms of niche theory.

8 Write an essay on the ecology of termites (suggested references Lee, K. E. & Wood, T. G., 1971. *Termites and Soil*, Academic Press, London, and Brian, M. V. (ed.), 1978. *Production Ecology of Ants and Termites*, Cambridge University Press, Cambridge are suggested references).

9 What are character displacements? By reference to Darwin's finches explain how they are believed to originate – and why theories as to their origin are difficult to prove. (Suggested references: Lack, D., 1947. *Darwin's Finches*, Cambridge University Press, Cambridge; and Lack, D., 1971, Chapter 11 *Ecological Isolation in Birds*, Blackwell: Oxford.)

10 Pages 53–61 deal with energy budgets for whole ecosystems, but it is also possible to construct an energy budget for an individual species. The essential quantities are those represented as C, A, R, P, F and U in Figure 3.9. Actual data for cattle and elephants are given by Golley, F.B. & Buechner, H. K. (eds) 1968. *A Practical Guide to the Study of the Productivity of Large Herbivores*, (p. 16), Blackwell: Oxford, and for young of impala and elephants by Delany & Happold, 1979, p. 310. Gill, F. B. & Wolf, L. L., 1975. *Ecology*, **56**, 33–45, give a very nice set of data for the golden-winged sunbird. Using one or more of these studies, discuss the concept of production efficiency.

11 Table 3.5 gives data on large mammals in two of six sample areas in a wildlife area in Cameroun. The duiker, giraffe and rhinoceros are browsers, whilst the diet of elephants can be considered as 25 per cent browse, with the rest mainly grass. The other species are grazers. Calculate the total biomass of grazers and browsers in each of the two areas. One of the two areas is well wooded, with *Terminalia, Isoberlinia* and other trees, and also contains many thickets. The other area is more open, having few trees, and is subject to more hunting. Which area is which? Comment on the differences in their faunas.

12 The Kwiambana Game Reserve in north-west Nigeria is mainly woodland, mostly dominated by species of *Isoberlinia*. Smaller areas of *Terminalia* wooded grasslands also occur, usually near to water-courses. Using indirect methods (such as recent footprints) as well as direct observation, counts were made of large mammals in both habitats; see Table 3.6. Calculate the densities (numbers of individual animals per km^2) for each species in each habitat, and the total density for each habitat. Also calculate H′, the information index of diversity, for each habitat (see Box 3.4). Comment on the differences in density and diversity of mammals in the two habitats.

Table 3.5 Wildlife in two areas of the Bouba Ndjida National Park, Cameroun

Species	Live weight (kg)	Number of animals km^{-2} Area I	Area IV
Elephant	1725	0.02	0.35
Rhinoceros	820	0.02	0.08
Warthog	45	0.51	0.22
Giraffe	750	0	0.19
Common duiker	10	0.37	0.40
Korrigum	100	0	0.33
Oribi	10	0.81	2.42
Reedbuck	40	0.65	1.85
Waterbuck	160	0.35	0.81

(Weights from Coe, M. J., *et al.*, 1976. *Oecologia* **22**, 341–54; animal numbers from van Lavieren, L. P. & Esser, J. D., 1980. *African Journal of Ecology*, **18**, 141–53.)

Table 3.6 Counts of animals in two habitats of Kwiambana Game Reserve, Nigeria

Species	Numbers of animals recorded *Isoberlinia* woodland	*Terminalia* wooded grassland
Porcupine	0	5
Baboon	115	10
Green monkey	21	0
Red patas monkey	15	0
Civet	5	0
Hyena	1	0
Aardvark	9	2
Elephant	4	23
Warthog	38	19
Bushbuck	1	0
Roan antelope	30	15
Kob	0	3
Hartebeest	72	24
Duiker	14	3
Oribi	1	3
Buffalo	1	0

The areas counted were as follows: *Isoberlinia* woodland, 213 km^2; *Terminalia* wooded grassland, 28 km^2. (Data from Ajayi, S. S. *et al.*, 1981. *African Journal of Ecology*, **19**, 295–98.)

References to suggested reading

Barnes, R. S. K. & Mann, K. H. (eds). 1980. *Fundamentals of Aquatic Ecosystems*, Blackwell: Oxford.

Beadle, L. C. 1981. *The Inland Waters of Tropical Africa*; (2nd edn), Longman, London.

Bourlière, F. (ed.). 1983. *Tropical Savannas*, Elsevier, Amsterdam.

Brafield, A. E. & Llewellyn, M. J. 1982. *Animal Energetics*, Blackie: Glasgow.

Cohen, J. E. 1978. *Food Webs and Niche Space*, Princeton University Press: New Jersey.

Ewer, D. W. & Hall, J. B. (eds). (1978) *Ecological Biology 2*, Longman: London.

Gorman, M. 1979. *Island Ecology*, Chapman & Hall: London.

Grodzinski, W., Klekowski, R. Z. & Duncan, A. (eds) 1975. *Methods for Ecological Bioenergetics*, Blackwell: Oxford.

Lack, D. 1976. *Island Biology*, Cambridge University Press: Cambridge.

Leith, H. & Whittaker, R. H. (eds). 1975. *Primary Productivity of the Biosphere*, Springer-Verlag: Berlin.

May, R. M. (ed.). 1981. *Theoretical Ecology – Principles and Application*, (2nd edn). Blackwell: Oxford.

Milner, C. & Hughes, R. E. 1968. *Methods for the Measurement of the Primary Production of Grassland*, IBP publication No. 6, Blackwell: Oxford.

Petrusewicz, K. & Macfadyen, A. 1970. *Productivity of Terrestrial Animals*, Blackwell: Oxford.

Phillipson, J. 1966. *Ecological Energetics*, Studies in Biology, No. 1, Edward Arnold: London.

Putman, R. J. & Wratten, S. D. 1984. *Principles of Ecology*, Croom Helm: Beckenham.

Richards, B. N. 1974. *Introduction to the Soil Ecosystem*, Longman: London.

Scientific American. 1970. *The Biosphere*, Freeman: San Francisco.

Sinclair, A. R. E. & Norton-Griffiths, M. (eds). 1979. *Serengeti: Dynamics of an Ecosystem*, Chicago University Press: Chicago.

Swift, M. J., Heal, O. W. & Anderson, M. J. 1979. *Decomposition in Terrestial Ecosystems*, Blackwell: Oxford.

Symoens, J. J., Burgis, M. & Gaudet, J. H. (eds). 1981. *The Ecology and Utilization of African Inland Waters*, UNEP: Nairobi.

UNESCO 1978. *Tropical Forest Ecosystems*, A state of knowledge report prepared by UNESCO/UNEP/FAO, UNESCO: Paris.

UNESCO 1979. *Tropical Grazing Land Ecosystems*, UNESCO: Paris

4 Aspects of plant ecology

We have introduced some of the main theories and ideas in ecology by using the large-scale approach of the ecosystem study. Now we begin an examination of the component parts of these large structural and functional units.

To start with, plants are examined in terms of their individual responses and adaptations to environmental change. This is followed by an investigation of the functioning of plants within populations, particularly in relation to evolutionary trends. The life strategies of plants are classified according to whether they invade new environments, compete in crowded environments or tolerate environmental extremes such as aridity or frost. We also consider the adaptations of plants that have apparently resulted from long periods of co-evolution with animals.

Following on from the introduction to communities in previous chapters, we look in detail at selected studies on plant communities in Africa, concentrating on the processes of primary production, phenology, succession and species diversity.

The ecologist is interested in whole organisms but, whether his background is primarily botanical or zoological, he can ignore neither plants nor animals. Both are integral components of ecosystems and each directly or indirectly influences the environment of the other. The separate treatment of plant ecology and animal ecology, like the many other subdivisions of the subject, assists in examining in detail the component parts of complex systems.

Plant ecology is the scientific study of the distribution and abundance of plants and their interactions with the environment. Within this definition of the subject there are two broad categories. **Plant autecology** which is the study of individuals or individual species, and is largely synonymous with the term **physiological plant ecology**, and **plant synecology** which is the study of populations, communities and ecosystems. This chapter broadly examines these different approaches in plant ecology, moving from the study of individuals to populations and then communities. Ecosystems were considered in chapter 3.

Plant autecology

Water relations of plants

The significance of rainfall as the main determinant of vegetation distribution in Africa was discussed in chapter 2. As we have seen, a large proportion of the land surface of Africa does not receive an abundance of rainfall, and in such areas plants have become adapted to compete for, and efficiently utilize, available water. Accordingly, plants can be classified into four groups as shown in Table 4.1.

Table 4.1 The categories of plants according to the water conditions to which they are best adapted

Hydrophytes Plants that live in water, either floating or submerged, e.g. bladderwort (*Utricularia*)

Hygrophytes Plants that live in damp situations such as forest floors or the edges of swamps, e.g. mosses and ferns.

Mesophytes Plants normally growing on well-drained soils that wilt when water is in short supply. Tend to be drought-evading, e.g. many tropical trees.

Xerophytes Drought-tolerating plants typical of deserts, dry grasslands and rocky places. Often succulent, spiny, with reduced leaves and deep roots, e.g. aloes, succulent euphorbias and ascelpiads

Because plants have not evolved cuticles that will allow a free passage of gases, they have to expose a large area of moist cell walls to the air to absorb sufficient CO_2 for photosynthesis. In higher plants this is the function of the stomata. However, water loss through transpiration is also one of the main ways by which plants keep themselves cool in hot environments (page 12). C_3 and C_4 plants with limited supplies of water and which grow in hot environments are faced with a dilemma: either they continue to photosynthesise which necessitates keeping their stomata open and thus losing water, or else they close their stomata, cease to photosynthesise and over-heat.

One solution to this problem has been the evolution of different photosynthetic strategies (Box 4.1). The energy from the sun required for photosynthesis is used as chemical energy in the form of ATP (adeno-

Box 4.1 The photosynthetic strategies that are used by different species of plants in tropical environments. Different biochemical processes lead to efficient photosynthesis according to temperature and moisture.

C_3 Photosynthesis

- Typical of temperate zone plants but widespread in wet equatorial areas and high altitudes in Africa
- Utilize the Calvin–Benson biochemical cycle, where phosphoglycerate (a 3C compound) is the first product of photosynthesis
- Examples found in many species, including *Agrostris* spp. *Hordeum vulgare, Helictotrichon* spp. (Gramineae) and coffee and cocoa plants.

C_4 Photosynthesis

- Found in low altitude tropical and semi-arid areas
- Oxaloacetic acid (a 4C compound) is the first product of photosynthesis. Higher temperature optima than C_3 plants and can reduce water loss by slightly closing stomata without limiting photosynthesis.
- Examples found in many tropical savanna grasses e.g. *Themeda triandra, Cymbopogon* spp., *Pennisetum* spp. and some tropical trees.

Crassulacean Acid Metabolism (CAM)

- Found in plants in arid zones
- Plants fix CO_2 at night by forming malic acid. Thus, during the day, stomata can be closed and water conserved
- Examples found in Crassulaceae

sine triphosphate). The majority of plants use this directly to fix carbon dioxide using a pentose phosphate (C_3) biochemical pathway, and are referred to as C_3 plants. An alternative strategy found in a number of tropical plants is to fix carbon dioxide as oxaloacetic acid, a dicarboxylic acid with four carbon atoms. This C_4 pathway is more efficient at higher temperatures, allowing photosynthesis to continue and therefore better at exploiting available sunlight. A third strategy, first discovered in succulent members of the family Crassulaceae, is known as Crassulacean Acid Metabolism (CAM). In these plants carbon dioxide is fixed at night and stored as dicarboxylic acids in vacuoles. This is further converted to sugars as ATP becomes available in sunlight the following day. The advantage of CAM metabolism is that the stomata can remain closed during the day to conserve water, allowing survival in extremely arid environments. Thus, C_4 plants have improved photosynthetic efficiency at higher temperatures and higher light intensities, compared to C_3 plants, but CAM plants are most efficient in extremely arid conditions. Because these environmental factors vary considerably through the year, the relative advantages of each strategy and the competitive abilities of the plants also vary. For this reason species using different strategies are found in quite close proximity to one another.

One study on the moorlands of the northern slopes of Mt Kenya found a zone of overlap of C_3 and C_4 grasses at altitudes between 2800 and 3200 m. C_3 grasses predominated at the higher elevation and in moist valley bottoms, whereas C_4 grasses were at lower elevations and in drier areas.

Useful information about the water relations of plants can be obtained using simple measurements of relative water content and water saturation deficit (Box 4.2). Water potential and the concept of the soil-plant-atmosphere continuum (SPAC) were discussed on page 18.

Box 4.2 Measurements of relative water content and water saturation deficit can be obtained using simple techniques that provide useful and original information on the water relations of plants. The method described below can be used to compare individuals of the same and different species from a variety of different habitats. It can also be used to examine diurnal or seasonal variations.

Method

1 Weigh the leaf or the whole plant (FW_1)
2 Float on water for 24 hours and reweigh (FW_2)
3 Oven dry to constant weight, at 80–105 °C for 24 hours, and reweigh (DW)

Calculations

$$\text{Relative water content (\%) } (RWC) = \frac{FW_1 - DW}{FW_2 - DW} \times 100$$

$$\text{water saturation deficit (\%)} = 100 - RWC$$

Results

The investigator will be able to comment on water availability, the competitive ability of different plants, their relative ability to conserve water and their drought tolerance.

Drought survival

Maize plants transpire some 98 per cent of the water they absorb; only 0.2 per cent is used in phosynthesis and 1.8 per cent is retained in the plant tissues. Even xerophytic desert plants transpire at least 50 per cent of the water they absorb. Obviously plants have evolved numerous adaptations, besides different photosynthetic strategies to conserve water and survive periods of drought.

The structure of plants in the tropics is influenced more by water availability than by any other environmental factor. Many of the structural adaptations to conserve water are quite well known, such as deeper rooting systems, reduced leaves and thicker cuticles. Many perennial plants shed their leaves in the dry season, whilst others lie dormant in the dry season, often as bulbs, corms or tubers. Succulent plants store water in their tissues, either in their roots (e.g. *Ceiba parvifolia*), stems (e.g. many members of the Euphorbiaceae) or leaves (e.g. many members of the family Agavaceae).

Many mesophytes can be classed as **drought-evaders**. For example, many annual plants complete their life-cycle during the rainy season and the early part of the dry season and then remain dormant as a propagule (page 79) during the remainder of the dry season. In the same way, many trees in the tropics are deciduous and shed their leaves during the dry season to conserve water (also see pages 86 and 87).

Other plants, the **drought-tolerators**, attempt to survive and even continue growing during the dry season. Adaptations to prevent excessive water loss include thick waxy cuticles on small, typically evergreen, leaves. Hairs on the plant surfaces are sometimes well developed to trap moist air close to the cuticle, and the leaf blades of some grasses roll up during drought to limit water loss from stomata. Drought-tolerant plants (xerophytes) tend to have low surface : volume ratios, being thick and fleshy, and their leaves are often reduced or absent altogether. They are often deeply rooted and widely spaced in arid environments in response to competition for limited water supplies. However, these classifications are largely artificial and many plants combine both xeromorphic and mesomorphic features.

The baobab tree which is widespread in tropical Africa remains leafless for up to eight months of the year in semi-arid areas. Trees of this species are believed to live as long as 800–1000 years and their fibrous wood is a favourite food of elephants (Figure 4.1). The baobab possesses a remarkable ability to conserve water. It has a very extensive root system, estimated at up to 0.6 ha for a single tree, and has extremely good control of stomatal and cuticular water loss, even though its leaves show almost no xeromorphic features. Seasonal and diurnal changes in girth take place (Figure 4.2) which have been calculated to represent a daily loss of as much as 367 litres of water, although this is only 2 per cent of the tree's total water. The bark of baobabs contains a distinct green layer below the outer waxy surface which may enable photosynthesis to take place even when the tree is not in leaf.

Figure 4.1 The baobab tree, one of the botanical 'big game' of Africa.
(Courtesy D. J. Thompson.)

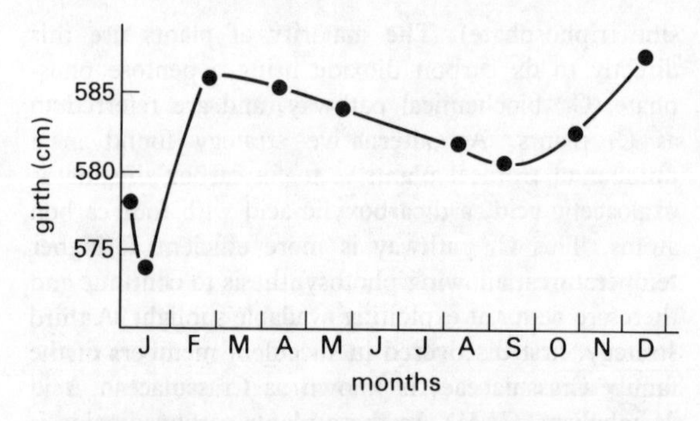

Figure 4.2 Mean girth of baobab trees throughout the year in Kibwezi, Kenya. These changes represent large volumes of water (see text).
(From M. Fenner, 1980, *Biotropica*, **12**; 205–09.)

Applied studies

The efficient utilization of water by plants is of particular importance in tropical agriculture and forestry. In contrast to developed nations, a majority of the rural and urban populace of Africa have practical experience of agriculture and are familiar with the water requirements of at least the common food crops. It is well known, for example, that pigeon peas and cowpeas will consistently outyield maize in many drier areas, although consumer preference often takes priority over the statistical chances of a high yield! Nevertheless, the merits of scientific studies of the water relations of crop plants are well founded. In Table 4.2 the results of an experiment comparing two varieties of sorghum show how planting of varieties most suited to the rainfall of different areas can increase yields. Irrigation is carried out whenever and wherever possible and its effectiveness could undoubtedly be improved with more scientific studies. For example, research has shown that costly irrigation to intensive coffee plantations is most effective at increasing yields when carried out just before and after the wet season rather than during the dry season. However, a period of moisture stress at some time before the start of the rains is important in promoting early and prolific flowering.

Table 4.2 Results of a ten-year experiment comparing the effects of planting density of two varieties of sorghum on yields. Variety Kafir grows as a single plant whereas variety Milo spreads vegetatively by producing tillers

	Variety Kafir		Variety Milo	
	2 driest years	2 wettest years	2 driest years	2 wettest years
Mean row spacing for maximum yields (cm)	91	7.5	65	62
Conclusions	Optimal planting density increases with increasing rainfall		Low planting density is optimal for all conditions	

A number of studies have also been carried out on the water relations of forest trees in Africa. It has been suggested that there may be an optimum leaf size for the most efficient utilization of water resources. Very large leaves are not common in tropical vegetation, nor are very small ones. In Ghanaian forests, for example, the predominant leaf size is 20–45 cm^2, the larger leaves being more frequent in wetter regions. Two exotic trees, teak and Gmelina occupy more than 60 per cent of the forest area and replace natural rain forest in southern Nigeria. Both species have very large leaves (500 cm^2 and 225 cm^2 respectively), with high rates of water conductance. Furthermore leaf area has been shown to be correlated with the amount of water conducting tissues (Figure 4.3). Are water resources being managed badly by planting these exotic species? Research of this type is starting to explain whether planting exotic trees affects hydrological cycles.

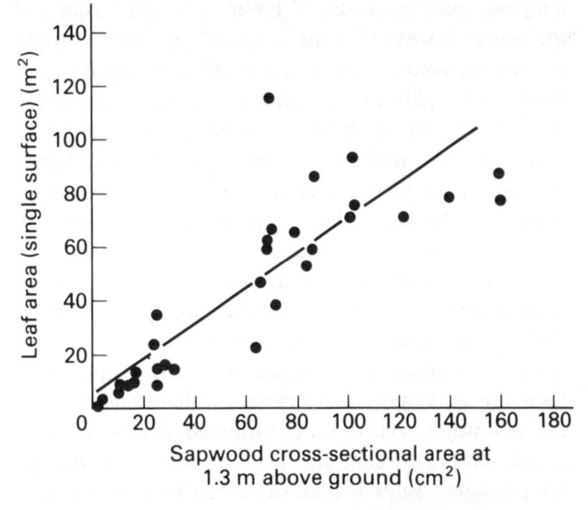

Figure 4.3 An autecological study of the water relations of teak trees in a plantation in Ibadan, Nigeria. Leaf area is positively correlated with the cross-sectional area of the water conducting tissues. The study showed that water conduction is very efficient in teak trees and water resistance from the roots to the leaves is low.
(From Whitehead, D. *et al.*, 1981. *Journal of Applied Ecology*, **18**, 571–87.)

Temperature

As we have seen, plants in the tropics have to be particularly concerned with keeping their temperature from rising too high (page 12). A range of morphological, physiological and behavioural adaptations maintain the plant at temperatures lower than ambient, with a few exceptions in high altitude areas. Even though temperature is an environmental factor of less significance in the tropics than in temperate zones (page 9), it affects a wide range of plant responses and has an interactive effect on a number of other environmental variables.

Temperature affects seed germination, plant growth rates, stomatal opening and closing, flowering and ripening of fruits. Physiological and biochemical processes are accelerated by rising temperatures, up to an optimum beyond which further temperature increases are detrimental.

In one study in Nigeria it was hypothesized that lower maize yields in the tropics compared to temperate regions are caused by higher night temperatures. Such temperatures, it was argued, would mean that night respiration rates are higher and thus also are the overall energy costs of metabolism. Nightly dry weight losses by whole plants averaged 40 per cent of the daily dry weight gain, but this was found to be only 3–5 per cent more than in areas where night temperatures are 10–20 °C lower. The conclusions of the research were that the extra energy costs of night respiration account for only a small proportion of the productivity differences, and that poorer yields are due to the limited success that has been achieved in breeding high-yielding varieties for the tropics. Yields in tropical maize varieties at best are in the range of 3000–6000 kg ha^{-1} (dry wt) compared to varieties in temperate areas which yield up to 18 000 kg ha^{-1}.

The way in which temperature interacts with other environmental variables is illustrated by considering how the temperature of a leaf affects its water balance. If leaf temperature is higher than air temperature, then the water potential gradient is high and the leaf can lose water even in water saturated air. Conversely, if the temperature of the leaf is lower than the air temperature, water will be deposited on the leaf, seen as dew in the early mornings. It has been suggested that water deposited in this way may be a particularly important moisture resource in arid and semi-arid areas.

a

b

Figure 4.4 High levels of soda have very definite effects on the distributions of both animals and plants in and around some East African Rift Valley Lakes. (a) In Lake Nakuru in Kenya, large numbers of lesser flamingoes (above) feed on the soda-tolerant alga *Spirulina platensis* which is present in great abundance. (b) The shoreline supports only sparse vegetation, dominated by two soda-tolerant grasses, *Sporobolus spicatus* and *Cynodon dactylon*. A recent rise in the level of the lake is believed to have increased soda concentrations in the soil and thus to have killed the *Acacia xanthophloea* trees, seen in the background. Note the light coloured substrate caused by the soda.

Mineral nutrition

The availability of nutrients differs considerably between soils (page 18), and plants have evolved mechanisms to tolerate deficiences and excesses of particular chemicals, and to maximize uptake in competition with other plants.

The response of individual species to soil nutrient status is well known for many agricultural plants. It has long been known that cassava and sisal will produce good yields on poor soils, whereas good yields of coffee and tomatoes need fertilizers. The modern fertilizer industry is based on a sound knowledge of plant mineral nutrition, and agricultural advisory services will test soils for farmers and recommend the types and quantities of fertilizers necessary for particular crops. Unfortunately there is little information on the mineral nutrition of non-agricultural plant species in the tropics.

In some cases the soil chemical status determines the distribution of plants. For example, high soda (sodium carbonate) levels in some of the East African Rift Valley lakes determine the distribution of plants in and around the lakes (Figure 4.4). Mineral prospectors can detect mineral deposits by recognizing the plants associated with particular minerals. One example is copper, which is an important plant nutrient but is toxic in high concentrations. In the copper belts of South–Central Africa, many of the plants growing near deposits contain concentrations of copper that would normally be considered to be toxic (Table 4.3). This is one of the few examples where detailed information is available for nutrients in natural plant communities in Africa.

Table 4.3 Copper concentrations in plants growing on soils of the copper belts of south–central Africa. Plant tissue concentrations in excess of 20 μg g^{-1} dry wt are normally considered to be toxic but in these areas plant populations have evolved to tolerate tissue concentrations as high as 13 700 μg g^{-1} dry wt. (1.4% of dry weight). These plants are referred to as hyperaccumulators of copper. (Data from Malaise, F. *et al.*, 1978. *Science*, **199**, 887–88, and Brooks, P. R., *et al.*, 1980. *Bulletin de la Société Belge* **113**, 166–71.)

Species	Family	Country	Copper concentration (μg g^{-1} dry wt)
Aeolanthus biformifolius	Labiatae	Zaire	13 700
Buchnera henriquesii	Scrophulariaceae	Zaire	3 500
Lindernia perennis	Scrophulariaceae	Zaire	6 000
Ipomoea alpina	Compositae	Zaire	5 000
Pandiaka metallorum	Amaranthaceae	Zaire	2 270
Silene cobalticola	Amaranthaceae	Zaire	1 660
Commelina zigzag	Commelinaceae	Zaire	1 210
Vigna dolomitica	Leguminoseae	Zaire	3 000
Fimbristylis exilis	Leguminoseae	Zimbabwe	420
Indigophora dyeri	Leguminoseae	Zimbabwe	890
Triumfetta digitata	Tiliaceae	Zaire	1 060

Plant populations

A population is the individuals of a species occupying a particular area of space at a particular period in time (cf. animal populations, pages 174 and 175). Thus a population of plants could mean a group of herbs that colonize a ploughed field or all of the trees of a particular species in a forest covering thousands of hectares. Also, as our definition infers, populations vary in time. The processes of dispersal of seeds and fruits are akin to emigration and immigration of animals from a population (pages 175 and 184), which also commonly involve young of the species. Similarly new individuals appear in the population as seeds germinate or propagules develop, this is a process of birth. And of course there are deaths among plants. Populations of plants will also be subjected to the effects of the physical and chemical environment, and to interactions with other plants and with animals. Population ecology, whether of plants or animals, attempts to quantify how the chemical, physical and biotic environment influences births, deaths and migration.

The life-cycle of plants in a population

Plants in a population undergo a series of stages during the course of their life-cycles, from the germination of seeds through growth, flowering and seed set (Figure 4.5).

Germination of seeds

In most habitats, the number of dormant propagules (mostly seeds but also including tubers and rhizomes) awaiting a stimulus for germination vastly exceeds the numbers present as growing plants. As many as 39 000 seeds m^{-2} belonging to forty seven different species, have been recorded from arable land soil in temperate areas, although numbers in tropical soils may be less. As many as 700 seeds m^{-2} have been found in West African soils, and seeds of thirty different species have been found in soils of mature forests in Ghana. Of these seeds, inevitably a certain proportion will be eaten or destroyed by animals and microorganisms. Whilst larger animals often aid in the dispersal of seeds, a proportion of the seeds will die, but others will survive until the right conditions for germination prevail. The environmental stimulus for germination is most often either moisture or light or both. Some seeds may not germinate until a gap in the forest canopy provides the light stimulus required for germination, or until they have been brought closer to the surface of the soil by cultivation. Of thirty seven weed species recorded in a study in East Africa, germination of fifteen was inhibited by darkness and sixteen by leaf shade. A high degree of dormancy was found in fourteen species.

Interactions with other plants

Seeds of many plants can remain dormant in the soil for months or even years, only germinating when the environment is favourable. They form a **seed bank**.

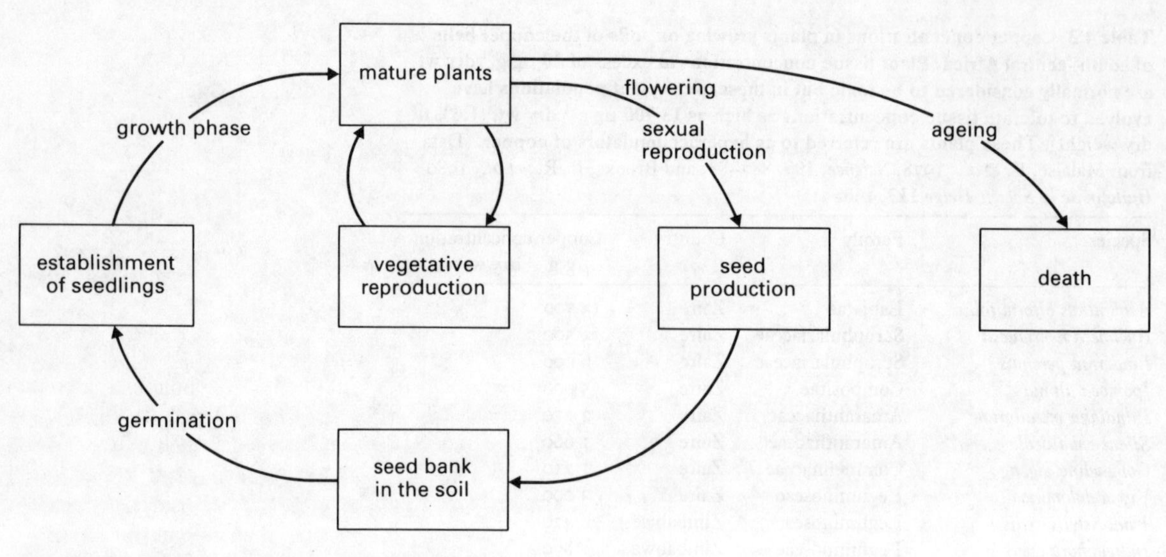

Figure 4.5 A model of the life-cycle of plant populations, illustrated as a flow diagram. During the course of its life-cycle the population will interact with its physical, chemical and biological environment. The seed bank is the population of dormant seeds in the soil.

Seedlings that establish from a seed bank will enter a growth phase, but as they increase in size there is an increasing chance that they will receive interference from other plants and begin to compete for resources. The more closely plants are crowded together the smaller each plant is liable to be, and as the population matures certain plants will be out-competed, predated (eaten) or become diseased and will die.

There is a simple mathematical relationship between density and the mean weight of surviving plants growing in a population that is progressively becoming 'thinner'. This is known as the $-\frac{3}{2}$(minus three over two) power law (Figure 4.6). In simple terms, this means that every change of 3 units in mean plant weight corresponds to a change of only 2 units in mean plant density, both units being expressed as logarithms. It appears that *all* plant populations can be described in these simple mathematical terms.

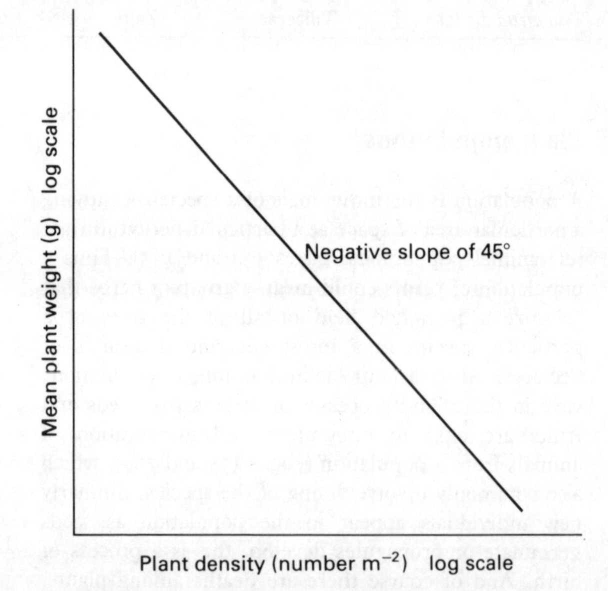

Figure 4.6 The $-\frac{3}{2}$ power law of the mean weight of plants growing in a crowded population. As the population matures individual plants move from the bottom right-hand corner to the top left-hand corner of the graph. The relationship appears to apply to all plant populations undergoing self-thinning.

Maturity, flowering and seed set

During its lifetime, a weedy annual plant commonly produces hundreds of seeds, and a tree or orchid produces thousands or millions. However, logic tells us that for each plant there is on average only one descendant; otherwise the earth would be covered with a single most successful species. Thus, populations maintain themselves through the survival of a small number of offspring. To propagate their genes, populations must maintain themselves and also attempt to expand into other areas. The latter is particularly true for species in successions (pages 87 and 89) which are doomed in their present habitat, their survival being dependent on reproduction and dispersal.

Dispersal mechanisms of seeds have developed in many and varied ways. Fleshy fruits are consumed by many mammals including bats and also by birds. The

Table 4.4 The types of seed dispersal and their relative frequencies (percentages) used by plants in a range of forests in Zaire. Animal dispersal is most common. (From Malaise, F., 1978. *The Miombo Ecosystem*, in *Tropical Forest Ecosystems*, UNESCO, Paris.)

Type of dispersal	Dense equatorial forest	Semi-deciduous forest	Transition forest	Mountain forest	Forest on waterlogged soil	Miombo
Self-dispersal	37.9	33.5	14	8	12.0	30.4
Animal dispersal	54.7	54.5	57	49	59.4	38.5
Wind dispersal	7.0	11.5	24	37	22.2	31.1
Water dispersal	0.4	0.4	5	6	6.4	0

seeds survive passage through the animal's gut; in fact this provides a pretreatment which has been shown to be an essential requirement for germination in several species. Other seeds rely on a passive dispersal by animals, and have hooks and barbs to attach themselves to animal's fur. Plumed and winged seeds rely on wind for dispersal, and a number of plants contain explosive mechanisms to disperse seeds, for example *Hura crepitans* (Family Euphorbiaceae).

Generally, wind dispersal is most common in open habitats and, for example, is used by most savanna grasses, whereas dispersal by animals is most often encountered in closed habitats such as tropical forests. Table 4.4 shows the proportions of species containing the different dispersal mechanisms in a number of sites during a succession in Zaire.

Timing is another important factor for flowering and seed dispersal. Flowering must be synchronized between members of the population to allow cross-fertilization and, if relying on animal pollinators, must also take place at a time when the animals are most abundant. In Ghana, it has been shown that species with very fleshy fruits mature their fruits during the rains, species with dry fruits during the dry season and species with wind-dispersed seeds near the end of the dry season when strong gusty winds prevail. Patterns of reproductive behaviour become more complex in parts of Africa close to the equator where there are two dry seasons and two wet seasons per year. A species that flowers towards the end of the rains may not mature its fleshy fruits until the end of the dry season. The role of animals in pollination and seed dispersal is discussed further on pages 83 and 84.

Eventually of course all plants die, although it could be argued that plants that reproduce vegetatively have discovered the secret of eternal life. Perennial plants, and particularly woody species, accumulate dead tissues as they get older and larger. They also acquire a progressively increasing respiratory burden with size. Thus, it seems likely that a combination of disease susceptibility and physiological stress limits the longevity of plants.

Plant strategies

With unlimited resources, a population increases geometrically (1, 2, 4, 8, 16 etc.). *r* is the intrinsic or innate rate of natural increase, which is the rate at which the population will increase in an ideal environment with no shortages. Consider populations of plants invading newly ploughed land or bare soil. To begin with, they grow with unlimited resources and without competition from other plants. Even at maturity, such plants tend to be small, having high reproductive capacity but a short generation time.

The parameter *K* can be defined as the carrying capacity of the habitat, and *K*- selection is typical of plants that spend a large proportion of their time under stress from neighbours. These plants are usually capable of becoming large and have long and complex life-cycles. Thus we can distinguish two broad categories of plants. Those which are *r*-selected are poor competitors and reproduce as soon as the population density becomes high (Figure 4.7). The *K*-selected

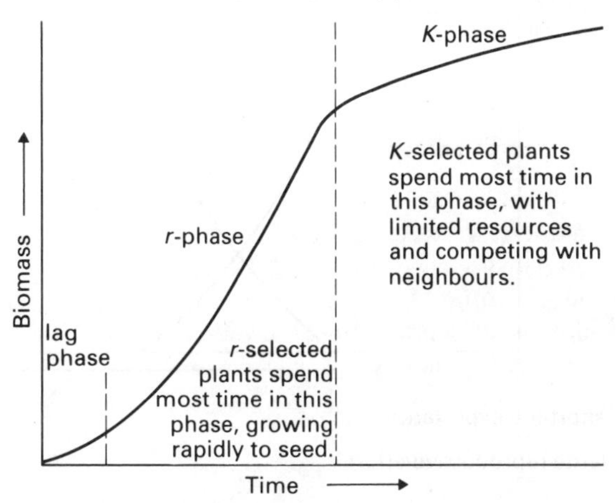

Figure 4.7 A hypothetical model of *r*- and *K*- phases during the growth of a population. Following an initial lag period, a population increases geometrically with unlimited resources. Eventually resources become limiting and plants begin to compete with neighbours of the same or different species. Plotting the number of plants on the *y*-axis would give the same shaped graph.

Box 4.3 The life strategies of plants
The life-cycles of plants follow one of three strategies according to intensity of stress and disturbance to which they are subjected. Few plants can tolerate a combination of high stress and high disturbance. The table illustrates these relationships and the characteristics of each type of strategy are shown below.

Intensity of disturbance	Intensity of stress	
	Low	High
Low	Competitors	Stress-tolerators
High	Ruderals	No viable strategy

Competitors
- Neighbouring plants competing for the same resource
- Tend to grow tall and maximize capture of resources
- In crowded vegetation

Examples Rangeland grasses, forest trees, lake algae

Stress-tolerators
- Tolerate shortages, but not disturbances
- Long lived, slow growth rates, slow turnover of nutrients and water
- Widely spaced

Examples Xerophytes, soda-tolerant grasses, metal-tolerant herbs, Afroalpine plants e.g. *Dendrosenecio* and *Lobelia*

Ruderals
- Annuals or short-lived perennials
- Rapid growth. Flower and set seed in a short time
- Invade recently disturbed fertile land

Examples Weeds of agricultural land, early invaders of cleared forests

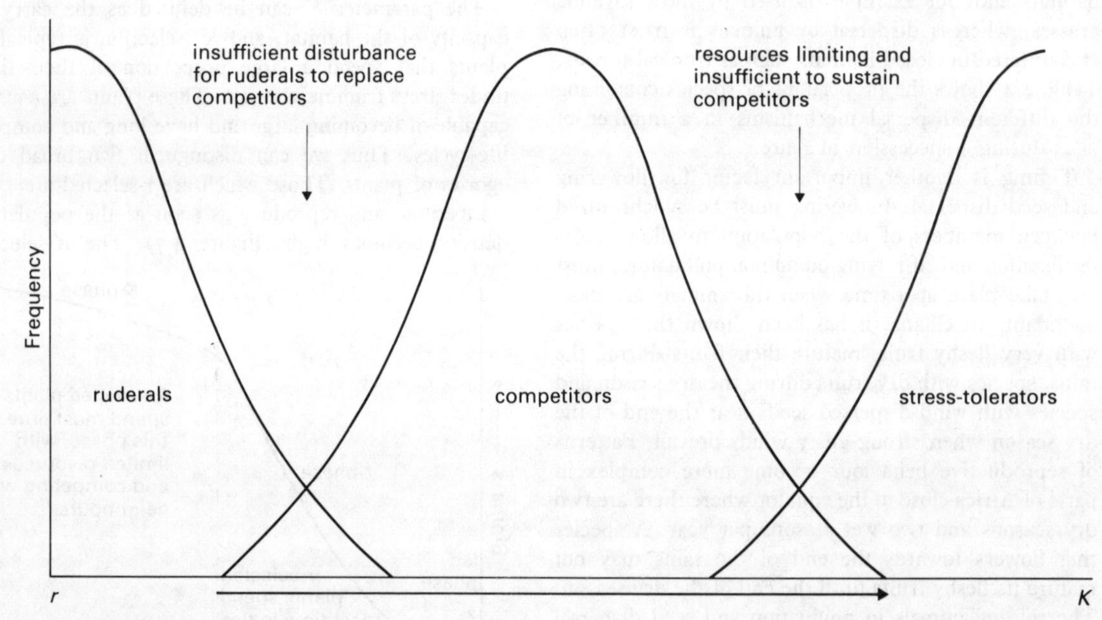

Figure 4.8 Distribution of ruderals, competitors and stress-tolerators along an $r-K$ continuum. As the life-form of plants becomes larger and longer competitors replace ruderals, but when resources become limiting stress-tolerators are best adapted. During succession the composition of plants in a community moves along this continuum.
(Adapted from Grime, 1981.)

species pass through a rapid-growing (*r*) phase but then enter a *K*-phase where, despite competition, they continue to grow (though more slowly) and may live for many years.

The ideas of *r*- and *K*-selection can be applied equally to animals, and are described on pages 199 and 200.

The distributions of plant species are limited in two ways. Firstly by **stresses**, such as shortages of light, water or nutrients and secondly by **disturbances**, such as destruction by herbivores, fire, cultivation or soil erosion. The intensity of stress and disturbance to which a population is subjected can vary enormously, and based on this plants can be described as conforming to one of three life strategies, as competitors, stress-tolerators or ruderals (Box 4.3). **Competitors** are the most common group, the competitive ability of plants ranging according to the suitability of the growing conditions; under optimal conditions of growth a plant will obviously compete better. In unproductive habitats **stress-tolerators** replace competitors, and **ruderals** are the plants involved in the early stages of secondary successions (pages 82, 87 and 88), for example weeds on arable land.

As in previous sections, an attempt is being made to describe ecological affinities between plants of different taxonomic groups. In this way, an understanding may be obtained of concepts such as **succession** (see pages 87 and 88). At this stage, it is worthwhile to draw together this section on plant populations by considering how plant strategies fit in with *r*- and *K*-selection (Box 7.5, page 199). In Figure 4.8 an *x*-axis has been drawn showing a range, or continuum, from typically *r*-selected plants through to typically *K*-selected plants. Superimposed on this continuum are the frequencies at which ruderals, competitors and stress-tolerators would be expected to be found. Once again, we see that these three strategies represent extremes of specialization and adaptation, and in between these extremes will be many plants experiencing intermediate levels of stress and disturbance. Species with strong ecological affinities will be found in close proximity to one another along this continuum.

Plants and animals

Pollination and seed dispersal

One of the characteristic features of tropical forests is the relatively large distances between individual trees of the same species (page 91), although a few species are gregarious. Although wind pollination is common in open habitats, such as grassland savannas, it would be extremely unreliable in forests where, in any case, wind is much reduced. Consequently many plants rely on animals to transfer their pollen in cross-fertilization.

In order to establish a successful rapport between a plant and its animal pollinator their life-cycles must be, to some degree, synchronised. The plant must also develop some characteristic by which it can be recognized by, or be attractive, to the pollinator. Plants with a large flower crop tend to be pollinated by large unspecialized pollinators, including birds, bats, moths, butterflies and bees, and noticeably tend to flower in synchrony over a short period. The environmental stimuli that these plants use to synchronize flowering through entire populations of such plants are generally unknown, although in many cases they seem to be triggered by some parameter related to moisture, for example a certain degree of soil moisture deficit. The alternative strategy is to produce a few flowers at a time over a long period, and to be pollinated by specific animal species; these tend to be birds, butterflies, large moths or large bees (Figure 4.9).

We briefly considered the dispersal mechanisms of seeds on pages 87–89. Seeds are a nutrient-rich food source for animals but, for dispersal, many seeds rely on being able to pass intact through the digestive system of a mammal or bird. Furthermore, they rely on the animal defaecating at a suitable site for germination and seedling establishment. One example is *Pimenta dioca* (allspice) which often grows beneath fences where birds perch and defaecate; germination of the seed is inhibited by its own seed coat (pericarp) which is removed during passage through the gut. Incidentally, such habitats are usually rich in the plant nutrients such as nitrogen and phosphorus from bird droppings, and are thus particularly favourable for the establishment of plants. In Africa *Acacia* seeds are frequently dispersed by cattle, for example in northern Nigeria cattle paddocks are often rich in *Acacia albida* and *A. nilotica*, the seeds having been voided in cattle faeces.

Fruits dispersed by birds are generally brightly coloured, such as the red berries of African mistletoe (*Loranthus*), but those dispersed by bats are usually green or yellow-green (wind-dispersed seeds tend to be the colour of dead plant tissue, e.g. *Khaya senegalensis* and *Terminalia* species, to camouflage themselves against herbivory). A range of animals also feed on fruits, including rodents, monkeys and elephants. Large crops of fruit that are produced during a single short period of time tend to be fed on by a whole range of animals, whereas small crops widely spaced over a long period of time attract the more specialist feeders. The flavours of the fruits (produced by secondary compounds, pages 84 and 85) are probably important in attracting particular species of animals.

(a)

(b)

(c)

Figure 4.9 The mutualistic relationship between plants and pollinators.
(a) *Poinsettia* (*Euphorbia pulcherrima*) with brightly coloured leaves to resemble large flowers.
(b) A sunbird uses it curved beak to reach down the corolla of the flower and feed on nectar. Pollen is transferred to the sunbird's head whilst feeding.
(c) *Thunbergia guerkeana*, presumably pollinated by a species of moth or butterfly whose proboscis can extend into the long thin corolla.

There are many examples of intricate plant and animal adaptations that illustrate these mutualistic (co-evolved) relationships. The fascinating story of the fig wasp is described in Box 5.3 (page 118).

Herbivores and plant defences

Habitats rich in plant species also tend to be rich in herbivorous species of animals. On average, herbivores consume about 10% or less of net primary production (pages 46–48), but it has been estimated that 21% of world food production is lost to plant-feeding insects alone. This higher figure probably results from the palatability of human food. Not surprisingly a whole range of plant adaptations have evolved to limit the damage caused by herbivores (Table 4.5).

Table 4.5 Morphological, chemical and behavioural adaptations that are used by plants to limit damage by herbivores.

1 *Physical protection* A hard covering which is difficult to penetrate, e.g. a nut-like seed-coat. Woody structures offer resistance to attack

2 *Physical defences* Thorns, spines, prickles, stings

3 *Chemical defences* Secondary compounds. Unpleasant tastes (repellent effect) or poisons (toxic effect), e.g. tannins, alkaloids, glycosides

4 *Buried perennating organs*, e.g. succulent tubers and rhizomes below ground

5 *A prostrate growth habit* Less liable to damage by large grazing herbivores if leaves are close to the ground surface. Many plants adopt this habit phenotypically in response to grazing and trampling

6 *Rapid regrowth following damage* Grasses are particularly tolerant to grazing and trampling. They contain a basal meristem close to the soil surface

7 *Timing* Events of the life-cycle when most susceptible to herbivore damage (e.g. flushing, fruiting) timed to avoid peaks of herbivore populations

8 *Intra-population spacing* Large distances between individuals of the same species reduce the likelihood of rapid discovery and proliferation of herbivores' populations

Plants produce a very wide range of substances. The **primary compounds** such as starch and cellulose are essential for metabolic or structural purposes. Others, termed **secondary compounds**, serve other purposes.

Chemical defences are particularly well developed in the tropics, indeed, many secondary compounds that plants evolved for this purpose have been exploited by man. Coffee and tea provide us with the stimulant effect of caffeine, and *Cassia* spp. are a source of laxatives (e.g. Senna from *Cassia senna*), together with a large number of other pharmaceutical and herbal products. The bitter taste of undercooked cassava is due to a poison which is a cyanogenic glycoside, while extracts from *Acokanthera* spp. provide a poisonous substance a cardiac glycoside to tip arrows. Diosgenin is an example of a saponin occurring in yams. Another common group of secondary compounds are the alkaloids, for example as produced by the widespread ordeal tree, which in the past was given to people accused of crimes. It was believed that if they were innocent they would survive the ordeal of eating food tainted with this poisonous tree, but few did!

We should remember, however, that plants and animals have evolved together over millions of years: flowering plants. It is hardly surprising that some sort of mutual interdependence will have developed between them over all these years. Consider, for example, the suggestion that high species diversity of trees in tropical forests, and the correspondingly low density of individuals (pages 90–92), may be at least partially maintained by insect and vertebrate seed eaters. Most species of herbivores consume a fairly narrow range of plant species, and the number of seeds consumed is inversely proportional to the distance of the seeds from the parent tree. Only seeds travelling large distances from the parent plant have a really good chance of avoiding predation, and eventually becoming established. Thus, we have the situation where herbivores apparently reduce the trees' evolutionary fitness by preventing any single tree species from becoming closely spaced and dominant within the plant community. It should be added that this is only an hypothesis, yet to be subjected to scientific scrutiny.

However, is it always correct to assume that the effects of herbivores on plants is to reduce their evolutionary fitness? With pollinators and seed-eaters this is certainly not the case, as we have already seen. Experiments carried out on grasses have shown that light grazing actually increases their productivity compared to when they are not grazed. Further more, many plagioclimax grasslands (see page 89) owe their continued existence to the presence of herbivores. In the absence of herbivores, bushland and forest encroach on to the land. Certainly the survival of some plant communities is benefited by herbivores.

Consider another example, of aphids feeding on a large tree. A single tree may support in excess of one million individuals, each of which sucks the sap of the leaves and deposits honeydew (sugar). This number of aphids deposits large amounts of honeydew (amounting to 1 kg m^{-2} yr^{-1} beneath some temperate trees), which falls to the soil surface underneath the tree. Now, it has been found that a 1 per cent addition of sugar to soil increases the number of nitrogen-fixing bacteria 1000 fold. Nitrogen is a nutrient which is often in limited supply (pages 18, 19 and 54), whereas sugar can be manufactured much more easily if sufficient water and sunlight are available. Here is a good example of a herbivore directly influencing nutrient cycling. However, this does not demonstrate that the fitness of the plant is increased. To do this it must be shown by experiment that a plant species produces more offspring in the presence of herbivores than without. It remains to be seen whether we can argue that plants benefit from being eaten by animals.

Plant communities

A plant community is an assemblage of two or more species living together as a result of their dependence on particular resources of energy, nutrients, water, air and space. Early plant ecologists were most concerned with the classification and naming of plant communities as related to different geographical regions, climates and soils. This we have dealt with on pages 21–32, and in chapter 3 we looked at the structure-ture and function of plant and animal communities. In this section, primary production, phenology, succession and species diversity of plant communities are considered in more detail.

Primary production

The productivity of plant communities (page 39–43) is amongst the most useful information that can be obtained about them. Primary production cannot, however, be estimated from harvest data alone. There is an important difference between yield and productivity, because during a growth period, from time t_0 to time t_1, some plant material will have died and some will have been eaten by herbivores (cf. Fig 3.4). The yield, $\triangle B$, merely represents a changing equilibrium between growth and losses. These additional factors must be taken into account to accurately determine net primary production (NPP):

$$\text{Yield}^{t_0-t_1} = \triangle B = \underset{\text{biomass at } t_1}{B_1} - \underset{\text{biomass at } t_0}{B_0}$$

$$\text{NPP}^{t_0-t_1} = \triangle B + \underset{\textit{loss to death and decomposition}}{D} + \underset{\textit{loss to grazing}}{G}$$

In grassland studies in Nairobi National Park in Kenya, it was found that failure to account for mortality (*D*) results in grass net primary productivity being underestimated by almost 50 per cent during the growing season of the long rains. Furthermore, grass production was found to occur during the dry season even though harvest samples alone indicated that there was no production. The conclusion was that most published estimates of NPP were in fact *underestimates*, because no account was taken of *D*.

In another study in a slightly drier part of Kenya (Kajiado) about 7.5 per cent of the harvestable yield was found to be removed by grazing (*G*) by large herbivores, which included both wild game and cattle, sheep and goats. In the northern Guinea savanna zone of Nigeria large mammals have been found to remove an exceptional 63 per cent of the harvestable yield of grasses, grazing most when the seasonal protein content of grasses is highest.

There will also be additional grazing losses caused by invertebrate herbivores. In the Serengeti grasslands in Tanzania invertebrates as a whole have been estimated to remove 7.6 per cent of NPP. Acrididae (short-horned grasshoppers and locusts) alone have been found to consume 0.8–3.2 per cent of NPP at different sites of semi-arid thorn bush scrubland in Marsabit in northern Kenya.

Studies of NPP in forests and woodlands are confronted with the problem of similar inaccuracies, although methodology is different. In tropical forests, losses to *D* and *G* are often substantial. In evergreen forests in the Ivory Coast litter-fall represents 61–70 per cent of NNP (Table 4.6). This means that only 30–39 per cent of NPP is retained as biomass, although this is probably an overestimate because no account was taken of *G* in the study. In tropical forests in other parts of the world, approximately 6–9 per cent of NPP is consumed by insect herbivores, the most predominant feeders.

Table 4.6 Results of a study on net primary production (NPP) in an Ivory Coast tropical evergreen forest. Leaf life is thought to be on average about one year, and the largest proportion of NPP is lost to the litter. Units are kg ha^{-1} yr^{-1}. (From Bernhard-Reversat, F. *et al.*, 1981, *Structure and Functioning of Evergreen Rain Forest Ecosystems at the Ivory Coast*, in *Tropical Forest Ecosystems*, UNESCO, Paris.)

Production	Sites Banco plateau	Banco valley	Yapo plateau
Growth of boles and branches	4 600	3 000	4 600
Growth of roots	700	500	700
Production of leaves and reproductive organs	11 700	10 100	10 200
Total annual production	17 000	13 600	15 500
Losses to litter			
Leaves	8 190	7 430	7 120
Flowers and fruits	1 100	660	1 050
Branches	2 580	1 090	1 450
Total losses to litter	11 870	9 180	9 620
Litter as a percentage of production	70%	68%	62%

Phenology

Phenology is the partitioning in time of the events of a plant's life-cycle: seasonal changes in leaf bearing, leaf shedding, growth processes, flowering and seed production. Studies of phenology provide a valuable insight into the functioning of plant communities as the following examples of studies in West and East Africa will serve to illustrate.

Dry tropical forest in Ghana

The vegetation of the Accra Plains in south–east Ghana contains patches of undisturbed dry tropical forests in vegetation that is described as a grassland/thicket mosaic. The region experiences two dry seasons with a mean annual rainfall of 1100 mm. Patterns of flowering, fruiting and flushing (production of new leaves) in the forest in relation to rainfall are shown in Figure 4.10.

These phenological patterns appear to reflect the rainfall pattern. The number of species flushing corresponds closely with peaks of rainfall: it was discovered that rain brings about a rapid response from many species. There was found to be a threshold level of rainfall of 35–40 mm over the preceding twenty one days before a large number of species began flushing. There was also a statistically significant correlation between the number of species in flower and the rainfall over the preceding twenty one days. If the species in the forest are grouped into different life forms, based on the habitat in which they are most commonly found, flowering patterns can be examined in more detail (Table 4.7). Grassland trees tend to flower in the dry season or soon after, while forest trees and shrubs usually flower and fruit for short well defined periods during the wet season. In contrast, most grassland thicket shrubs flower almost continuously, and climbers show an equal combination of all three patterns of flowering.

With respect to fruiting (Figure 4.10), fruits of fleshy-fruited species were most abundant in wet seasons, whereas those of dry-fruited species were most abundant in dry seasons. Fleshy-fruited species were less successful in setting fruit than were dry-

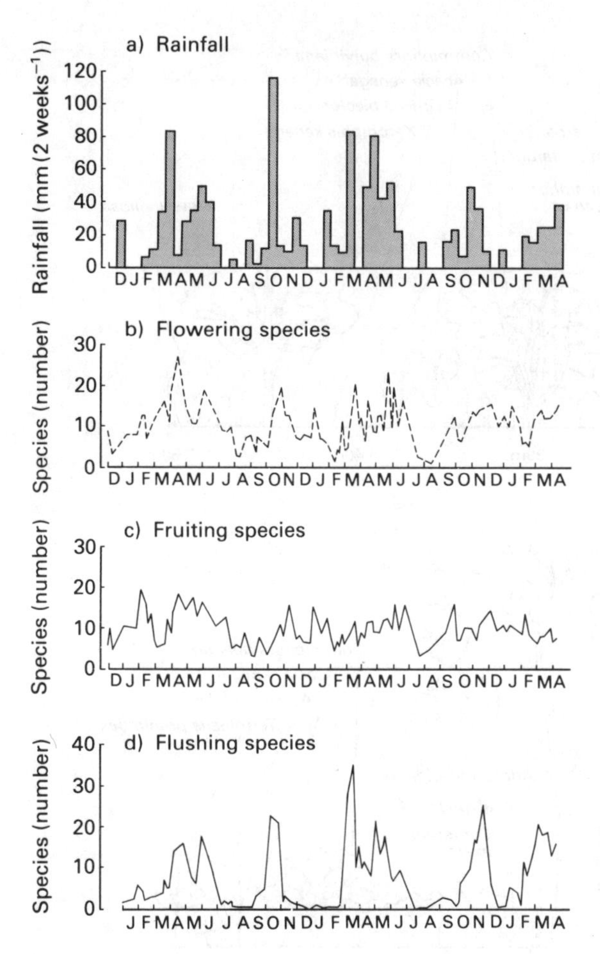

Figure 4.10 Dry tropical forest in Ghana. Seasonal patterns of (b) flowering, (c) fruiting (d) flushing reflect (a) the seasonal pattern of rainfall, mm per 2 weeks (From Lieberman, D., 1982. *Journal of Ecology*, **70**; 791–806.)

Table 4.7 The percentage of species, grouped into four life-forms depending on their usual habitat, flowering at different seasons in dry tropical forest in Ghana. (From Lieberman, D., 1982. *Journal of Ecology*, **70**, 791–806.)

Life-form	Flowering season			Total number of species
	Dry	Wet	Continuous	
Grassland/ thicket trees	67	22	11	9
Forest trees and shrubs	11	78	11	9
Grassland/ thicket shrubs	8	33	58	12
Climbers	27	36	36	11
Total	27	41	32	41

fruited species, probably because of a greater dependence on water supply. Amongst the dry-fruited species those that are wind-dispersed bear fruit near the end of the dry season when there are strong gusty winds. Flowers and fruits in deciduous species were typically produced while the plant was leafless.

The phenological patterns examined in this study illustrate the way in which the plant community has become adapted to the relatively low seasonal rainfall of the area. Most phenological patterns appeared to be moisture dependent.

Semi-arid bushland in Kenya

The vegetation in the Kibwezi area in Kenya varies from woodland thicket to sparse bushland, receiving a mean annual rainfall of 490 mm, 80 per cent of which falls in two wet seasons. Profile diagrams of the vegetation, together with phenology of leaves, flowers and fruits are shown in Figure 4.11.

The majority of species lose their leaves in the long dry season. The *Commiphora* spp. lose their leaves earliest, but *Combretum exalatum* and *Grewia bicolor* roll their leaves up and retain them until September. *Adansonia digitata* (the baobab tree) is in leaf for only four months of the year (see page 76, Figure 4.1). Only three common species are evergreen.

A large number of plants bear their flowers towards the end of the long dry season, often before much rain has fallen. Surprisingly individuals of the same species often flower at different times, whilst seven of the thirty species did not flower at all during the months of the study, and the pattern of others appeared to be rather unpredictable. Fruiting was also very variable, although a wide range of fruits were available throughout the year which probably accounts for the rich variety of bird life in the area (Figure 3.25(a)).

The phenological patterns of the vegetation are adaptations to the main climatic characteristic of the area, which is low and erratic rainfall. Regular and predictable phenological patterns would not favour survival of species, nor necessarily would a direct response to rainfall – because of its usually brief duration. Slightly erratic phenological rhythms ensure the continued existence of the plant community.

Succession

We have briefly come across the term **succession** in a previous section (page 83) in relation to the development of populations. Consider an abandoned agricultural field, left alone for nature to take its own course. To begin with, small herbaceous ruderals colonize the open land, rapidly reproduce and proliferate. During the course of time, perennial herbs will establish themselves to be followed by woody shrubs

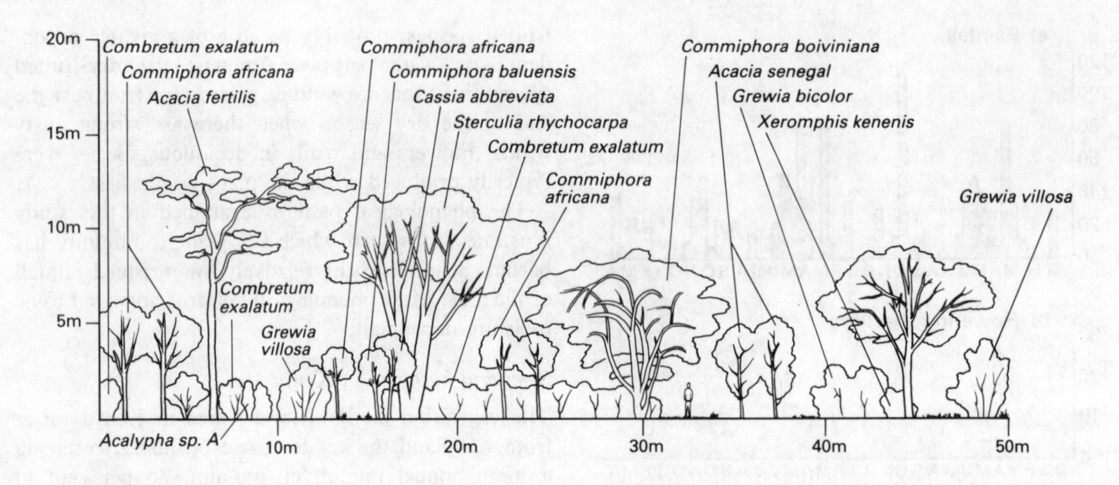

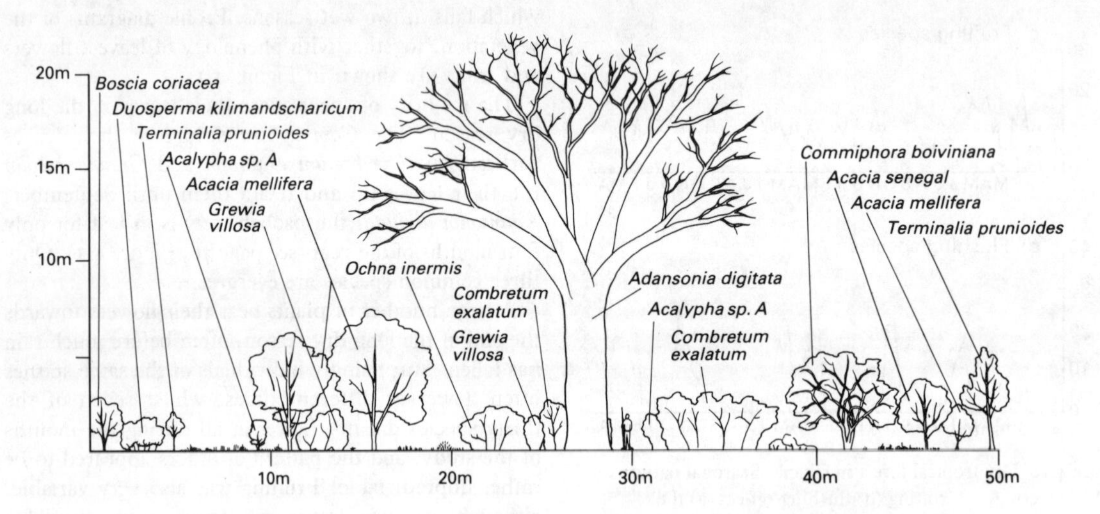

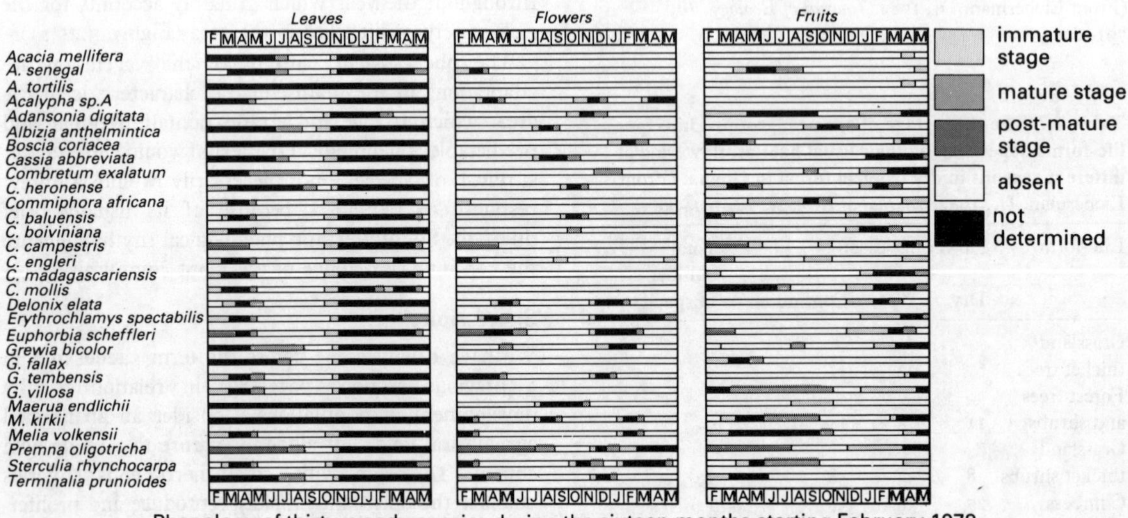

Phenology of thirty woody species during the sixteen months starting February 1978.

and then probably trees. The process of increasing community complexity is what is referred to as a succession. This particular example which took place on disturbed land that had previously supported a plant community is called a **secondary succession**. If the process had started on a virgin substrate (a bare area), such as on the debris of volcanic ash following an eruption, or in a newly created water body, it would be referred to as a **primary succession**. The early colonizers of primary successions are usually the lower plants, the algae, mosses and lichens.

The end-point of a succession is the **climax**, which may be of one of three types. A **climatic climax** is determined by the climate of the region: ranging from humid rainforest in lowland areas of West and Central Africa to poor grassland in arid areas of northern and eastern Africa (pages 21–32). In other instances the nature of the substrate results in an **edaphic climax**, typified by swamp vegetation and vegetation surrounding soda-rich lakes in the Rift Valley. However, in by far the largest part of Africa, the activities of man have interfered with the progress of succession by felling trees, grazing cattle, burning and developing land for agriculture. Such an artificially maintained end-point of succession is called a **plagioclimax**.

During the process of succession, plant communities pass through a series of stages or **seres**, and there is a characteristic development and change in the structure and function of the plant community, as well as in the animal community, the soil and the microclimate (Table 4.8).

Table 4.8 Changes that occur in plant communities during succession. Points 1 to 7 concern changes in the plant community, although succession is an orchestrated response affecting the whole ecosystem (points 8–10). (Adapted from Grime, 1981.)

1 Height and size of the plant community increases

2 Physiognomy becomes more complex. Usually an increased stratification within the vegetation

3 Small short-lived species are replaced by larger longer-lived species

4 Productivity of the community increases

5 Species diversity increases (see pages 90 and 91)

6 Changes in the stability of the community usually occur (see page 92)

7 More nutrients are contained within the vegetation

8 Microclimate (pages 13 and 14) is increasingly determined by the community itself

9 Soil develops, having increasing depth with more organic matter

10 Corresponding successional changes occur in the animal community

There are many examples of terrestrial and aquatic successions, for example transects through a fresh water swamp, as shown in Figure 2.15 and through a mangrove swamp. Here we have chosen a secondary succession following shifting cultivation in Nigeria.

Secondary succession following shifting cultivation in Nigeria

In many African countries, shifting cultivation is widespread. Although seemingly rather primitive, it is probably one of the most successful forms of agriculture in the tropics. Forest is cleared and high crop yields are obtained from the land for a few years, after which the site is abandoned in the knowledge that fallow vegetation restores fertility to the soil. Problems have been caused in recent years because human population pressures are tending to reduce the period of time before the land is brought back into cultivation.

A study in south-west Nigeria compared fallow plant communities of different ages with mature forest (Table 4.9). After seven to ten years fallow, vegetation is similar to mature forest with respect to the number of species, tree density and species diversity. Organic matter in the soil has reached 78 per cent of that in mature forest after ten years and nitrogen in the top soil has significantly increased. However, mature forest is obviously much more complex in vertical and horizontal development. Tree height, diameter and the ground cover of shoots and trunks are substantially less in ten-year fallows than in mature forest.

The results of this study, which is one of the few of its kind that have been carried out in Africa, indicate that at least seven years of fallow should be allowed between agricultural crops in forest areas. Further studies should provide information as to the extent to which man can exploit tropical forests. Shifting cultivation appears to be a viable and long-term method of agriculture, if properly managed. Scientific studies are necessary to prevent land lying 'idle' for longer than necessary, and also to prevent over exploitation of the valuable tropical forest resource by increasing human populations.

Figure 4.11 Semi-arid bushland in Kenya. Profiles of the vegetation along two transects, together with the phenology of thirty woody species. The wet seasons are indicated below. (From Fenner, M., 1982. *Journal of the East African Natural History Society and National Museum*, no. **175**; 1–12.)

Table 4.9 A comparison of fallow communities of different ages with mature forest in south-west Nigeria (From Aweto, A. O., 1981. *Journal of Ecology*, **69**, 601–11.)

Successional stage (year)	Trees				Species diversity			Soil		
	Density (ha⁻¹)	Mean height (m)	Mean crown diameter (m)	Mean basal cover of shoots or trunks (%)	Diversity Index	Number of species	Number of tree species	% organic matter	% N	Water-holding capacity (%)
1	56	1.3	0.7	1.5	0.37	39.2	2.2	2.5	0.19	36
3	512	2.5	1.8	1.6	0.43	42.5	7.0	2.2	0.19	35
7	2270	4.7	4.6	2.7	0.71	56.6	14.5	3.1	0.31	48
10	2670	5.8	7.0	5.1	0.71	53.5	19.0	4.2	0.29	47
Mature forest (c.80 years)	2260	10.4	20.0	10.3	0.71	60.6	24.6	5.4	0.49	56

Species diversity

In the example of a fallow succession in the previous section (Table 4.9) the results list both the number of species and species diversity. The number of species in the ten-year fallow was not as high as in the mature forest, although species diversity in a seven-year fallow was the same as in mature forest. As we have seen in chapter 3, species diversity is an index which incorporates measurements of the number of species and also the number of individuals of each species.

Diversity of plants in Africa

Plant and animal communities are generally more diverse in the tropics than in temperate regions (pages 65 and 66), although African forests are floristically relatively poor compared with forests of tropical America and Indo-Malaysia (Figure 4.12).

Nevertheless, African tropical forests contain an abundance of species. In West Africa, 7000 species of higher plants have been recorded in forests, and even this is probably an underestimate because a number of species probably remain undiscovered or unidentified.

Table 4.10 The number of species of flowering plants recorded in tropical rainforests in four African countries. (Anon, 1978, *Inventory and Survey: International Activities*, in *Tropical Forest Ecosystems*, UNESCO Paris.)

Country	Number of species	Number of genera	Number of families
Ivory Coast	600 (trees > 10 cm GBH* only)	275	60
Nigeria	4 500	–	–
Cameroun	8 000	1 800	220
Zaire	11 00	–	–

* GBH = Girth measured at breast height.

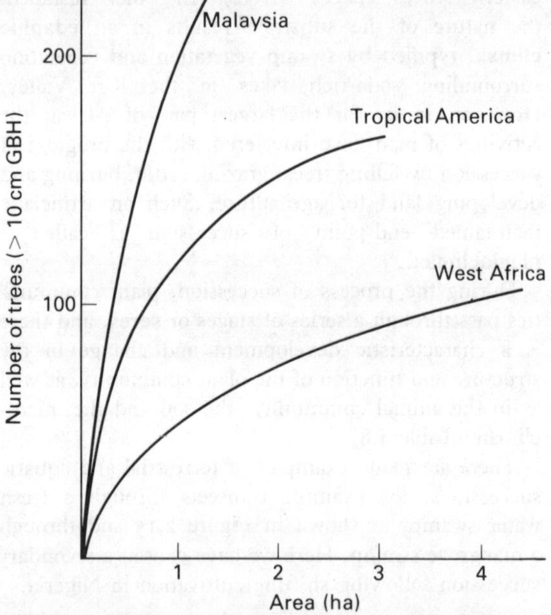

Figure 4.12 Species/area curves comparing West African forests with those outside Africa. African forests contain relatively few species compared with forests in tropical America and Malaysia. (c.f. Figure 3.29). (From Bernhard-Reversat, J. *et al.*, 1978, pp. 557–74, in *Tropical Forest Ecosystems*, UNESCO, Paris.)

Some examples of the numbers of higher plants recorded in African forests are shown in Table 4.10. Note that numbers in this table refer to much larger areas than does Figure 4.12. Examples of the numbers that may be found in smaller areas, more comparable to most ecological research projects are shown in Table 4.11. The notes on equitability (the relative number of individuals of each species) in this table show that although there are many species, a large proportion are represented by only one or a few individuals.

Table 4.11 The number of species and notes on the equitability of communities in smaller study areas in tropical rainforests in three African countries. (From Anon, 1978-reference as in Table 4.10.)

Country	Area of study	Number of species	Notes on equitability
Nigeria	18 ha	170 (trees > 10 cm GBH)	18 species each contained > 100 individuals 90 species each contained < 10 individuals
Cameroun	1 ha	230 (trees)	1 species contained 150 individuals 125 species contained 1 individual
Gabon	8 ha	122 (shrubs and trees > 3 m in height)	26 species of lianes were recorded amongst 96 individual plants

Intra-population spacing is considerable.

Forests with an extremely large number of individuals of one species are rare. Some forests *appear* to contain only a few species, as for example the dry forests of *Gilbertiodendron dewevrei* in Zaire, whereas in fact they contain more than 300 species. The same is true in forests in other parts of Africa where species of the family Caesalpiniaceae appear to be dominant. Exceptions occur where there are peculiar topographic or edaphic influences, as for example in mangrove swamp and *Raphia* swamp communities, and in non-woody papyrus swamps. It seems that these plants have successfully occupied an extremely specialized habitat, where a lack of competitors results in colonization by virtually just one species, or closely related plants, rather reminiscent of agricultural monocultures. Mangrove swamps are particularly interesting in this respect. Several unrelated species of mangrove trees have evolved the same unique morphological adaptations to conditions in the marine intertidal zone: a good example of convergent evolution in plants.

Moist tropical forests undoubtedly contain the largest number of species, particularly compared to temperate forests which are commonly dominated by only one or a few tree species and rarely if ever, contain more than about twenty five tree species. Compared to this, forests in drier parts of Africa also contain many species: for instance East African forests have in the order of 100 species of trees. Similarly, much other tropical vegetation contains a large number of species. In 80 ha of dry woodland and wooded grassland vegetation in north–east Uganda, 400 species of vascular plants have been recorded. Generally with increasing aridity species richness and diversity declines.

Diversity and stability

Tropical rainforests are vegetation climaxes where plants and animals are diverse and abundant, they are places of considerable biological complexity. At the other end of the spectrum, there is intensive agriculture which is plagued with the pests, diseases and erodable soils associated with monoculture crop production. Is it correct to conclude that high diversity means high stability and that low diversity means low stability? Perhaps so, but first we should clearly define what we mean by these terms.

Diversity, as we have seen, can mean the number and richness of species, vertical and horizontal stratification or even the range of different habitats. **Stability** introduces rather more of a problem. Most obviously we imagine a tropical rainforest or similar complex habitat as places of high stability, meaning there is little change and low levels of stress and disturbance. But where do the natural fire-dependent grasslands fit into this scheme? Although they have been subjected to regular and severe disturbance by fire they have stood the test of time, as shown by the many species of large mammals inhabiting them, and which must have evolved in the grasslands. Furthermore, there are marked diurnal and seasonal changes in most tropical rainforests, and also a large amount of disturbance usually takes place in the form of a continuous series of 'secondary successions' as gaps are created by the death and collapse of older trees.

To clarify the meaning of stability it helps to introduce another term, **resilience**, which is the ability of a plant community to cope with a disturbance by returning quickly to a steady state. Indeed if the disturbance is removed, for example, if the grassland is not burned, the resilient plant community is soon replaced by disturbance-sensitive competitors, including woody plants. Recent research has shown that, in semi-arid grasslands in southern and eastern Africa, the resilience of plant communities decreases as their stability increases; in other words they become more fragile. Tropical forests may be very stable but generally they are not resilient. Thus it appears that relatively stable environments allow highly diverse but fragile communities to exist, while unstable or unpredictable environments will support only a resilient and relatively simple ecosystem.

Suggested reading

There are many excellent texts dealing with plant ecology. Kershaw's book, 1975, is recommended particularly since it is available in an ELBS low-price edition. Fitter & Hay, 1981, and Etherington, 1982, are good textbooks on physiological plant ecology, while Harper, 1977, Grime, 1981, and Silvertown, 1982, give useful accounts of plant population ecology, and Whittaker, 1975, writes on communities and ecosystems. Edwards & Wratten, 1980, and Hodkinson & Hughes, 1982, serve as useful introductions to plant/insect interactions.

Amongst those dealing with tropical plant ecology Janzen, 1975, is one of the better books, although unfortunately it is primarily concerned with the American tropics. Ewusie, 1980, provides an introduction. Lind & Morrison, 1974, is essential reading for East African ecologists and Hopkins, 1979, and Longman & Jenik, 1974, for those in West Africa. The recent books of Mabberley (1983) and Vickery (1984) on the ecology of tropical rain forests are particularly useful.

Essays and problems

1 Do you consider that a knowledge of the water relations of plant species would be useful to our exploitation of forest resources? If so, how?

2 What is the significance of vegetative propagation in the population biology of flowering plants? List examples of plants that you know spread vegetatively. Use references in the Suggested reading section to illustrate your answer.

3 Do the common tree species in the area where you live tend to flower in synchrony during certain seasons or at different times of the year? Why do you think this is?

4 What is the impact of increasing human populations on secondary forests used for shifting cultivation? How do secondary forests in the area where you live differ from primary (unexploited) forests.

5 What is the relationship between plant zonation and succession in mangrove swamps?

6 How do two rainy seasons per year affect leaf shedding, flowering and seed set of deciduous trees in the tropics?

7 How do plants respond to herbivores?

8 Draw histograms of the data in Table 4.9 and comment on features of interest.

References to suggested reading

Edwards, P. J. & Wratten, S. D. 1980. *Ecology of Insect–Plant Interactions*, Studies in Biology, No. 121, Edward Arnold: London.

Etherington, J. R. 1982. *Environment and Plant Ecology*, (2nd edn). Wiley: London.

Ewusie, J. Y. 1980. *Elements of Tropical Ecology*, Heinemann: London.

Fitter, A. H. & Hay, R. K. M. 1981. *Environmental Physiology of Plants*, Academic Press: London.

Grime, J. P. 1981. *Plant Strategies and Vegetation Processes*, Wiley: Chichester.

Harper, J. L. 1977. *Population Biology of Plants*, Academic Press: London.

Hodkinson, I. D. & Hughes, M. K. 1982. *Insect Herbivory*, Chapman & Hall: London.

Hopkins, B. 1979. *Forest and Savanna*, (2nd edn). Heinemann: London.

Janzen, D. H. 1975. *Ecology of Plants in the Tropics*, Edward Arnold: London.

Kershaw, K. A. 1975. *Quantitative and Dynamic Plant Ecology*, (2nd edn). Griffin: London.

Lind, C. M. & Morrison, M. E. S. 1974. *East African Vegetation*, Longman: London.

Longman, K. A. & Jenik, J. 1974. *Tropical Forest and its Environment*, Longman: London.

Mabberley, D. J. 1983. *Tropical Rain Forest Ecology*, Blackie: Glasgow.

Silvertown, J. W. 1982. *Introduction to Plant Population Ecology*, Longman: London.

Vickery, M. L. 1984. *Ecology of Tropical Plants*, Wiley: Chichester.

Whittaker, R. H. 1975. *Communities and Ecosystems*, (2nd edn). Macmillan: New York.

5 Animals' environments

In chapter 4 we discussed factors affecting the populations of plants. Now we turn our attention to animals' environments, and consider ways in which these environments affect the abilities of animals to survive and reproduce themselves. In many ways, animals' distributions reflect those of plants, since all animals depend, directly or indirectly, upon plants. But animals are more mobile than plants, and can respond to unfavourable environments by moving away.

We look at four major aspects or components of animals' environments and consider how they influence distribution and abundance. Firstly the physical environment, including especially water, temperature and light; secondly resources, principally food and shelter; thirdly other organisms, which can be individuals of the same species, or of different species. In the latter case, they may be either animals or plants. Furthermore, interactions between individuals can be positive, benefiting one or both of them, or negative, as when they compete. Finally we need to consider **man** *and the impact of his many activities on the environments of animals other than himself (man is the subject of chapter 8). It often happens that animals' environments are unfavourable to them, and we conclude by considering their responses to those circumstances, which often lead to inactivity or loss of 'condition'.*

Animal distributions

Nobody knows how many species of plants and animals are to be found in Africa. Indeed, innumerable species have yet to be named and no doubt many interesting discoveries are still to come. Nevertheless, there could well be over a million species of animals in Africa with more than half of them being insects; and there are at least 30 000 species of flowering plants. Some examples from groups which have been studied relatively well give an idea of the numbers involved. Over 200 different species of beetle have been found in the dung of elephants in Tsavo National Park, Kenya. Lake Malaŵi contains some 250 species of fish, whilst the Zaire River has more than 400. There are about 100 species of snakes to be found in Kenya, and in Nigeria more than 1000 species of birds have been recorded.

Practically nothing is known about the ecology of most of these species. But we did point out on pages 64–67 that habitats vary greatly in the numbers of species they support. We also know that many species have rather restricted distributions – that is, they only occur in a few places.

The **range** or **distribution** of a species is the geographical area within which it occurs. Some species have much more extensive distributions than others (Figure 5.1). At the same time, almost all species are

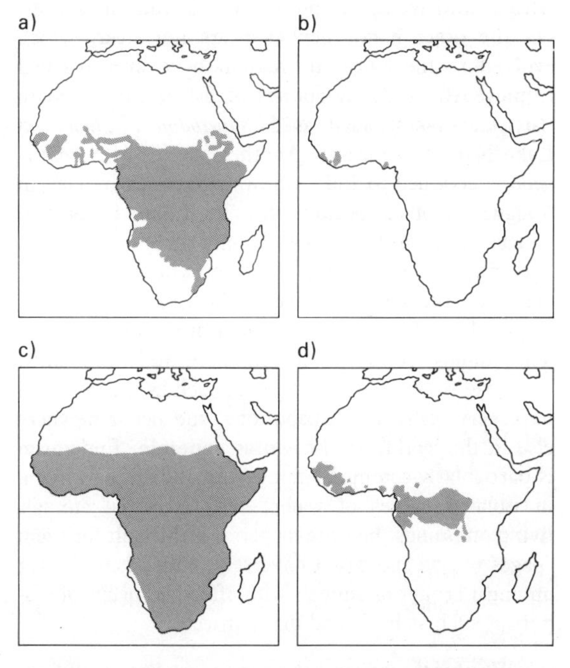

Figure 5.1 Distributions of some African mammals. The hippopotamus (a) is widely distributed in tropical Africa but its close relative the pygmy hippopotamus (b) has a very restricted distribution. The aardvark or antbear (c) has a continuous distribution in all favourable habitats over a vast area, whereas the distribution map of the chimpanzee (d) shows it to be found in at least eight separate areas, between which it is absent.
(Based on Dorst & Dandelot, 1970.)

patchily distributed within their overall range. For example, the aardvark (Figure 5.1c) is not present in every square kilometre of tropical Africa, as the map might seem to imply. Detailed studies would be needed to show exactly where it does and does not occur within tropical Africa. There are very few species for which this has been done (and the aardvark is not one of them). The distributions of many species, probably most, are fragmented, i.e. divided into a number of separate populations, as in the case of the chimpanzee (Figure 5.1(d)). Interchanges of individuals between the separate populations happen only rarely; if they cease completely, then the different populations could eventually become separate species (see chapter 1). This process is reflected in the occurrence of different forms of species (i.e. races, strains or subspecies) in different places, or sometimes in the same place. A familiar example is our own species.

The present-day distributions of animals can be explained in terms of various factors, thus:

(a) *Where they evolved* The chimpanzees, zebras, ostriches, guinea fowls and many others evolved in Africa and have never spread much beyond its shores. On the other hand, neither bears nor deer, which evolved in the Palearctic region, have ever reached tropical Africa. Many species of fish are confined to particular lakes, such as *Sarotherodon alcalicum* in Lake Natron, Tanzania. Another species of *Sarotherodon* is confined to Lake Chilwa and a nearby lake in Malaŵi; another occurs only in Lake Bosomtwi, Ghana, whilst in Lake Victoria only three out of about 130 species of cichlid fish occur anywhere else. There are some 180 species of fish in the Lake Chad basin, and at least twenty five of them are confined to it. They almost certainly evolved there and have failed to spread to other waters. Species such as these, which have very restricted distributions, and occur nowhere else in the world, are known as **endemic**. Endemism is particularly common amongst aquatic animals living in isolated bodies of water, and terrestrial animals living on islands, because dispersal is difficult for them (pages 67, 70 and 71). Conversely, endemism is rarer amongst groups of animals with effective means of dispersal, such as birds and many insects.

(b) *Whether the species' habitat is present* You would not expect to find a chimpanzee in the desert, nor a cattle egret in a forest. Each species has particular tolerances and requirements. Discovering what these are helps us to know where to look for it.

(c) *Human activities* Man has had a profound effect on the distributions of many species, especially during this century. As a result, the ranges of some species, such as most pests, have been greatly extended, whilst many other species, such as lions and ostriches, have disappeared from areas where they used to occur. Some have become extinct already, and it seems almost inevitable that many more will suffer the same fate (page 128 and chapter 8).

The second and third of these factors affecting distribution are generally similar to those affecting abundance (chapter 7). Elephants provide a good example. Their preferred diet consists largely of grasses and herbs, but they need trees too, for browsing and shade, and water for drinking and bathing. They are also prime targets for poachers. Hence their present distribution in Africa coincides largely with areas of low human population and moderate or high rainfall: they are absent from deserts (Figure 5.2). Within their overall distribution they are most abundant where there are least people (e.g. in National Parks) and where grass, suitable trees and water are also plentiful – that is in wetter areas or near to permanent water.

Components of environment

A major concern of population ecologists is to discover why particular species are common or abundant in some places, but rare or absent in others. The ecologist can approach his study in several ways. One way is to imagine the world as it might seem to the animal being studied, and to ask oneself, if I were that animal, what would be most important to me? This philosophy led the Australian ecologist H.G. Andrewartha to develop the idea that **an animal's environment is anything which may influence its chance to survive and reproduce**. He restricted the idea of environment to those things which were likely to affect the animal directly. Andrewartha recognized several different aspects or **components of environment**: the scheme we follow is a modification of his. Taken together, the components of any given species' environment determine its survival and reproductive rate, and hence its population size. Since some of the components vary from place to place, the populations in different places would be expected to vary too, in size or in composition.

We can group the components of environment which an ecologist needs to consider under five main headings, namely:

(a) *The physical environment*, which in turn has various aspects, different for terrestrial and aquatic species;

(b) *Resources*, these are the animals' requirements from their environment;

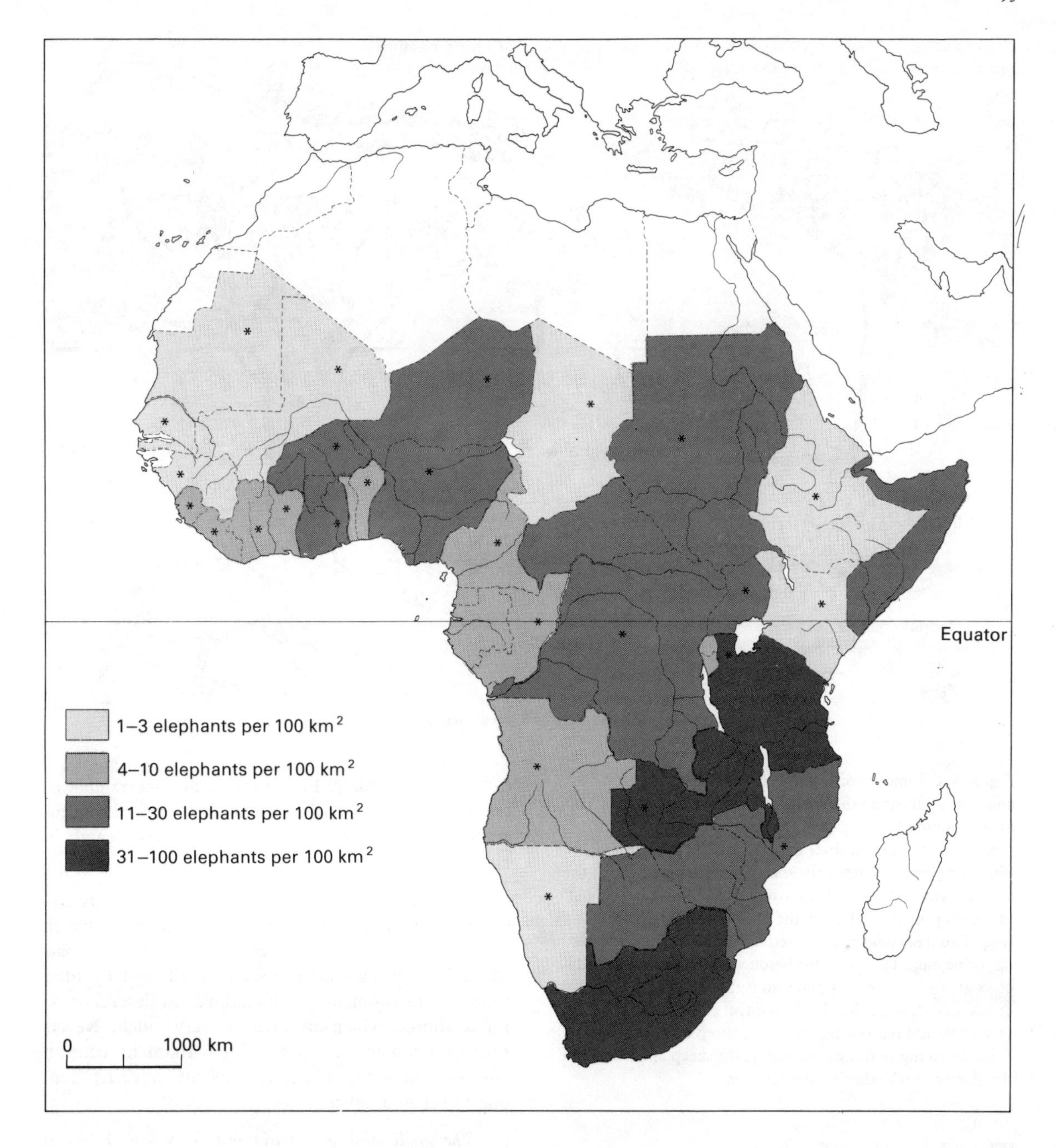

Figure 5.2 The distribution and abundance (numbers 100 km^{-2}) of elephants in Africa in 1980; compare with map of rainfall (Figure 2.3). In 1980, elephants occupied an area of about seven million square kilometres, approximately 23 per cent of Africa, but this is declining rapidly especially in some countries.*
(After Douglas-Hamilton, O., 1980. *National Geographic*, **158**, 568–603.)

(c) *Other individuals of the same species*, such as mates and competitors;

(d) *Individuals of other species*, including predators and parasites;

(e) *Man* and his various activities.

Figure 5.3 illustrates this idea of components of environment, which we shall now examine in some detail by taking each in turn.

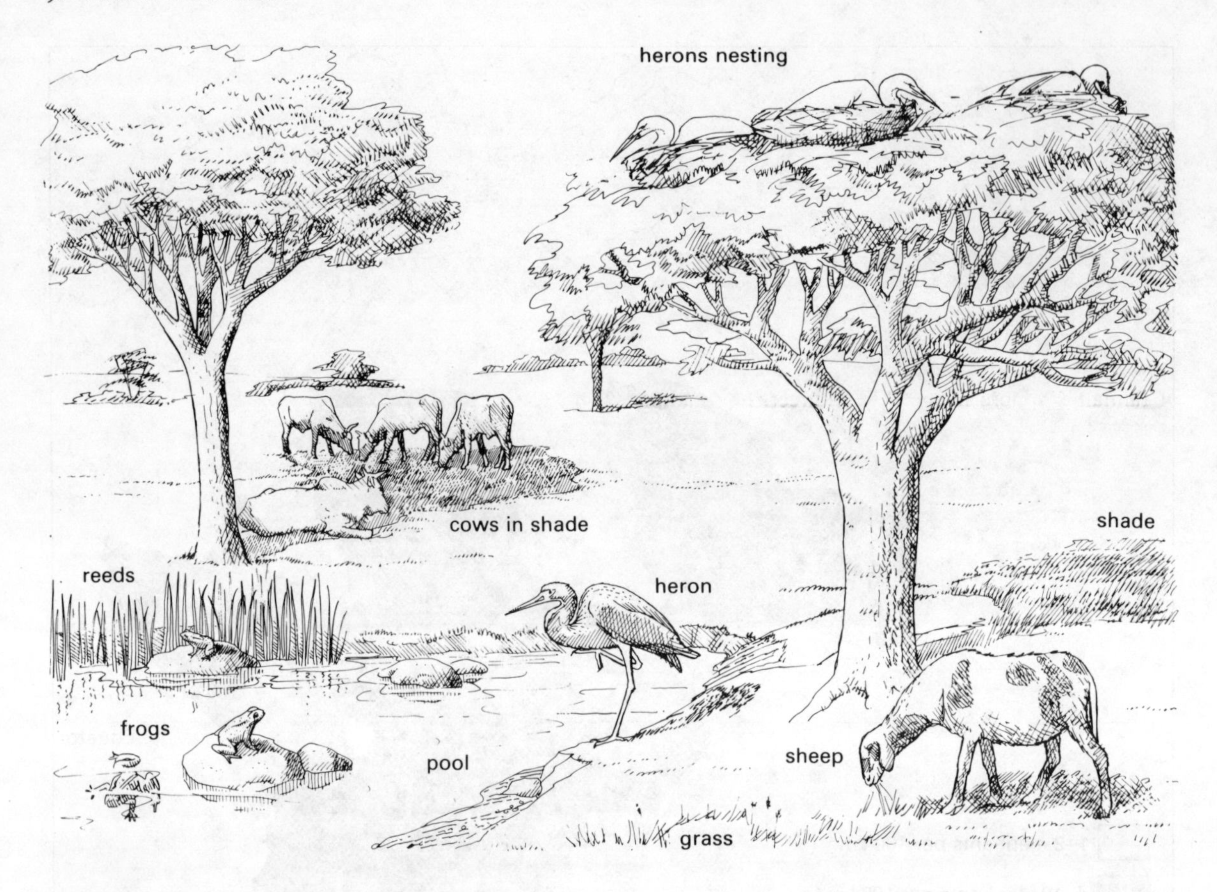

Figure 5.3 Some aspects of the environments of terrestrial animals, to illustrate the special ecological sense of the term **environment**. The frogs and the heron are parts of each others' environments, since the heron might eat a frog. Similarly, grass is part of the sheep's environment whilst the trees provide shelter (in the form of shade) for the cows and the reeds provide shelter (in the form of a hiding place) for a frog. The tree could also be used by the heron for perching on, or nesting. However, the heron and also the sheep and cows are unlikely to be significant components of the tree's environment, since they have no direct effect on its chances to survive and reproduce, unless the sheep eats the tree's fruits or young seedlings. Similarly, the sheep and heron are not parts of each other's environments.

The physical environment

The physical environments of aquatic and terrestrial animals differ so fundamentally that we need to start by considering them separately. However, there are also similarities: the physiological effects of temperature are much the same for all animals, and their responses to light are comparable.

There are three aspects of any particular variable which determine its significance to animals. These are:

(a) *Its average value*. For instance, average rainfall or windspeed in terrestrial environments, oxygen concentration in aquatic ones, temperature or day-length in either.

(b) *How much it varies*. In some places the physical environment is practically constant – for example in rainforest soil, or in deep lakes or oceans. Animals living in such habitats are scarcely affected by their physical environment, unlike animals in deserts or on rocky shores, which are affected very much. Nevertheless, although the physical environment beneath any one rain forest is nearly constant, it varies from one forest to another.

(c) *The predictability of variations*. Day length varies in a highly predictable way, and so does the interval between high tides. Other variables, such as rainfall, temperature and water currents can change suddenly, although an underlying pattern is usually discernible. These less predictable variables are often the most important in terms of survival.

The sum of the physical conditions at a particular place is its **climate**: whilst the conditions at a particular time are referred to as the **weather**. In chapter 2 we considered the importance of microclimates for plant survival, and they can be equally significant for animals, particularly small ones. If we want to understand the effects of changing temperatures on termites, for example, we should measure the temperature where the termites actually are, which is not, as a rule, inside a Stevenson's screen (which is where meteorologists place their instruments).

There are many ways in which animals are affected by their physical environments. Rising temperatures have a direct physiological effect which in turn can directly change an animal's behaviour. But the effects of rainfall are more likely to be indirect, for example by changing the food supply. Changing day length provides a signal to some animals, initiating their breeding cycles, whilst for diurnal predators, day length determines how much time they have for hunting. What all of these examples have in common is their influence upon animal's chances to survive and reproduce, which was how we defined components of environment.

Aquatic environments

In this section we shall consider the major features of the physical environment as they affect aquatic animals, except for temperature, which is of greater significance to terrestrial animals and hence comes under 'Terrestrial environments' (pages 99–105).

Water and its solutes

Living cells contain a great variety of substances, many of them in solution. A few of these substances are also found in the sea, rivers and lakes, but almost always in concentrations differing from those of living cells. So for life to continue, organisms need to have some means of keeping their internal environment distinct from that outside. They do this largely by surrounding themselves with an **integument**, in the form of cell membrane, cuticle or skin, which in aquatic organisms reduces the flow of dissolved substances (solutes) in and out. Even the best integuments are not perfectly impermeable, so animals have to dispose of solutes entering in excess of their needs, or make good losses of solutes that leak out.

Aquatic environments show considerable variations in their physical properties, including salinity and the concentration of various solutes (Table 5.1). Traditionally, the adjective marine is applied to seas, oceans and coastal waters, and freshwater to rivers and lakes. But the water in some African lakes has a very high salinity, so we shall refer to them as being inland waters rather than freshwater.

Animal species vary in their abilities to cope with differing environments. Thus the distribution of the

Table 5.1 Physical characteristics of a range of African lakes, and for comparison, the open sea. Note that the figures are only approximate; fluctuations occur, and can be considerable in small and/or saline lakes. The majority of rivers have low ionic concentrations. Phosphorus can also be important as a limiting resource (see pages 51 and 52)

Lake	Location	Salinity† (g litre^{-1})	pH	Ionic concentrations (milli-equivalents per litre)		
				Na^+	Ca^{2+}	Cl^-
Tumba	Zaire	0.02	4.8	–	0.03	–
Victoria	East Africa	0.09	7.5	0.4	0.3	0.1
Chad	Chad	0.17	8.2	0.5	0.8	0.0
Malaŵi	Central Africa	0.19	8.5	0.09	1.0	0.1
Tanganyika	Central Africa	0.53	8.5	2.5	0.5	0.8
Turkana	East Africa	2.48	8.6	35.3	0.3	13.5
Chilwa*	Malaŵi	16.7	10.8	142	1.0	88
Nakuru	Kenya	45	10.3	217	4.0	34
Magadi	Kenya	74	10.0	1650	0.4	640
Sea water (average)		15	7.9	500	5.1	550

(Mainly adapted from Beadle, 1981)
* maximum values, after prolonged drought
† total solutes

less tolerant species can be restricted to one, or a few, lakes of a particular salinity. Likewise, almost all marine species are confined to the sea, very few inhabiting inland waters too. Generally speaking, tolerance to changes in the concentration of solutes in their environment is low in animals with a relatively permeable integument, such as protozoans and cnidarians. Animals with less permeable integuments, such as fish, have a greater tolerance, but they still gain or lose ions through their gills and digestive systems. However, there are fish, such as the Atlantic salmon, which can survive in the sea as well as in rivers.

Insects living in water are secondarily aquatic; that is, they have evolved from terrestrial ancestors which, in turn, had aquatic ancestors. Some insects, such as dragonflies, mayflies and mosquitoes, are only aquatic in their immature stages. Only a few insects (e.g. some hemipterans and beetles) have adult stages that are totally or partially aquatic. The insect integument or cuticle is relatively impermeable, and in most species is coated with a waxy monomolecular layer which greatly reduces water loss. Many aquatic insect larvae, however, have a permeable cuticle lacking the waxy layer, and their anal papillae (so-called gills) are usually even more permeable than the rest of the cuticle. The anal papillae of mosquito larvae play an important part in osmoregulation and the uptake of salts from the water. Larvae in water where the concentration of salts is very low develop larger papillae than those in water of higher ionic concentration. Since most aquatic insects live in waters with low salinity, they suffer from water gain rather than water loss, and they absorb very little water from the excreta in the rectum. This is in marked contrast to most terrestrial insects.

Aquatic animals living in temporary habitats such as small rivers, pools and dams, have an additional problem to face – the risk that the water in which they live may evaporate. They sometimes tackle this risk of desiccation very successfully, as we shall see on pages 129 and 130.

Figure 5.4 Larvae of simuliid blackflies live in flowing waters. They are sedentary for much of the time, attaching themselves firmly to submerged rocks, stones and vegetation by the circlet of small hooks at the posterior end (a). Using these and a smaller circlet of hooks on the proleg, they move in a looping fashion (b). Larvae have very large salivary glands that produce silk-like threads with which they spin their cocoons and attach them to a rock or submerged vegetation (c). The silk is also used by larvae moving downstream. The silk is sticky, and the larvae having produced a blob of it on a stone can let go, and be carried downstream in the water current on a fine silken line; they can then reattach themselves to a stone further downstream. This is rather reminiscent of small spiders being dispersed by wind whilst attached to long silky threads.
(Based on Service, M. W., 1980. *A Guide to Medical Entomology*, Macmillan: London.)

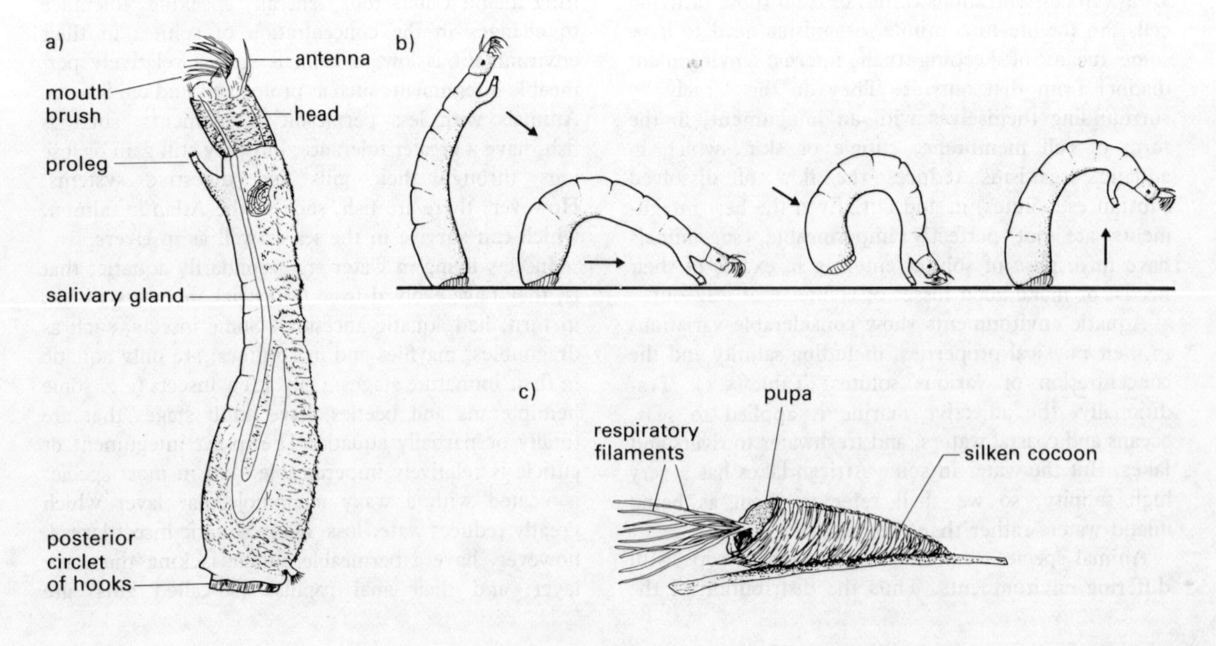

Oxygen

Just as the concentration of solutes can vary considerably, so can the availability of oxygen. Oxygen is used up in respiration by both plants and animals, but green plants can produce more by photosynthesis than they use in respiration. Oxygen also enters water directly from the atmosphere, particularly in fast-flowing rivers and on lakes and seas with waves. The sources of oxygen are near the surface in both cases, and consequently oxygen concentration drops rapidly with depth. As a result, all plants, and most animals are confined to the upper layers of the water. However, there are animals adapted to survival on minute amounts of oxygen and thus able to live at great depths. There are also some important aquatic animals that breathe air – various arthropods, adult amphibians, sundry reptiles including turtles and crocodiles, and many species of birds and mammals. Many of these swim on the surface, but others dive. Some show striking physiological adaptations, allowing them to remain submerged for long periods: the sperm whale can submerge for more than an hour, and reach depths of 1000 m.

Water currents

The density of water is some 800 times greater than that of air, and water currents can exert proportionately greater forces than winds. The distributions of many aquatic animals are related to their ability to survive water currents. Thus fish such as barbels (Family Cyprinidae) are characteristic of fast-running streams, whereas most cichlid fish are confined to still or slow-moving waters.

Animals inhabiting fast-flowing rivers are likely to show special adaptations. A few species are virtually sessile, such as the larvae of blackflies, *Simulium* spp., which adhere to submerged stones or vegetation by means of a posterior ring of hooks (Figure 5.4).

More active animals, such as dragonfly nymphs and many fish are able to walk or swim against a strong current although they avoid doing so if they can. Weak swimmers are inevitably carried downstream whenever they move into the current. The larvae of some insects, including caddisflies and mayflies, are subject to this **drift** but the adults, when they emerge, fly back upstream and lay their eggs near to where they themselves hatched.

Even more rigorous as a habitat is a rocky shoreline exposed to breaking waves. But whilst some species are adapted to water currents or to wave action, many are not, and are confined to sheltered waters. Thus schistosomiasis (bilharzia) is much commoner in places near to slow-flowing rivers or ditches than along fast-flowing rivers, because the host snails survive best in water with little or no current.

Light

Water is not only denser than air, it is also much less transparent. Even when pure, the maximum depth at which light can support photosynthesis (the **euphotic zone**) is about 50 m. Most water contains suspended particles, however, and light penetration can be reduced to 0.5 m or even less. Soil erosion leads to increased silt loads in rivers, and this can interfere with the functioning of fish gills as well as reducing light. The effects of light on plants, added to those of oxygen, affect animals. The vertical movements of zooplankton provide a good example of this. Some of the phytoplankton, upon which they feed, move up and down in response to the depth of penetration of sunlight, and the zooplankton follow (Figure 5.5).

Although not a physical component of water, its chemical composition can be very important. The data in Table 5.1 show that great differences exist between lakes. The effects on animals of the varying concentrations of individual solutes are mainly indirect, influencing physiological activities such as nervous conduction which depends upon movements of ions. These functions are affected if the ions are in short supply. Increasingly, however, pollution (pages 219 and 220) is adversely affecting aquatic animals, either through poisons or by causing eutrophication (Figure 3.16).

Terrestrial environments

Water

As a medium for supporting life, air is very different

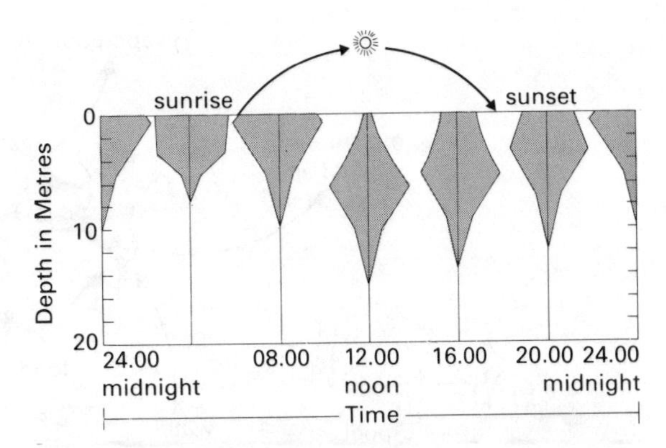

Figure 5.5 A good example of a 'component of environment' which acts indirectly. The diagram shows the proportions of freshwater planktonic animals at various depths and at different times of day. The width of each polygon represents the number of animals, which move up and down following the phytoplankton on which they feed. These in turn are responding to sunlight, going deeper when it is brighter. The correlation between the numbers of animals and plants is so good that an ecologist studying the animal would be justified in recording the amount of light, which acts *indirectly* but is easily measured, rather than plant numbers, which affect the animals *directly*, but are harder to measure.
(From Whittaker, 1975.)

from water. Unlike aquatic animals, those on land never experience a disappearance of the medium in which they live; but they still need water, and its conservation can be a major problem. Death from desiccation can happen very quickly. For example, it will kill a termite worker exposed to the heat of the sun for more than a few minutes. Terrestrial animals lose water through their general body surface, from respiratory organs, and in the disposal of wastes (Figures 5.6). In most animals, loss of water through the skin occurs when the relative humidity of the air is less than about 98 per cent, although the exact value depends upon the species and the temperature. At humidities near to saturation water is absorbed by the body. Under certain circumstances, a few arthropods (including some ticks) can absorb water from air at relative humidities as low as about 50 per cent, but only by expending a considerable amount of energy.

Like aquatic forms, terrestrial animals need to maintain a relatively constant amount of water in their bodies if they are to survive. To do so, the water gained in various ways must equal that lost in others (Figure 5.6). Most food contains some water; in living grass, for example, the water content is usually between 65 and and 85 per cent, although it may fall below 20 per cent in dry grass on a hot dry day. In addition, water is produced within the animals' bodies as a result of normal metabolic processes, particularly

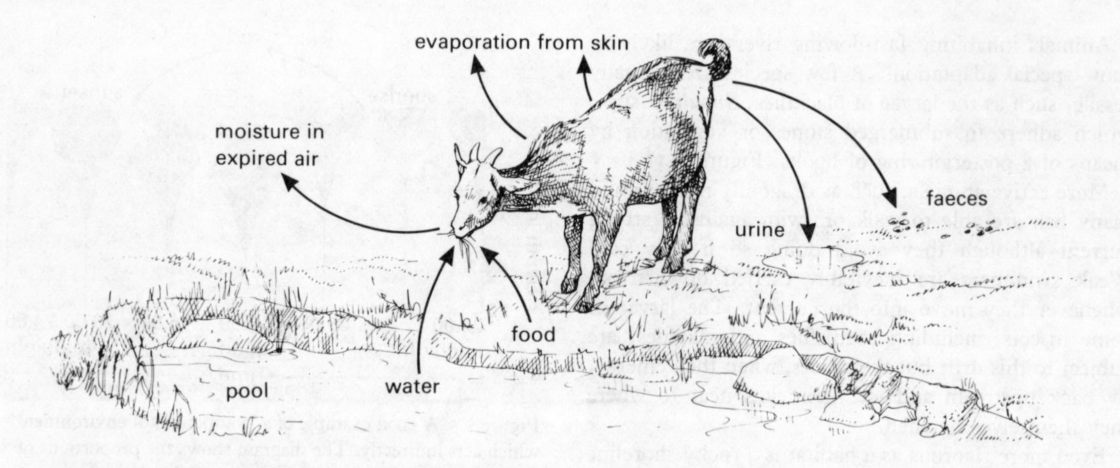

Figure 5.6 There are several ways in which water enters and leaves a terrestrial animal, such as this goat. For the composition of its body fluids to remain constant, the water gained from food and drink must equal the total lost through expiration, faeces, urine, and from the skin. The last includes both sweating and losses through epidermal cells.

the oxidation of fats and carbohydrates. For example, every molecule of glucose yields six molecules of water during cellular respiration. These two sources, food and metabolism, provide enough water for many terrestrial animals, but others, especially some birds and mammals, also need to drink. Their distributions are then restricted by the availability of water.

Arthropods have various sources of water, apart from food. Some drink from droplets of dew, or the exudations of plants, and certain spiders can suck capillary water from soil. Butterflies in drier environments may be seen to congregate where a mammal has defaecated or urinated.

The problems associated with conserving the body's water are greatest in arid environments, and reach extremes in deserts. Camels, and some other ungulates such as Dorcas' and Grant's gazelles, and the two species of oryx, can survive indefinitely without drinking at all, although if water is available they take it. The same is true of various birds in dry regions. Domestic livestock, in comparison, are very water-dependent; most breeds of cattle and sheep for example need to drink at least once in two days.

Reducing water loss from the body minimizes the amount of water an animal needs to take in. Some land animals, notably annelids, nematodes and molluscs, have comparatively permeable integuments, and so their activities are largely confined to times and places where there is moisture. This severely restricts their distributions.

Almost all successful terrestrial animals belong to two of some twenty five phyla in the animal kingdom, the Arthropoda and Chordata. These two phyla contain about twelve classes, but almost all of the terrestrial species are contained within the five classes Arachnida, Insecta, Reptilia, Aves and Mammalia. What these groups have in common, unlike most of the rest, is a very waterproof integument, constructed in such a way that evaporative loss of water is reduced to an extremely low level. Other adaptations include reduction of the water content of urine and faeces.

Animals adapted for life in dry places produce dry faeces; most of the water in their guts is absorbed by the rectum. Many reptiles and birds, as well as the majority of terrestrial arthropods, excrete insoluble uric acid, thus disposing of their toxic nitrogenous wastes with very little loss of water. Although descended from reptiles, mammals produce urea, not uric acid. Consequently they cannot avoid losing water when they excrete because urea is highly soluble; it cannot be precipitated and has to be voided in solution. Nevertheless, mammals from hot and dry climates minimize the loss by producing very concentrated urine.

Arthropods living in dry habitats possess cuticles which are highly impermeable to water. The main cuticular layers have a compact molecular structure, and water loss is further restricted by the electrochemical properties of the outermost lipid layer. The net effect is to reduce water-loss to an extremely low rate: less than 0.1 mg per square centimetre of cuticle per hour at 25 °C. (The loss from a free water surface at the same temperature would be several hundred times greater, in still air and a humidity of 50 per cent). But at higher temperatures (30 °C for cockroaches and 48 °C for locusts, for example) the structure of the epicuticle changes, water loss increases rapidly (Figure 5.7), and the insect soon dies from desiccation. Insects also lose water from their tracheal systems. Thus a

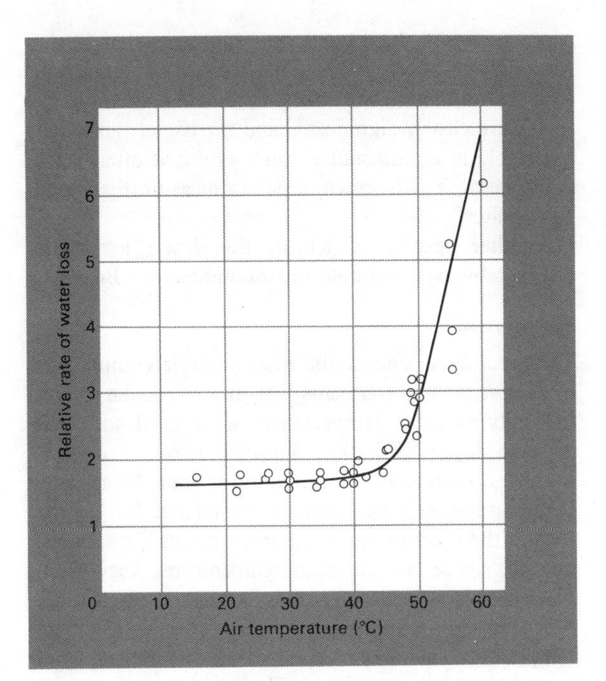

Figure 5.7 The rate of water loss through the cuticle of the migratory locust. In all insects, the rate increases steeply above a certain critical temperature, which in this case is about 40 °C. In hot weather, locusts try to avoid places with high temperatures (see Figure 5.11). (The unit of water loss is a complex derived one, and is not given here.)
(From Loveridge, J. P., 1968. *Journal of Experimental Biology*, **49**, 1–13.)

locust losing water at a rate of about 5 mg cm^{-2} h^{-1} can cut this to less than 1 mg cm^{-2} h^{-1} by reducing the ventilation of its tracheal system, and partly closing its spiracles. Water resorption by the rectum is very effective in many terrestrial animals. In tsetse flies, for example, the water content of the faeces is around 75 per cent when the ambient (i.e. surrounding) air is humid, but reduced to about 35 per cent when the air is dry.

The opposite problem – too much water – can happen at times. A locust feeding on green grass can take in 2 ml of water a day, but the total volume of its body fluid is less than a quarter of this. The excess water is removed through the Malpighian tubules, under the control of a diuretic hormone.

Biologists have long been interested in how desert animals are able to survive their harsh environment (such species are sometimes called **xeric**, in contrast to **hygric** ones living in moist places, and **mesic** species of intermediate habitats). The kangaroo-rats or jerboas have been studied extensively as examples of xeric species. One North American species (there are others in Africa) has been found capable of indefinite survival without drinking, obtaining water only from its food of air-dried barley seeds. Each 100 g of seeds yielded 54 g of metabolic water. About three-quarters of this was lost by evaporation, mostly in the expired air from the lungs. This rate of respiratory loss is considerably lower than in mesic species of comparable size, because jerboas have long nasal passages where a heat exchange process enables up to 80 per cent of the moisture in the expired air to be condensed and resorbed. Furthermore, the jerboa's urine is highly concentrated and its faeces almost dry.

A more famous example of a xeric animal is the camel (Figure 5.8). It, too, has an enlarged system of nasal passages which are thought to act as heat sinks where water is resorbed. Camels deprived of water have been found to exhale air with a relative humidity of only 75 per cent – the figure would be close to 100 per cent in mesic or hygric mammals. These and other adaptations allow camels to survive for up to two weeks without drinking, despite living in hot deserts. A camel can tolerate a loss of 25–30 per cent of its body water, whereas mesic species, such as dogs or humans, would die before losing 15 per cent. Camels can drink unusually fast, and do so when water is available. Contrary to popular belief they cannot store water, but instead are very efficient at conserving what they have. Camels are also helped by their matted fur, which reduces the amount of heat reaching the body surface.

A further adaptation of camels, and some other xeric species of birds and mammals, is their ability to tolerate fluctuations of the body temperature. By

Figure 5.8 Camels are extraordinarily well adapted to life in hot, dry places. Their hump, however, contains fat, not water.
(D. K. Jones.)

contrast, mesic species sweat profusely on hot days to prevent their body temperatures from rising; and in so doing they inevitably lose water. The camel, instead of sweating, allows its temperature to rise, sometimes to above 40 °C, which would soon be fatal to humans and most other mesic species. At night, the camel's temperature may drop as low as 34 °C. These fluctuations can reduce water loss in a camel of 500 kg by about 4 litres a day. Such adaptations to life in dry environments help to explain the surprising range of animal species that inhabit them. But the conservation of water, for example by its resorption in the rectum and kidneys, uses considerable quantities of energy, which in turn requires a high rate of food consumption.

Survival in terrestrial animals is often linked to rainfall. An example is shown in Figure 5.9 which compares the number of red locusts with rainfall in the same locality some 18 months previously. Rain increased the growth of vegetation and two important

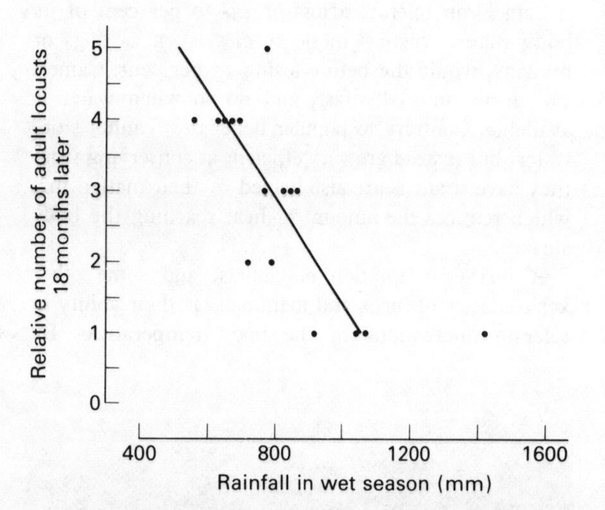

Figure 5.9 Populations of the red locust are influenced, probably indirectly, by rainfall. The graph shows a measure of locust densities near Lake Rukwa, Tanzania, plotted against the rainfall one-and-a-half years previously. Biological data often show a wide scattering of points, as here. In this case the scattering suggests that rainfall, though important, is not the only factor involved.

Nevertheless, the abundance of locusts one and a half years after a wet season is significantly correlated with that season's rainfall ($r = -0.811$, $P < 0.001$). The calculated regression can be used to predict the likely level of future locust populations. For example, when there is a wet season with rather low rainfall, an outbreak of locusts some 18 months later can be expected, and the planning of control measures would be justified.

(Redrawn from Symmons, P., 1959. *Bulletin of Entomological Research*, **50**, 507–21.)

components of the locusts' environment, that is the resources of food and shelter were thus increased. However, although the effects of rainfall were indirect, they were clearly significant, and the use of rainfall as a variable in circumstances such as this is often justifiable because it is much easier to measure than food and shelter.

Another species of locust, the desert locust, is affected by both rainfall and humidity: see Box 6.5.

Temperature

Temperature is one of the most variable components of terrestrial environments. At any particular place, the daily range of temperature may exceed 50 °C, as in some desert habitats, although it is only 1 or 2 °C in others, such as in litter on a forest floor. Mean daily temperatures can also change seasonally, but in the tropics the daily range commonly exceeds the seasonal one (cf. page 9). Average temperatures vary from place to place, and are closely correlated to altitude. Since data on altitude are often easier to collect than temperature data, many ecological studies in the tropics record altitude as a variable. There is, however, no evidence that animals can respond to altitude (although some insects respond to changes in barometric pressure).

Temperature affects animals in various ways, and these effects are often more pronounced in exotherms than in endotherms (Figure 5.10). The body temperature of an exotherm largely reflects the **ambient** temperature, i.e. that of the surroundings. All of its activities are directly affected by temperature changes, since the rates of all metabolic processes are determined by temperature. Whenever an exotherm's body temperature is below the optimum, its speed of movement is reduced, and this in turn affects its ability to search for food, or escape from predators.

The specific heat of air is very much lower than that of water, and air temperatures can therefore change much faster. Hence temperature changes are usually of greater importance to terrestrial animals than to aquatic ones. But water at high altitudes in the tropics sometimes freezes, and this requires special adaptations in the few aquatic species inhabiting such places.

As one would expect, most species are adapted to the mean temperatures of their normal habitats, which are usually close to their optimal temperature at the times when they are active. Species vary in their tolerance to temperature variation and in their ability to evade extremes of temperature by moving to more favourable places.

The effects of temperature on insects have been studied extensively, and they provide many examples

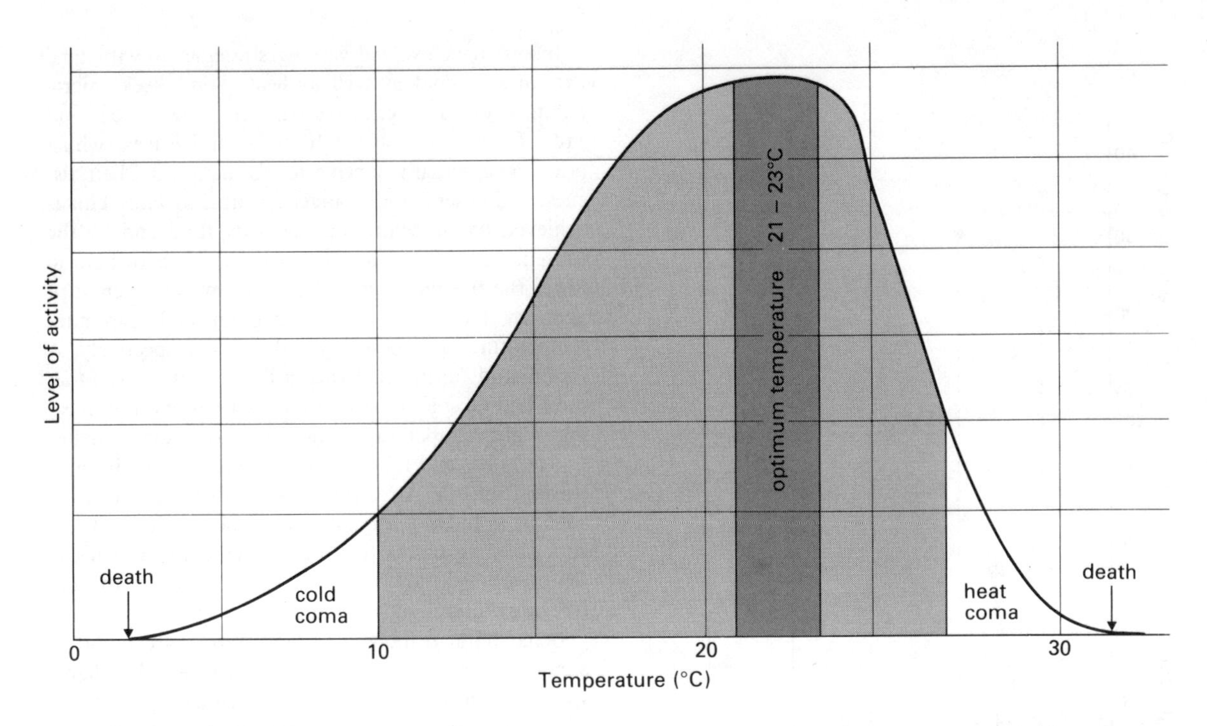

Figure 5.10 A diagrammatic representation of the effects of temperature upon a representative exotherm. The temperature values vary considerably between species, but in this imaginary animal the optimum temperature is between 21 and 23 °C. It would be lower for a species living in a generally cooler climate. Activity decreases outside this range. Below 10 °C the animal ceases moving: it is in a **cold coma**, and would die if the temperature fell below 2 °C, the **lower lethal limit**. Similarly, it would go into a **heat coma** above 27 °C, and its upper lethal limit is 32 °C.

to illustrate the generalizations in the preceding paragraphs. For instance, the time taken by a female mosquito to digest a blood meal is two days at 27 °C, but a whole week at 15–18 °C. Growth rates (often referred to as speed of development in invertebrates) are affected too. Cotton-stainers (which are species of Pyrrhocoridae bugs) can complete their life-cycle, from egg to egg, in about twenty five days at 30 °C, but take thirty five days at 25 °C, and may need as many as forty days if temperatures drop to 20 °C. Below 20 °C, the life-cycle is unlikely to be completed. This prolongation of development times also increases the time that the insect is exposed to predation.

Many animals respond to changes in temperature by moving. The favourable temperature range for the desert locust is from 17 °C to about 40 °C, with an optimum in the low thirties. Below 17 °C it is in a cold coma, whilst between 17 and 20 °C, it can move only slowly, and orientates itself at right angles to the sun, to maximize absorption of radiant heat (Figure 5.11). Flight is only possible if the ambient temperature is above 22 °C, and temperatures of about 30 °C are necessary for sustained flight. A locust can increase the temperature of its flight muscles by vibrating its wings, which it does when the ambient temperature is low. On the other hand, if exposed to high temperatures (and the bare ground in a desert can reach 80 °C on a hot afternoon) the locust would go into a heat coma if it remained sitting on the ground. However, if the ground temperature exceeds 38 °C, it raises its body by extending its legs, whilst ground temperature above 43 °C cause it to climb grass stems, to a height where it is cooler (Figure 5.11).

Flight muscles function effectively within a narrower range of temperatures than those used in walking and swimming. During flight, the muscles generate heat, and honey bees, for example, can maintain thoracic temperatures of about 30 °C when the ambient temperature is as low as 8–10 °C; when it is colder than that, they do not fly. High temperatures also inhibit insect flight; 38 °C is the maximum for sustained flight by locusts (See also Box 6.5).

These examples provide a further illustration of the need to be aware of microclimatic differences. The temperatures to which a locust on bare ground is exposed are quite different from those which a meteorologist would measure in a standard Stevenson's screen, in shade at a height of 1.5 m. Many plants and animals survive in small but favoured spots where

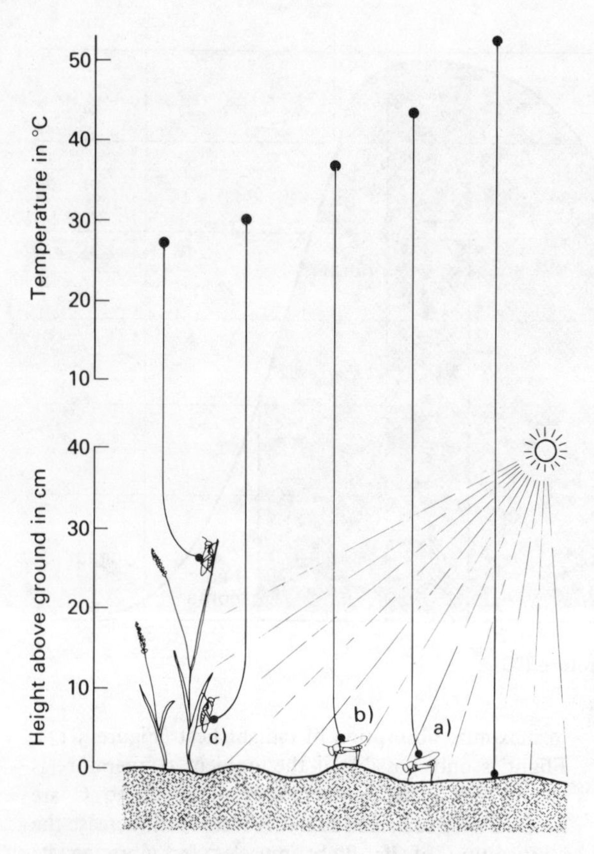

Figure 5.11 A small animal can sometimes change its physical environment by moving a very short distance, e.g. climbing a blade of grass, which will take it from a warm place to a cooler one. Locusts, as shown in this figure, respond to the midday heat of their surroundings by (a) orientating themselves parallel to the sun's rays, thus minimizing the area exposed to direct solar radiation; (b) raising their bodies on extended legs; or (c) climbing the vegetation.
(Data from Hardy, 1972.)

the microclimate meets their needs (see pages 13 and 14). Another good example is the tsetse fly, *Glossina morsitans*, which survives best in the laboratory at 30 °C and at relative humidities of between 19 and 44 per cent. Under these conditions, flies ingest more blood during feeding, and produce more offspring, than at other temperatures and humidities. Observations suggested that environmental conditions at Gadau, northern Nigeria, would only allow flies to survive for about five days during the heat of July. But in fact they survive much longer. They do so by avoiding the lethal temperatures, and seeking refuge in cooler microhabitats, such as by resting beneath large branches of trees.

Insects have evolved varying strategies to withstand extremes of cold as well as heat. Some seek microhabitats offering effective shelter (pages 109 and 110), for example, beneath stones and leaves, whilst others can actually survive ice formation in their tissues, and others evade danger by supercooling. This is achieved by concentrating the body fluid and by the presence of glycerol in the blood, which both help to lower the freezing point. On Mt Kenya at a height of 4200 m, the mean daily maximum and minimum temperatures are about +5 and −4 °C respectively. A recent study of insects living at this height showed that most had no special adaptations to withstand the freezing night temperatures, but instead sought shelter under stones and fallen plants, or amongst the litter of leaves from the Afroalpine plants. But a few insects, such as aphids, springtails and some beetles had low super-cooling points; two beetles were able to survive temperatures as low as −7 °C. Others could withstand freezing without supercooling.

Behavioural responses to temperature are found in almost all animals, including ourselves. Elephants spend the heat of the day in the shade (Figure 5.12). Their large size prevents them from behaving like many smaller mammals and reptiles which can escape the greatest heat by sheltering in a hole in the ground or a deep cleft in a rock. Hippopotamuses have a unique response for a mammal. Although their food is grass, they spend most of the day in water, grazing mainly at night (Figure 5.13). This of course restricts their distribution considerably, since they can only graze within walking distance of water, and few hippos travel more than three or four kilometres from water.

Figure 5.12 Shade is an important resource to elephants in hot weather. In places where there are few trees; they crowd into the available shade. Their relatively small surface area-to-volume ratio makes heat loss slow; they augment it by flapping their highly vascularized ears.
(D. K. Jones.)

Figure 5.13 Another method of heat disposal is to submerge the body in water, and hippopotamuses do so for much of the day; at night they leave the water in search of grass. (D. K. Jones.)

Elephants also cool themselves by bathing, and buffaloes and warthogs are fond of wallowing in mud during hot weather.

The direct effects of temperature on endotherms are much less than on exotherms – this, of course, is a major advantage of being an endotherm. Endotherms can control their internal temperatures, and thus remain active over the full range of environmental temperatures that they normally encounter. But, as we saw on page 45, the cost of endothermy in terms of energy is considerable, so it benefits a bird or mammal to avoid the coldest places. On the other hand, when the ambient temperature increases, so does the rate of evaporation of water, increasing water loss through epidermal cells of the skin as well as the loss through sweating.

A good example of the influence of temperature on an animal, and of the interaction between the effects of moisture and temperature, is found in the mosquito *Aedes vittatus*. The larvae inhabit rock pools, especially those left by seasonal rivers as they dry up. The species is widely distributed in Africa, as well as occurring in parts of Europe and Asia. Its northern extent appears to be controlled by temperature (Figure 5.14), since it is virtually confined to places with winter temperatures of at least 10 °C. Within these limits, the mosquito's seasonal abundance and behaviour is dependent upon rainfall, and the availability of rock pools, which in turn depends upon the distribution of suitable rocks.

In areas of southern Nigeria receiving at least 2500 mm of rain in a year, *Aedes vittatus* breeds throughout the year, because there are always some rock pools with water. In northern Nigeria, however, most pools are dry from October to April, and the species survives as dry eggs which only hatch when the pools fill. But not all of the eggs hatch on their first soaking at the beginning of the rains, some needing up to six soakings to stimulate hatching. The reason for the staggered pattern of egg hatching is that the first rains may fill the pools only partially, and if the temperatures are high they dry up, killing the larvae before they can complete their development. With each successive soaking the risks of desiccation decrease. Similar adaptations have evolved in other inhabitants of temporary pools.

Winds and other physical factors

The effects of wind can likewise be direct or indirect. Strong winds are a major hazard to flight, especially for fragile insects. Many insects avoid the risks of being blown over water or other unfavourable places by not flying at all when it is windy. Insects found in exposed places, such as deserts and mountains, are either comparatively strong fliers or have reduced wings. A measure of the risk of being blown off a mountain is the complete loss of flight by a number of insects, including various flies, which inhabit the higher parts of Mts Kilimanjaro and Kenya. The most important advantages of flight are easy dispersal and escape from predators. Presumably, for flightless insects, these advantages are outweighed by the risk involved in attempting to fly in a persistently windy environment. On the other hand, moderate winds can assist in the dispersal of insects which are weak fliers. This sometimes frustrates attempts at eradicating pests, which are brought into controlled areas from elsewhere by the prevailing winds.

Air currents, including winds, are important in the survival of many African birds. On a warm day, the reflected heat from different surfaces, such as forests and open ground, causes upcurrents in the air. These thermals enable vultures to soar for several hours a day, searching for food whilst using very little energy. Indeed, they are so adapted to soaring that they rarely attempt to use flapping flight for more than a short distance and during cool periods merely perch in trees waiting for the weather to improve.

We do not intend to offer an exhaustive list of physical factors, because it would be lengthy. We have not, for example, mentioned the risks of being hit by a falling rock, or struck by lightning, or drowned in a flood. These and other such hazards do happen, and may occasionally be of more than local significance. The important thing for the ecologist to bear in mind is that many things can affect the chances of animals to survive and reproduce, and that a study of any species' population has to identify the particular components that apply to it. For most species, the headings we have used in this chapter will provide a good initial framework.

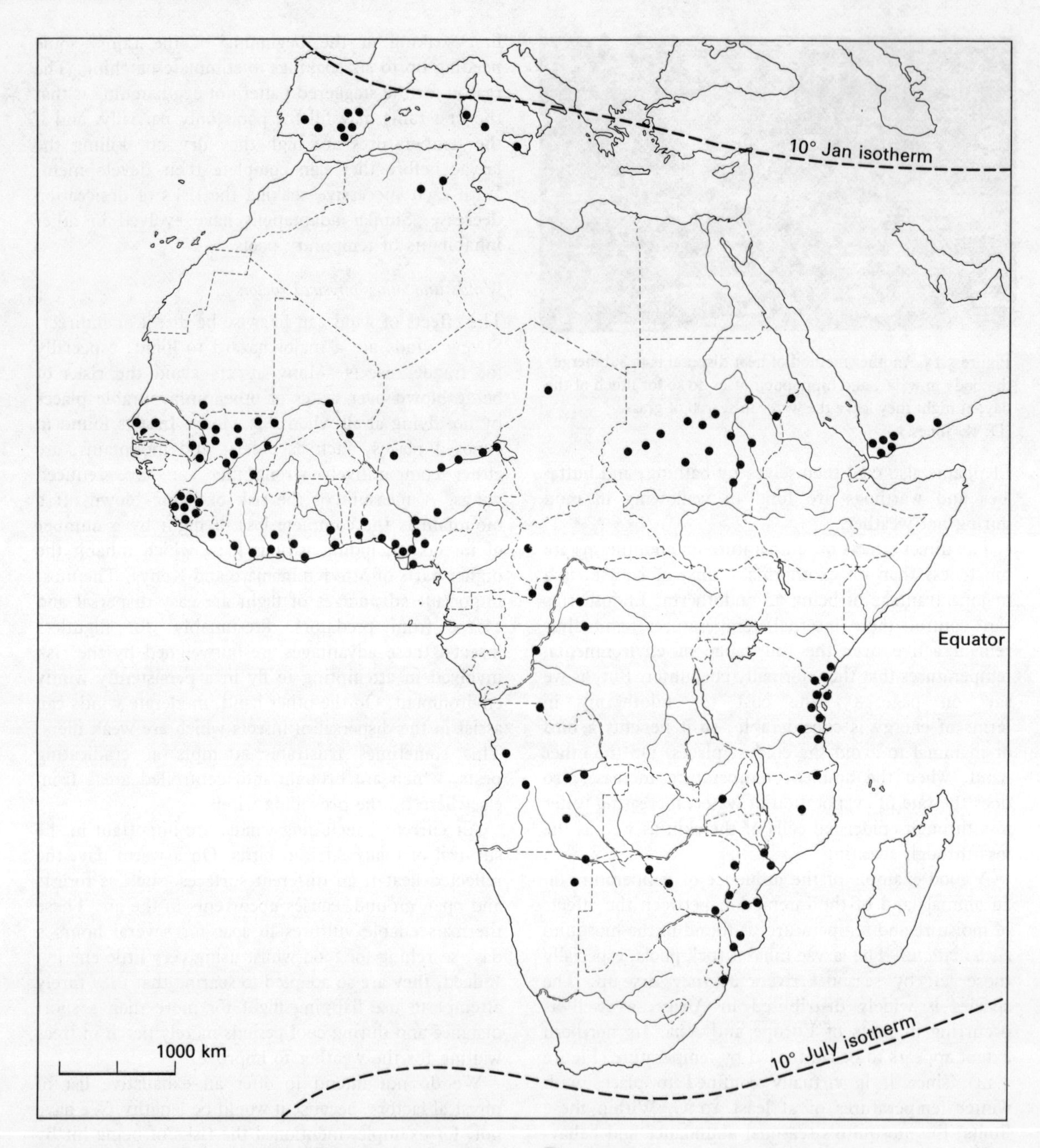

Figure 5.14 The distribution of the mosquito *Aedes vittatus*. Each point on the map represents a locality where this species has been collected. All of the localities lie within the mid-winter isotherm of 10 °C. There are doubtless many more places where this species occurs than those shown on the map, which reflects to some extent the distribution of scientists who collect mosquitoes.
(Service, M. W., 1970. *Transactions of the Royal Entomological Society of London* 122, 101–43.)

Resources

We defined resources as animals' requirements from their environment. Food is a resource for all animals, and the majority also require some form of shelter, at least when breeding. Oxygen is a resource too but only becomes significant ecologically when it is scarce, as in some aquatic habitats. Water for drinking is vital to some terrestrial species, but like oxygen was considered in the previous section. With the exception of a comparatively small number of species which breed asexually, mates are also essential; but they will be considered on pages 112, 113, 139–149, 145 and 146.

Food

Food is a major component of the environment of all animals and its availability often limits their distribution and sometimes their abundance. Food is one of the harder components to measure. As we shall see, the notion that 'cows eat grass', and comparable statements for other species, are gross over-simplifications. One reason for this is that, for most species, the quality of the food can be as important as the quantity. This is why many species are highly selective in what they eat (see for example pages 43, 44, 122 and 123).

Table 5.2 Approximate composition of various forage grasses when dry, as judged by the criteria commonly used in agriculture. The *starch equivalent* is a measure of the energy value of the grass; whilst *digestible crude protein* is a measure of the usable protein content: notice its low values

	Composition, as % dry weight		
	Total digestible matter	Starch equiv-alent	Digestible crude protein
Kikuyu grass (*Pennisetum clandestinum*)	15	7.5	2.4
Elephant (Napier) grass (*P. purpureum*)	24	8.5	0.6
Rhodes grass (*Chloris gayana*)	28	5.5	1.5
Sudan grass (*Sorghum arundinaceum*)	31	23	1.8
Guinea grass (*Panicum maximum*)	26	11	0.5

(Based on Ngugi, D. *et al.*, 1978. *East African Agriculture*, Macmillan: London.)

A cow provided with unlimited supplies of fresh, green grass, will produce good quantities of milk, it will maintain its weight, and its calf will grow quickly. With the same dry weight of grass at the end of the dry season, a cow in a marginal or semi-arid area will produce little milk, and is almost certain to lose weight, and moreover its calf will grow slowly, or not at all. In bad seasons, the calf may die, partly because dry grass contains insufficient protein (Table 5.2). Their diet will also have contained a smaller proportion of the more easily digested carbohydrates, and relatively more fibre (Figure 5.15). Under these

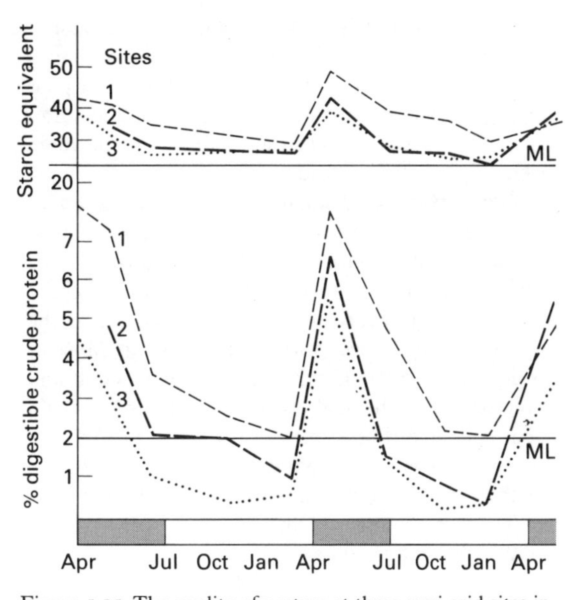

Figure 5.15 The quality of pasture at three semi-arid sites in Karamoja district, northern Uganda, over a two-year period. ▒ = months normally receiving > 50 mm of rain. Two measures are shown: the starch equivalent (above) is a measure of the energy value of the grass, whilst the digestible crude protein (below) is an indication of the usable protein content. The amounts of energy available remained above the minimum levels (ML) required by cattle for maintenance, but at sites 2 and 3 there were marked seasonal deficiencies in protein. Under these circumstances, the cattle would lose weight. As well as seasonal variations, there are considerable specific differences in the nutritive values of grasses (Table 5.2). Based on Pratt, D.J. and Gwynne, M.D. (eds) 1977, *Rangeland Management and Ecology in East Africa*, Hodder and Stoughton; London)

circumstances, the amount of energy needed to digest the grass may approach the amount of energy obtained from it. Since the cow also needs energy for its basic metabolic activities, and for locomotion, it suffers a net loss, even when the food is plentiful; it is said to lose 'condition' (pages 125 and 130).

The relationship between the quality and the quantity of food on the one hand, and the response of the animal on the other, is even more complex. When eating fresh green grass, a cow will derive more benefit from a given amount than a horse; similarly a buffalo will do better than a zebra. This is because the cow and the buffalo are ruminants, and their digestive systems are exceptionally efficient at extracting nutrients from forage of good quality; but rumination is slow. Non-ruminants like the horse and zebra compensate for their coarse diet by being able to process approximately twice as much food in a given time. So, if the non-ruminant is only two-thirds as efficient at extracting protein from grass, but can process twice as much as the ruminant, it will assimilate four-thirds as much (i.e. one third more) in a given time. Hence, when grass is plentiful but of poor quality, the non-ruminants do better than the ruminants. The ruminants' answer to this is to be more selective in what they take.

Another example of quality being more important to animals than quantity is that of plant-sucking bugs, such as aphids. Once an aphid has inserted its proboscis into a sieve tube of the plant, it is literally force-fed, the pressure of the sap ensuring that the liquid enters the insect without its sucking. Most of this liquid passes straight through the insect's gut, and is collected from the anal aperture by ants who obtain much of their carbohydrates in this way. But this, of course, is not why the aphids do it; their problem is that the concentration of amino acids in the sap is so low that a large total volume has to be processed to extract sufficient amino acid.

Even when the quantity of food in the animal's environment is adequate, and of a suitable quality for the animal's physiological needs, it still has to be found. This may be comparatively easy, as with a cow in a field and a wood-eating termite living in a piece of dead wood. Other animals spend considerable amounts of time searching for food. An extreme case is that of scavengers, such as vultures and hyenas. Some predators also spend much of their time searching. For example, lions commonly hunt for several hours without catching anything. Birds which prey on insects take much more time searching for food than actually eating it. In contrast, elephants feed for up to eighteen hours a day. The question of how an animal can best use its time to obtain as much food as possible is discussed further on pages 156 and 160.

Blood-sucking insects, once they have located a suitable host, appear to have an unlimited supply of food. But finding the host can sometimes be difficult. It is possible to identify the hosts of blood-sucking insects by using immunological tests on the blood in the insects' stomachs. These tests show that many blood-sucking insects are very selective. For instance, a study in Nigeria of the tsetse fly *Glossina morsitans* showed that it favoured warthogs, even though various wild bovids (especially antelopes) were more numerous. Similarly, in Tanzania and Kenya, *G. swynnertoni* preferred warthogs to impala, zebra or wildebeest, even when these were abundant. In these cases, the tsetse flies may well have been suffering from a **relative shortage** of food; the total amount of warthog blood in the area would have supported many more flies. Female tsetse need to feed every three days for successful reproduction and survival. So when warthogs were sparse, many flies will have failed to find any food, or will not have been able to feed as frequently as every three days. Those that succeeded in locating a warthog, on the other hand, will have had far more food available than they could possibly take. Contrast this with **absolute shortages** of food, when the total amount of food available is inadequate, regardless of the animal's ability to find it.

Amongst birds and mammals, and probably in other groups too, larger species tend to occupy larger territories or home ranges (Box 5.1). This is taken as evidence that their populations are limited by the availability of food, the argument being that a larger species requires more food and consequently needs a bigger area from which to obtain it.

Variations in territory size of a particular species from place to place probably reflect differences in food supply too, and hence also suggest that food is a limiting factor in such cases. Thus troops of baboons (typically forty to eighty animals) in forested areas of western Uganda occupy territories of about 5 km^2. In southern Kenya, troops of a closely related species require 15 km^2 or more in the Amboseli area, presumably because their food is widely dispersed.

When animals feed they are obviously affecting their own future food supply by eating it. How a particular species interacts with its food depends partly upon whether or not the food is composed of living organisms. If it is, then the chances of the food organism surviving and multiplying are affected by its being eaten. But there is no simple relationship between, for example, the populations of predators and the populations of their prey (see pages 193–195). One reason for this is that most predators can eat more than one species of prey, but they do not always select the commonest prey species, rather as the tsetse flies above were highly selective of their hosts. Again we are impinging on aspects of animals' behaviour, and will return to this point in chapter 6.

The effects of herbivores on the plants they eat may not be straightforward either. The productivity of most species of grass is increased by moderate levels of grazing, but declines when the grass is ungrazed. It also declines with overgrazing, which can, in turn, affect the survival of the grazers, since it diminishes their food supply. In extreme cases, the grazers themselves die as a result of killing their food plants. An interesting contrast to this is the case of animals whose food is not living – such as dung-beetles, or termites that feed on wood. No matter how much dung the dung-beetles eat, they will not affect the future supply of dung.

Mineral salts are an essential part of animals' diets, and mammals will sometimes go to remarkable extremes to obtain them, often by licking soils or

Box 5.1 Home ranges of some birds of prey (raptors)

Most African raptors are non-migratory. With careful observations, the area over which individual birds hunt can be mapped and its area calculated – this is its home range (see pages 139 and 142). Usually a pair of birds share the same range, pairs of most species remaining mated until one partner dies. The size of home ranges in raptors increases exponentially with their body weight, as shown in the Figure (notice that both scales are logarithmic). The weights are those of the female, who is larger than the male in most raptors. Result of different studies on the same species are joined by vertical lines.

The solid regression line was calculated from these data on raptors; interestingly, it is almost parallel to the dashed line, which is for carnivorous mammals. (The mammalian data are from various parts of the world.) On average, mammals required less than half the area occupied by birds of prey of the same weight (the birds' home ranges are for pairs). This may be because birds have higher metabolic rates and hence require a greater intake of food.

For both birds and mammals, home ranges were larger for larger species. The simplest hypothesis to explain this is that the size of the home range was determined by food availability and since larger animals need more food, they have to hunt over a wider area than smaller species. However, although this hypothesis provides a simple explanation for the observations, it does not *prove* that food was limiting – there were no observations on feeding as such.

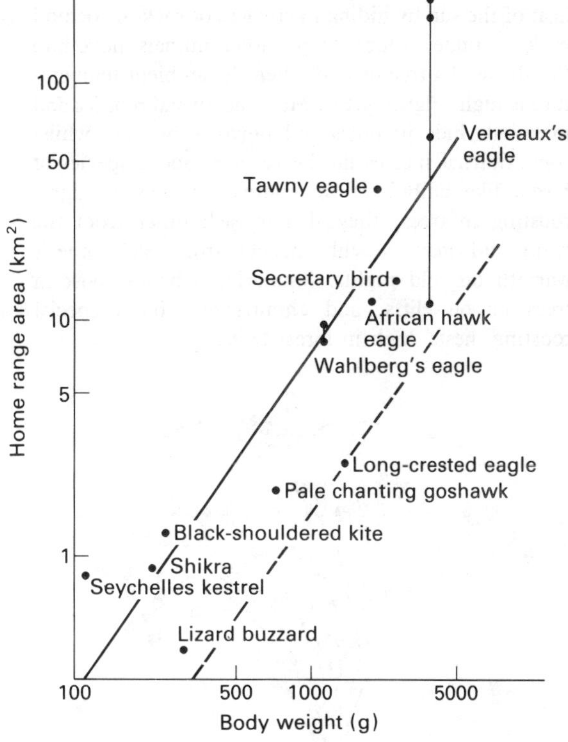

The home ranges of pairs of raptors (birds of prey) in relation to the weights of the birds.
(Bird data from Newton, I., 1979. *Population Ecology of Raptors*, Poyser, Berkhamsted; mammal line from A. S. Harestead and F. L. Bunnell, 1979. *Ecology*, 60: 389–402.)

rocks with a high salt content. There are caves in the slopes of Mt Elgon, on the borders of Kenya and Uganda, containing salt-bearing strata which are visited regularly by elephants and other species. They penetrate deep into the caves, more than 100 m from the outside and sometimes remain there for several hours in total darkness. In the Yankari Game Reserve in northern Nigeria, natural salt-licks which occur along the River Gaji basin are favourite haunts of elephants, western hartebeests, roan antelopes and baboons. The urine and faeces which these animals pass out during the long periods they spend licking soil at the salt-licks help to maintain the mineral contents of the soil at the sites. This is an unusual example of the cycling of nutrient materials.

Shelter

Shelters take many forms but they have in common the fact that they offer a degree of protection from some adverse component of the animal's environment – heat, or predators, for example. Few animals are active throughout the day *and* night, and at times of inactivity they rest. So nocturnal species need somewhere to spend the day, and diurnal species need a place to pass the night (sometimes called a roost, especially in birds). Many animals also use special places for breeding, and some build nests for this purpose. We are using the word shelter for these and similar resources, but it has to be recognized that the term does cover a diversity of situations. In some cases, the protection may not be obvious. Thus when pelicans or other water birds roost on exposed sand spits in the middle of a lake or river, they are protected in the sense that predators cannot stalk them and take them unawares.

We mentioned earlier that small animals often seek protection from extremes of temperature by taking advantage of microhabitats where the temperature is more favourable. Thus many small animals avoid the heat of the sun by hiding in crevices of rock, or behind bark, or under stones. Some larger animals move into the shade of a tree or cliff when the ambient temperature is high (Figure 5.12). Many nocturnal reptiles and mammals hide in holes and burrows by day, whilst some diurnal species do the reverse. Social species of birds, like bulbuls, form compact groups at night, roosting in trees; they shelter each other from the wind and may possibly benefit from each other's warmth on cold nights. Similarly, baboons roost in trees or on cliffs, and chimpanzees build special roosting 'nests' high in forest trees.

Figure 5.16 Weaver birds sometimes nest in the same tree as an eagle, whose presence presumably deters other predators. The eagles themselves do not eat weavers.

Shelters associated specifically with breeding are diverse too. An eagle's nest on top of a tree gives it safety from most mammalian predators, whilst in turn other birds may seek to be protected by the eagle (Figure 5.16). Birds' nests are mainly seasonal; only a few, such as some weavers, use nests outside the breeding season, which is usually only a few weeks or months each year (pages 161 and 162). Most birds lay their eggs in nests, but another way of sheltering eggs is to bury them, as happens in many reptiles, including crocodiles and turtles.

When a female hartebeest is about to give birth, she seeks dense vegetation; but this shelter is used for a period of an hour or so only. Giraffes hide their young in thick bushes for several weeks; at times the mother leaves the young for several hours so that she can feed.

The most complex shelters in the animal kingdom are those constructed by the social insects. Many of these belong to the order Hymenoptera, which includes the ants, bees and wasps. The well known nests of honey bees (the introduced domestic species is *Apis mellifera*) are excellent examples of structures providing shelter for breeding, as well as a food store. The shelters built by ants are very varied, though usually less complex than those of social bees. The cocktail ants, members of the genus *Crematogaster*, inhabit shelters which consist of spherical growths, or galls, on the whistling thorn shrub *Acacia drepanolobium*. The ants use the galls primarily as nurseries, although the nest containing the queen is underground. Shortly after the eggs are laid, worker ants carry them to the galls, where they remain throughout their larval life, tended by the workers. The ants vigorously attack any herbivore which attempts to browse on the *Acacia*, which is thus protected. This relationship, of mutual advantage to both ant and plant, is a nice example of symbiosis (pages 116 and 117). Other species of *Crematogaster* chew plant fibres to produce a brownish mixture from which they construct large nests on the trunks and branches of trees.

The weaver ants, found throughout the tropical regions of the world, build nests from leaves. The

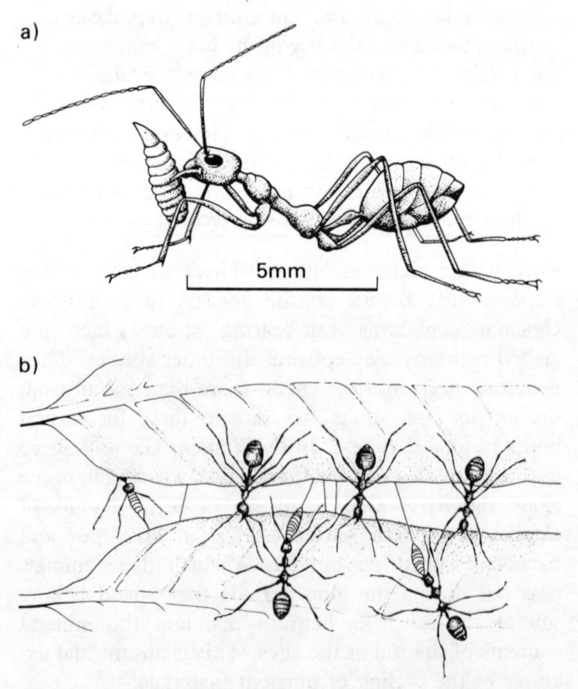

Figure 5.17 The construction of a nest from leaves by weaver ants, using silk threads produced by their larvae to sew the leaf-edges together.
(Ewer & Hall, 1978.)

worker ants pull leaves together by forming a living chain, each ant gripping the next with jaws and legs. Meanwhile, other workers collect some of their larvae and stimulate them to produce silk from their salivary glands. By holding the larvae in their mouths, the workers shuttle them from one leaf edge to the other and back, until the edges of the nest are held firmly by the silk threads (Figure 5.17). They then make their nest inside the leaf shelter.

The mound-building termites have succeeded in combining all the functions of a shelter into one amazing structure. The most sophisticated nests are those constructed by members of the genus *Macrotermes* (Figure 5.18). Their mounds consist of anything up to fifty tonnes of soil, and they provide protection from most predators and, equally important, shelter from the outside weather. Termites have thin cuticles and dehydrate rapidly once outside their mound or its passages. In the hive, the part of the mound where the majority of the colony lives, the relative humidity is maintained at a level just below saturation, and the temperature is kept at a favourable 28 to 30 °C regardless of outside conditions. Carbon dioxide levels in the nest exceed those outside but are usually below 2 per cent and termites are unaffected by this. The climate inside the mound is thus almost constant and at or near the optimum all the time, and it is also favourable for the symbiotic fungi on which the termites depend.

In order to be able to construct this system, the termites need suitable soil (they prefer loams but can use soils with a clay content as low as 10 or as high as 60 per cent). They also need to keep the air humid throughout the dry season, which they apparently do by collecting moist soil from beneath the mound, sometimes from considerable depths. Although the mound provides a food resource (the fungus combs), as well as a favourable climate, and protection from most predators, some predators have specialized in finding a way in. The aardvark does this frequently and army ants of the genus *Dorylus* sometimes enter *Macrotermes* mounds. This causes the workers to leave the mound and wander helplessly over its surface, where they fall easy prey to other ants or to birds.

A few other examples will serve to emphasize the great variety of animal shelters. Hermit crabs, each time they moult, need to find an empty shell of a particular shape and size. In contrast, larvae of caddisflies build their own protective cases using mineral particles or fragments of plants (Figure 5.19).

Some shelters are provided, although unintentionally, by man himself. Houses and other buildings provide shelter for a wide variety of animals, including many pests like rats and cockroaches. Some birds apparently regard tall buildings as cliffs, a common

Figure 5.18 A diagrammatic representation of a mound of the fungus-growing termite, *Macrotermes michaelseni*. There are thousands of foraging holes connected to each mound by means of an extensive, branching system of underground passages, but only a few are shown. When not in use, the foraging-holes are closed by a plug of cemented soil particles. The association of fungus and insect is a good example of symbiosis (pages 161 and 162).

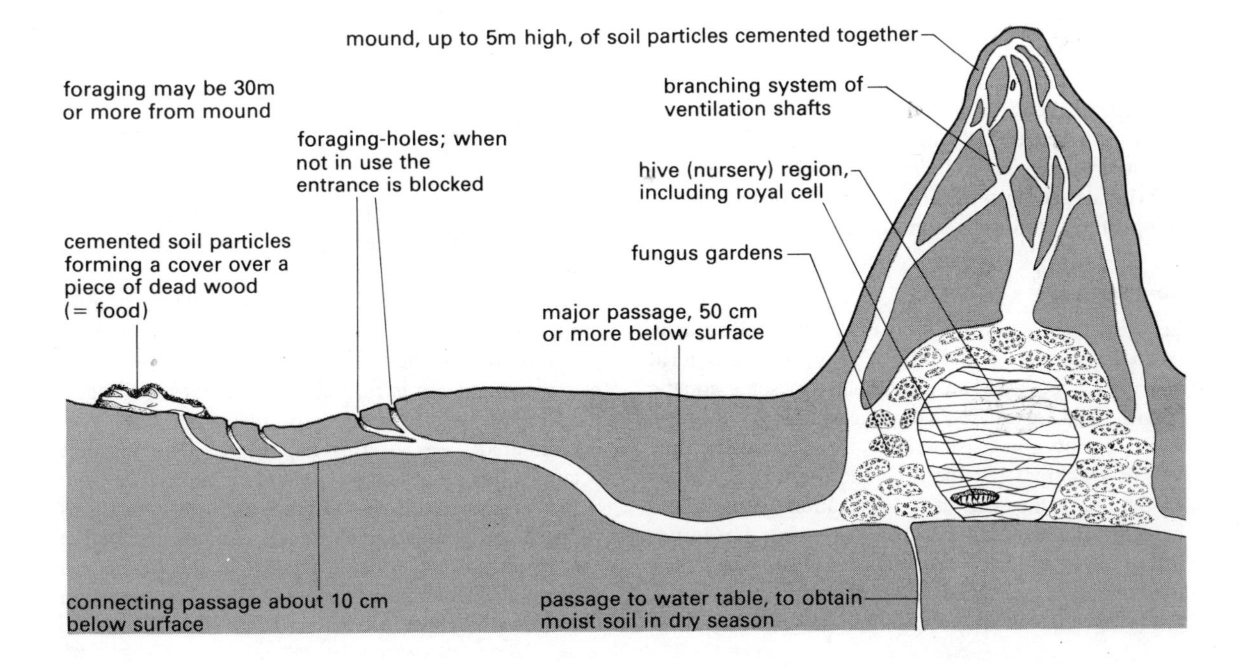

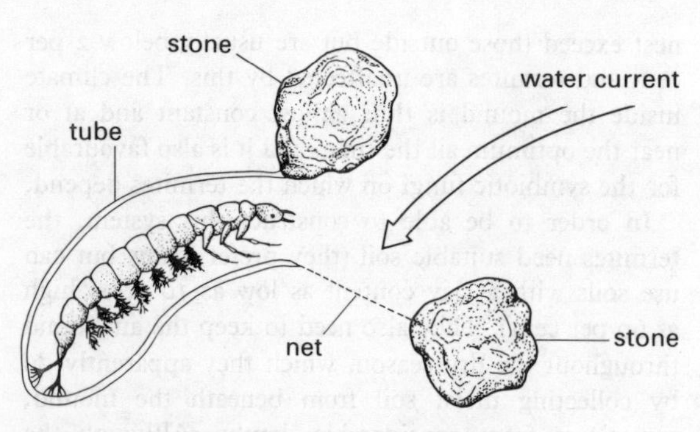

Figure 5.19 Caddisfly larvae construct a shelter from materials collected from the stream where they live; consequently they are difficult to see. Some species (e.g. *Hydropsyche*) spin a net to trap food particles in the water. (From Ewer & Hall, 1972, reference in chapter 1.)

example being the little swift, a common species in many African cities. These birds build their nests on the buildings which provide a resource that was previously missing – there are no natural cliffs in most cities. Other birds which treat buildings as cliffs include feral pigeons, and one of their main predators, the peregrine falcon. The latter nests every year on the Parliament Buildings in Nairobi.

Animals which are cryptic evade predators by being difficult to see – in some cases almost impossible (Figure 6.16). These species sit in the open, but their colour and patterns resemble closely their background. Hence they do not need shelters.

Other individuals of the same species

Interactions between individuals of the same species can be divided broadly into two kinds – those from which both benefit and those in which one or both individuals are harmed. The first kind is usually called co-operation, and the second competition.

Co-operation

Ensuring fertilization is clearly essential for all sexually reproducing species and this requires a special form of co-operation between individuals. In some aquatic species, both sexes liberate gametes directly into the water, but even in these animals there is usually some way of synchronizing the time and place of release so as to increase the chances of fertilization. Most species have specific breeding seasons. Thus many freshwater fish breed at times of rising flood waters, when their food supply is enhanced. In most free-living animals, prospective mates seek each other actively.

Sometimes there is a shortage of mates, and in species which are uncommon, special adaptations may be needed to locate a mate. For instance, the black-throated honey-guide, a bird which is widespread in Africa and well known in many areas for its habit of leading people to bees' nests, has a loud and distinctive call which carries for several kilometres. The species is never common, but by its call, it can locate a mate even in wooded country where trees prevent the birds from seeing each other if they are more than a few metres apart. Amongst animals which locate mates by sound are various mammals, most frogs, a variety of insects (including cicadas and grasshoppers) and many birds. To be effective of course, each species requires a different call. A good naturalist also knows the calls, and uses them when looking for animals.

Sexual pheromones (members of a group of substances secreted by animals, and collectively known as seriochemicals) play an important part in the mating behaviour of various species, and in the location of mates in some of them. Amongst insects, it is usually the females that produce pheromones, and they are known to occur in a wide variety of species. Female termites, such as those of the genus *Trinervitermes*, attract males in this way, as do several species of *Diparopsis*, lepidopteran bollworms which often damage cotton. Cocoa capsids (Hemiptera), such as *Sahlbergella singularis*, provide another example of a pest species using pheromones, a fact which has sometimes been exploited as a means of control.

Amongst the most remarkable instances of attraction by pheromones are those found in moths. For instance, a male silkworm moth is able to detect a female's pheromone when its concentration is as low as 10 million millionths (10^{-12}) of a gram per millilitre of air! Cases are known of male moths locating a female at a distance of several kilometres. In other species the pheromones are only effective at close range. Thus, the sexual pheromone produced by a female tsetse fly induces mating behaviour only if the male touches her.

Despite such adaptations, it seems likely that individuals do sometimes fail to locate mates, especially when the population is at a low density. However, the fact that a species is rare obviously makes it difficult to observe, and consequently there are few well-documented cases of this happening. One instance occurred during attempts to eradicate tsetse flies from a part of Zambia. Bush clearance and other measures reduced the population of flies to such an extent that the proportion of unmated females in the population increased considerably. Soon after that the population became extinct and it is believed that this was due, in part at least, to most of the females dying unmated. In a quite different situation, wardens in Tsavo National Park, Kenya, are concerned about the survival of the black rhinoceros. Poachers have killed so many in recent years that the few remaining individuals might fail to locate mates, and thus become extinct despite vigorous anti-poaching programmes.

When mates are plentiful, individuals are presented with a choice. The Darwinian theory of evolution postulates that natural selection operates partly through individuals selecting the fittest mate, that is, the one which will enable them to produce the largest number of surviving offspring. The selective advantages of doing this are thought to be very great, and explain the amazing variety of mating systems to be found in living organisms (chapter 6).

There are many ways in which the individual members of a species benefit from each other, apart from the special case of mates. Co-operation within a species is analogous to symbiosis between species – both individuals benefit, at least on a long-term basis. For example, many species of birds have distinctive alarm calls. The first individual to spot a predator gives the alarm, and all take cover, thus increasing their chances of survival. Antelopes are generally silent, but they also signal to each other when alarmed. In their case it is often by means of distinctive patterns on their hindquarters, which are white (Figure 5.20). When one individual sees a predator and begins to flee, these patterns attract the attention of others in the group, and they are also warned.

Quelea quelea, a serious bird pest of cereals in some countries, illustrates the sort of co-operation which may take place in relation to feeding. The birds roost communally, often in enormous numbers. The same group of trees or shrubs may serve as a roost for several days, or even weeks, if food in the area is plentiful. Flocks set out from the roost in various directions to feed each day, returning in the evening. Flocks that have been successful in finding food one day leave early the next morning, returning to the same place. A flock which found little food in an area on one day will not return to it the next, but will follow one of the early departing flocks instead.

The well-known dances of the highly social honey bee in its hives give precise information on the location and quality of food sources, and this undoubtedly benefits the colony as a whole.

The feeding behaviour of vultures, which are essentially solitary birds, is also beneficial to the majority, but not to the individual that spots the food first (Figure 5.21), since it has no choice but to share the food with many others that it has attracted to the carcase. Indeed, the co-operation in this case, although effective, is presumably unintentional, and leads to intense competition as the vulture party grows in size (Figure 6.4). The inevitable progression from co-operation to competition illustrates the inherent difficulty of attempting to categorize animal behaviour.

A less widespread but very interesting form of co-operative behaviour is mutual preening or grooming (Figure 5.22), which is seen in a number of social species of birds and mammals, and is particularly common in primates and members of the cat family. Such behaviour is important in establishing the attitudes of individuals to each other (see pages 135 and 138).

(a)

(b)

Figure 5.20 The patterns on the rumps of some antelopes readily catch the eye if they are disturbed and start to run, and in this way other members of the group are warned. (a) impalas, which are common in many wooded areas of eastern and southern Africa, and (b) Grant's gazelles, a dry country species of eastern Africa
((a) D. K. Jones; (b) D. E. Pomeroy.)

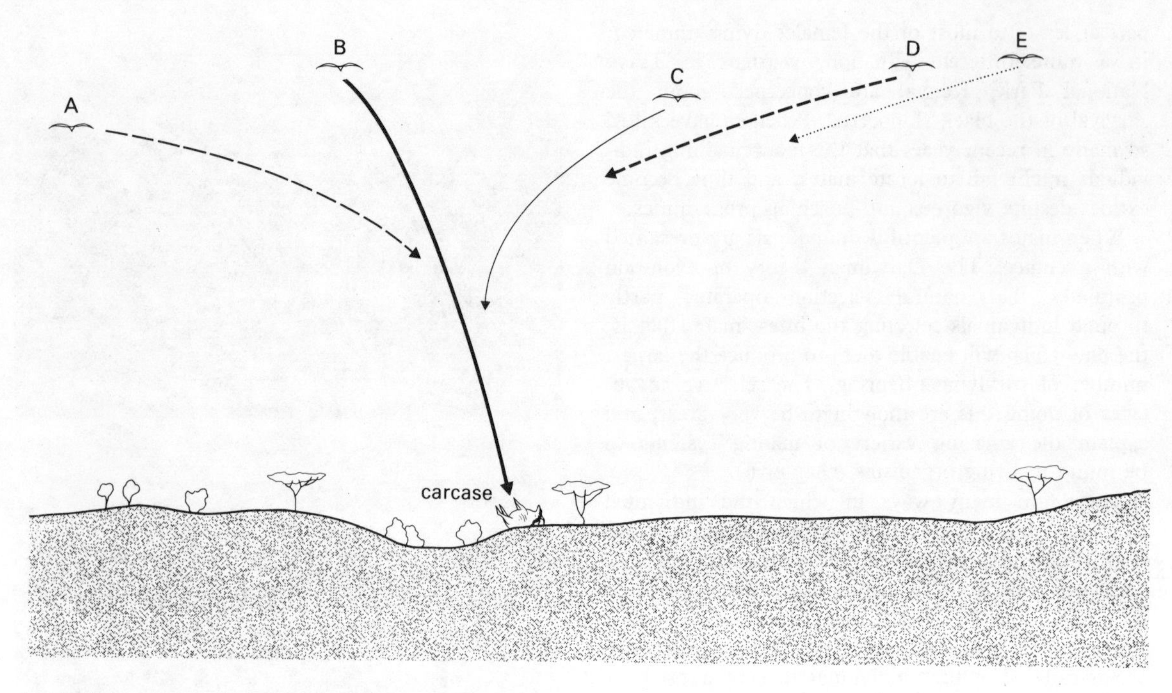

Figure 5.21 Vultures soaring on a warm day over open savanna. The horizontal scale is condensed; birds A to E are 10 to 20 km apart. As well as watching the ground, each bird keeps an eye on the others within range. So when bird B spots the carcase and drops from its soaring height of 4–500 m, A and C immediately notice its fall and give chase. Vulture D sees C set off in haste, E sees D, and so on all round. Within 15 minutes of B reaching the carcase it is likely to have 50–100 companions. The chain-reaction affects all species of vulture – there can be as many as six species at the same carcase (cf. Figure 6.4).

The birds drop very fast and are conspicuous from several kilometres away, and as they drop, any lions and hyenas in the vicinity spot them and come running in too.

Figure 5.22 Mutual grooming by monkeys – here vervets (or grivets). They are searching for ectoparasites in each other's coats. Grooming amongst adults typically involves males grooming females, or dominant males grooming sub-dominant males. (D. E. Pomeroy.)

Intraspecific competition

We defined competition in relation to plants on page 82; the essential features are shortage of shared resources leading to a reduction in the chances of at least some of the individuals to survive and multiply. If the individuals are all of the same species, then the competition is **intraspecific**, but when two or more species are involved it is **interspecific**. These two words are so similar as to be confusing, and 'within-species' or 'between-species' are good alternatives. Unfortunately not all ecologists agree on the meaning of the word competition and some have even advocated that its use should be avoided; but it is used so widely that attempts to abandon it seem unrealistic. Nevertheless, each time you see the word competition in a book or article, it is wise to check the author's definition. In the sense of our definition, the ecologist needs to be able to show that some individuals are not obtaining sufficient of at least one of their resources.

Intraspecific competition is often the means of determining which individuals survive (see, for example, page 200). In some cases, intraspecific competition also determines how many survive. Suppose, for example, that 100 eggs of a blowfly are placed on a piece of meat weighing 5 g, and that 0.2 g of meat are needed for one larva to complete its development. It is clear that, on average, each larva will have only 0.05 g, and as a result none could complete its development and all would die. In nature, however, such situations rarely arise; a female would not lay so many eggs in a single batch, unless the quantity of food was greater. If the female fly had laid only twenty five eggs, there would have been enough food for them all. Nevertheless, some degree of overcrowding does sometimes occur, resulting in the production of smaller flies.

A more frequent result of intraspecific competition is to reduce individual growth-rates, which again can lead to smaller adults, who may therefore have a reduced chance of survival. This effect is often seen in plants, but is less marked amongst animals, many of which have determinate adult sizes: that is, the size of adults only varies within a fairly narrow range. Nevertheless, stray cats, for example, often grow to less than half the size of well fed cats kept as household pets. In nature, individuals suffering from a severe shortage of resources usually die, but domestic stock are protected and it is common on overgrazed pastures to see livestock in poor condition which would certainly die of disease or predation were it not for the herdsmen tending them. Unfortunately, keeping them alive only aggravates the problem of overgrazing.

Under certain conditions, natural selection favours behaviour that reduces intraspecific competition. A good example of this is seen in the egg-laying habits of the large and colourful mosquitoes of the genus *Toxorhynchites*. Unlike most other mosquitoes, female *Toxorhynchites* lay very few eggs in any one place. This is probably because their larvae are predatory on those of other mosquitoes, and most of their oviposition sites (e.g. tree-holes, leaf axils, snail shells) are too small to support many prey. If there were more than a few *Toxorhynchites* larvae they would be unlikely to find enough food, and they would then eat each other, as in fact they sometimes do.

Individuals of other species

Generally speaking, animals depend upon other living things for their food (but some eat dead plants and animals), and they also associate with other members of their own species as we discussed in the last section. We now consider ways in which the distribution and abundance of animals can be affected by animals other than those that they may eat. These effects may be positive, involving interspecific cooperation, or negative where one or both species is harmed by their interaction.

Most ecosystems contain many different species of plants and animals, so that the number of possible interactions between them is enormous. Box 5.2 summarizes the kinds of interactions that can occur between any two species, although it can happen that three or more species are involved. (And, of course, each species interacts with only some of the others in its ecosystem – cf. Figure 5.3). The different kinds of interactions shown in Box 5.2 form the basis for the following sections.

Co-operation between species

When cattle or other large mammals walk through grass, they frequently disturb grasshoppers. This is why flocks of cattle egrets are often seen walking between them, since the disturbed grasshoppers are easy prey (Figure 5.23). If, as they move through the grass, the birds see a lion, they fly up, thus warning the mammals. This is an example of **facultative mutualism**, where both species benefit. But the relationship between the species is temporary. We find many instances amongst birds. The black-throated honey-guide has developed an extraordinary relationship with man, and its distinctive call is known to honey-gatherers right across Africa, and they know too that by following the bird as it moves and calls through the bush, they will be led to a bee's nest. Once the men have taken what they want, the bird gorges itself from the exposed remains of the nest, eating comb, grubs and honey. Without the men to open the nest, it would not have had its meal. In this instance, neither species depends for its survival upon the other but men show a strong liking for honey, and honey-guides for wax, and in this way they both benefit. Without man, the honey-guide may try to lead another species of mammal, such as the honey badger, or rely on its staple food of insects.

Figure 5.23 Cattle egrets walking close to a party of elephants. As they move, the elephants disturb grasshoppers on which the egrets feed.
(D. E. Pomeroy.)

Box 5.2 Interactions between different species of animals and plants

The various species which compose an ecosystem can interact with each other in many ways, some of which are of fascinating complexity (see, for example, Box 5.6). Here we present the commoner sorts of interaction; this classification forms the basis of pages 115 and 127.

The terms in bold type are defined in the text, where examples of most of them are given.

The terms *symbiosis* and *parasitism* need further comment because their use varies between writers. The literal meaning of symbiosis is 'living together' and the word is sometimes used to cover any form of intimate and enduring relationship between species, including parasitism. Other writers, especially from Britain, restrict the word symbiosis to relationships where both benefit, as shown here.

The sorts of benefit obtained by one or both of the interacting species can take many forms, but the most frequent are food and shelter.

- **Positive interactions** one or both species benefit, neither is harmed
  - **Mutualism** – both species benefit
    - **Facultative mutualism** – the relationship exists for only part of the time
    - **Symbiosis** (or obligate mutualism) – The two species have evolved together to such an extent that one cannot exist without the other
  - **Commensalism** – one species benefits while other is unaffected
    - **Commensalism** – usually involves the benefiting species obtaining food (or more food than otherwise)
    - **Phoresy** – one species uses another for transportation but does not derive food from it
- **Negative interactions** one or both species harmed and may die
  - Only one species harmed, the other benefits
    - Harmed species dies
      - **Predation** – one animal consumes another; or consumes plants; or vice versa
      - **Parasitoids** and some other parasites kill their hosts in completing their life-cycles
      - **Competition** – between species commonly ends in the death of the 'loser'
    - Harmed species normally survives
      - 'True' **parasitism**, where host is not killed
      - **Grazing** and **browsing** by animals on plants
  - Both species harmed – some cases of interspecific competition

Tree-snakes prey on birds and their eggs. One bird will mob a snake in a tree by flying around, calling loudly, but being careful to keep out of range. Other birds, often of several species, join in. The snake usually responds by retreating rapidly, since the noise might attract a snake-eating eagle and the snake would itself become the prey! There is a close and mutually beneficial relationship between tick-birds and the cattle, giraffes and other mammals whose ticks they eat, but again it is not obligatory for either.

In contrast, some forms of mutualism are obligatory – the two species cannot exist without each other – and these are usually called **symbiotic** (Box 5.2). Symbiotic relationships are often between animals and plants, and must have taken a very long time to evolve. They have resulted in some remarkable situations.

The dependence of two species on each other comes about through processes of **co-evolution**. A familiar example is the association between mound-building termites, and fungi of the genus *Termitomyces* which they cultivate in their fungus gardens (Figure 5.18). The details of this relationship are still being studied, but it is clear that both fungus and animal benefit, and that the termites obtain an essential part of their food by eating the fungus. The fungus, which is a basidiomycete and occasionally produces mushrooms around the mound, thrives in the warm, moist microclimate inside the mound. It never grows anywhere else, and has to be introduced in each new termite colony by the termites themselves. Other species of termite (but not the mound-builders) have symbiotic Protozoa in their guts, without which they could not digest the wood which they eat.

Like termites, ants are often involved in symbiotic relationships, suggesting that social behaviour favours symbiosis. *Crematogaster*, the ant associated with gall acacias (described on page 110) is an example. Another is the interdependence of ants and caterpillars of the small blue lycaenid butterflies, which are often common in forests. The caterpillars of these butterflies feed on lichens, and they produce sugary secretions that are an important food source for the ants, which protect the caterpillars by attacking would be predators.

Most flowering plants in the tropics are pollinated by animals, the main exceptions being savanna plants such as grasses. Many plants have become dependent, through co-evolution, on particular species of animals for pollination. Whilst the members of some families such as the Compositae can be pollinated by any of a wide variety of insects, particularly beetles or flies, many others rely on the behaviour of a limited number of specialized insects for pollination and sometimes on only a single species. For example, amongst the Labiatae and Orchidaceae, many species can be pollinated only by certain bees. The most extreme examples are probably those of the figs, described in Box 5.3.

Other species of plants are pollinated exclusively by one or a few species of bat, or by sunbirds. The latter, with their brilliant iridescent colours, also provide an example of a related phenomenon, **parallel evolution**. The humming-birds of tropical America, and the honey-eaters of Australia, are strikingly similar in appearance, behaviour, and as pollinators, to the predominantly African sunbirds. But each of the three families have evolved independently of the others and each is endemic to its particular region.

Another important form of plant–animal interaction, with both benefiting, is seed dispersal. During the course of its life, a tree may produce fruits equal in total to its own weight, an interesting example of the enormous demands, and consequent selective pressures, involved in the process of self-perpetuation. All of this fruit is potential food for animals, some of which are totally dependent upon it for at least part of the year, whilst the plant will benefit if the occasional seed is carried by the animal to a place where it may germinate and become established. The fruit bats are almost exclusively frugivorous (fruit-eating); and fruit is a major part of the diets of many primates and a wide variety of birds.

Some seeds are much more likely to germinate after passing through the alimentary canal of an animal. In this process, the seed coat is softened, and its permeability increased. But for *Acacia tortilis* seeds, the digestive juices may have an even more important effect. This tree is one of the commoner species in the drier parts of Africa and its pods are very rich in protein. Goats, gazelles, and even elephants seek them and eat them in large quantities. *A. tortilis* seeds collected from the faeces of a variety of large mammals showed germination rates of 11 to 28 per cent. This rate seems low, but it is much higher than the 3 per cent rate for seeds which were treated experimentally to soften their coats, but not eaten by mammals. The explanation is believed to be that most seeds are infested by beetles of the family Bruchidae, whose females lay their eggs in the young pods. The beetle larvae continue to grow after the pods fall to the ground, and eventually eat the embryo of the seed – thus preventing germination. But if the seeds are eaten by mammals before the beetle larvae have had time to eat the seed embryo, the mammalian digestive juices kill the larvae, and hence those seeds can germinate. This fascinating situation was discovered only as a result of very careful observations, and illustrates how complex the interactions between species can be.

In contrast to symbiosis, where both species benefit, are commensalism and phoresy, where only one benefits whilst the other is unaffected (Box 5.2). **Commensalism** literally means eating together at the same table, and this is particularly common in social insects. So many kinds of animals make their homes inside various parts of termite mounds that they have been given a special name – termitophiles. These snakes, mongooses, spiders, flies, beetles and many other species obtain shelter from the termite mound, but the termites themselves are largely unaffected.

The term **phoresy** describes an association between two animals in which one is transported by the other, but does not directly obtain food from it. Larvae of certain blackflies, e.g. *Simulium neavei*, are phoretic in East African streams on freshwater crabs of the genus *Potomonautes*. Certain bird lice (Order Mallophaga) cling to the bodies of peculiar flies called

Box 5.3 Fig-trees and fig-wasps

Fig-trees are widespread in the tropics; there are numerous species in Africa where they grow well in areas of higher rainfall. Every form of woody growth is represented: deciduous and evergreen trees, shrubs, creepers, vines, epiphytes and stranglers. Seedlings of strangling figs are epiphytic on other trees, but gradually they outgrow the host tree, and in many cases contribute to its death. Their fruits – figs – are remarkable structures, highly adapted to their associated fig-wasps. The fruit, in fact, is derived from an inverted head of flowers (see Figure), forming a hollow structure called a syconium which has only a narrow opening to the outside, the ostiole, and is guarded by overlapping teeth.

The remarkable degree of coevolution of tree and wasp is shown by the fact that every species of fig-tree is pollinated by a single species of fig-wasp. Furthermore, closely related species of tree have wasps which are closely related too.

Female fig-wasps enter the syconium by squeezing through the ostiole, losing their wings in the process. Once a wasp is inside, she lays eggs in some of the female flowers. She also pollinates other female flowers with pollen brought in her pollen sacs from the syconium where she was born, and in this way cross-pollinates the tree. The female wasp dies, but the female flowers containing eggs enlarge into galls, providing food and shelter for the next generation of wasp larvae. After pupation the adult wasps emerge, first the wingless males, and then the females. They mate, and the male then makes a hole through the wall of the syconium, after which he dies. The hole enables the female to escape, but before she goes she collects pollen from stamens which have ripened later than those previously visited by her mother.

Female fig-wasps pollinate more flowers than they lay eggs in, so the syconia produce seeds and their flesh ripens. The process of pollination stimulates the ostia to close, thus excluding more wasps from entering, whilst the presence of eggs in the female flowers inhibits the fig fruit from ripening too quickly or dropping. Eventually however it does ripen, but not before most of the wasps have emerged. The ripe fruit is sweet and often colourful, attracting parrots, hornbills, bats and primates, including man.

Different individual fig trees of each species in a forest produce their syconia at different times of year, and consequently both wasps and fruit-eaters can rely on them all year. Interestingly, species of figs that are typical of small islands, and other habitats where fig trees are sparse, produce syconia that ripen at varying times on the same tree. (Figure based on sketches by M. G. Murray (pers comm.), and Janzen, D. H., 1979 *Annual Review of Ecology & Systematics* **10**, 13–51.)

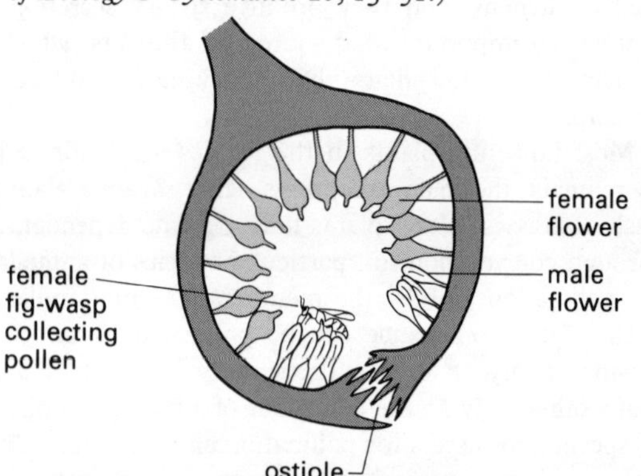

Cross-section of a fig fruit (syconium) showing the ostiole through which the female fig-wasp squeezes in. Whilst inside, she lays eggs in some of the female flowers, and pollinates others with pollen from another syconium where she was born.

hippoboscids which parasitize birds, and thus the lice are dispersed amongst their bird hosts. Many inhabitants of small and temporary water bodies rely upon other animals to transport them to new sites. The eggs of water snails are sticky, and they can be carried for considerable distances on the feet of migrating birds, and in this way new ponds and dams are quickly colonised by snails. Temporary pools are often inhabited by fairy shrimps of the genus *Chirocephalis*, whose sticky eggs similarly increase their chances of being transported to other pools.

Interspecific competition

The question as to whether species compete with each other, and if so how often, is one of great interest to ecologists. Our approach in this chapter is through the individual species; however, many ecologists consider that interspecific competition is of major importance to understanding the structure of ecosystems. In particular it helps explain niche separation (pages 61–63).

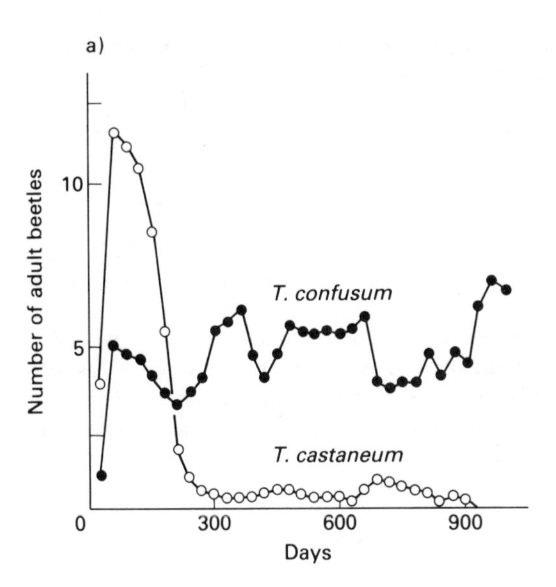

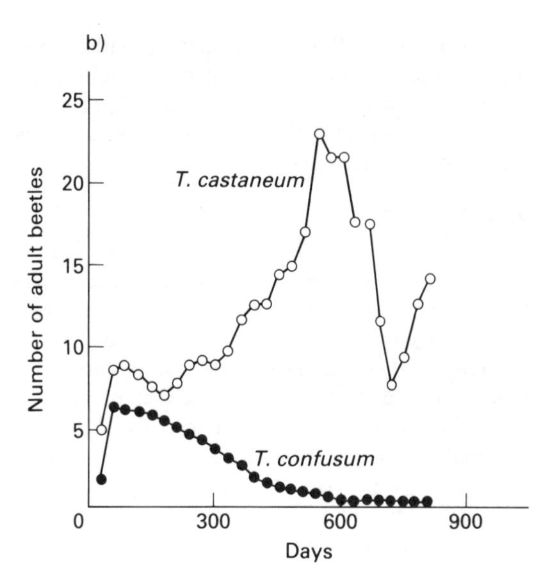

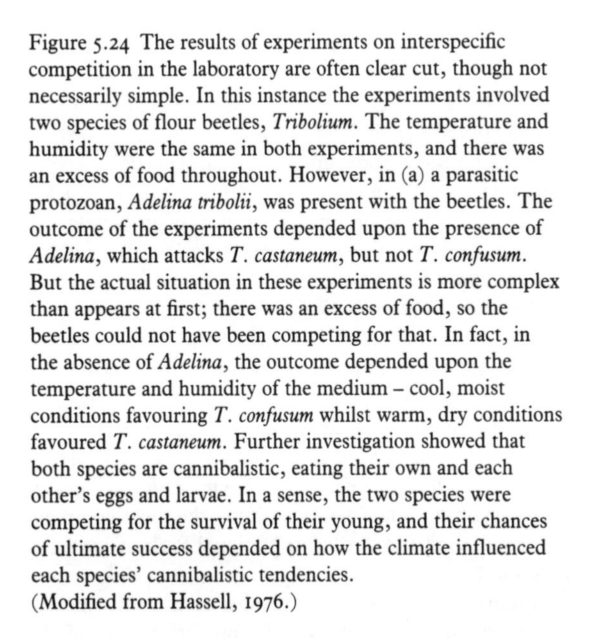

Figure 5.24 The results of experiments on interspecific competition in the laboratory are often clear cut, though not necessarily simple. In this instance the experiments involved two species of flour beetles, *Tribolium*. The temperature and humidity were the same in both experiments, and there was an excess of food throughout. However, in (a) a parasitic protozoan, *Adelina tribolii*, was present with the beetles. The outcome of the experiments depended upon the presence of *Adelina*, which attacks *T. castaneum*, but not *T. confusum*. But the actual situation in these experiments is more complex than appears at first; there was an excess of food, so the beetles could not have been competing for that. In fact, in the absence of *Adelina*, the outcome depended upon the temperature and humidity of the medium – cool, moist conditions favouring *T. confusum* whilst warm, dry conditions favoured *T. castaneum*. Further investigation showed that both species are cannibalistic, eating their own and each other's eggs and larvae. In a sense, the two species were competing for the survival of their young, and their chances of ultimate success depended on how the climate influenced each species' cannibalistic tendencies.
(Modified from Hassell, 1976.)

Laboratory experiments

A classic approach to the study of interspecific competition is to observe closely related species reared together in the laboratory. When this is done, the result almost invariably is that one 'wins' and the other becomes extinct. The early experimenter G. F. Gause created simple ecosystems using various species of the protozoan genus *Paramoecium*. When he kept *P. caudatum* and *P. aurelia* together in laboratory cultures, he found that *P. caudatum* populations survived whilst *P. aurelia* became extinct. But when he kept *P. aurelia* with another species, *P. bursaria*, they both survived. The reason is that *P. aurelia* lived in the upper parts of the culture jars and *P. bursaria* at the bottom, and as a result they rarely came into contact with each other. These observations illustrate the care needed in interpreting experiments on competition.

More extensive experiments have been performed on flour beetles of the genus *Tribolium* kept in uniform jars of flour. Here the final result depends upon the physical environment. For example, when cultures of *T. confusum* and *T. castaneum* were kept at low temperatures and relative humidities (24 °C, 30% RH), it was *T. castaneum* which died out. But at higher temperatures and humidities (34 °C, 70% RH), the result was reversed. However, the addition of a parasitic protozoan, *Adelina tribolii*, changed the results, since it attacked *T. castaneum* but not *T. confusum* (Figure 5.24). When the parasite was present, it was always the population of *T. castaneum* which became extinct, even in cases where it would have been the 'winner' in the absence of the parasite.

Experiments such as these led Gause and others to conclude that two species with the same requirements – in all these cases they shared a single food source – inevitably compete, and one is exterminated. A neat version of what came to be known as Gause's principle is that 'Complete competitors cannot coexist'. In a simple man-made ecosystem such as a culture jar the principle holds true, but already we have to add one qualification, namely that to predict the winner the physical conditions must be specified. This suggests that in a more complex environment, where conditions vary, it might be possible for the species to co-exist, as happened with *Paramoecium bursaria* and *P. aurelia*.

In effect, the culture jar situation, with constant temperature and humidity, is like a one-dimensional niche. In nature, niches are multidimensional, allowing considerable overlap between species' requirements (pages 61–63).

Species replacement

Rabbits were introduced into Australia by European farmers in the last century. Although for some years their attempts were unsuccessful, the farmers eventually managed to establish the rabbit, and it began to breed in the wild. Despite this slow start, rabbits soon increased to such an extent that within a few years they were major pests in extensive areas, and eventually throughout the grazing lands of Australia. As they spread, rabbits displaced several native mammals, most of which were marsupials. Some of these were exterminated in large parts of the continent. Whilst proof is lacking, it seems reasonable to suppose that the marsupials, which had evolved in the absence of rabbits, were less efficient in some way and hence were displaced by rabbits that competed successfully for the same resource – mainly grass.

In antimalarial campaigns in East Africa during the late 1950s the insides of houses were sprayed with residual deposits of dieldrin to kill the malaria vector, *Anopheles funestus*, which habitually rests indoors. The spraying was successful, but at the same time there was an increase in the population of a closely related mosquito, *An. rivulorum*. These two species coexist in larval habitats, but while adults of *An. funestus* rest indoors those of *An. rivulorum* rest outside houses. It seems that the reduction in the *An. funestus* population, caused by house spraying, reduced the larval competition between the two species. This allowed the *An. rivulorum* population to increase greatly, and because their adults rest outside, they were unaffected by house spraying.

Another interesting example of species replacement seems to have occurred in Nigeria, and most likely in many other areas of West Africa. Prior to the 1940s the mosquito *Culex quinquefasciatus* (= *fatigans*), a vector of bancroftian filariasis, was rare in urban areas, whereas *Cx. nebulosus*, a bird-biting mosquito, was common. Larvae of both species occur in polluted groundwaters. From the mid-1940s such larval habitats were regularly sprayed with DDT and this seems to have caused the virtual disappearance of *Cx. nebulosus* and its replacement by large populations of *Cx. quinquefasciatus*. The explanation seems to be that *Cx. nebulosus* is extremely susceptible to DDT whereas *Cx. quinquefasciatus* is not only less susceptible, but quickly evolves insecticide-resistant populations. *Cx. nebulosus* is still common in rural areas and the few towns where there has been little if any insecticidal spraying against mosquito larvae. So, here we have two examples of species replacements being caused by insecticides. There are probably many other undetected cases.

Interspecific competition in natural ecosystems

The replacement of one species by another nearly always seems to be associated with human activities – it was man who introduced rabbits to Australia. The successful species had not coevolved with the one it replaced, and this led to one species 'winning', as with the laboratory cultures of *Paramoecium* and *Tribolium*. But in natural ecosystems, many species appear to coexist and an obvious conclusion is that they have evolved together in such a way as to avoid, or at least minimize competition (see also pages 61–64). If this is the case, then interspecific competition has had an important influence on the course of evolution, but how significant is it over shorter periods of time? We have already mentioned on page 93 the fact that over 200 different species of beetles have been found in elephant dung – is it really possible that they are all doing something different? Or are they all competing with each other? Some answers to these questions were revealed by a fascinating study from which it appears that they do have important differences in behaviour, but that competition, both intra- and interspecific, can be intense (Box 5.4).

Another situation where, at first sight, competition might be thought likely, is that of the large mammals of the African plains. In many areas these have been replaced by domestic stock, but where they survive, one may see four or five species of large herbivores, all in the same general area, and all apparently eating grass. Detailed studies have been made in several places where such situations occur, the most complete being those of the Serengeti National Park in northern Tanzania.

Figure 5.25 shows the results of a study in one region of the National Park. It concerned three of the most important grazers of the Serengeti, all of which undergo extensive annual migrations (pages 170–172). The heaviest rains during the study period were in March and April, and they stimulated the growth of grasses. Populations of the common zebra peaked at the time of greatest growth. They are coarse grazers, whose need is primarily for a large quantity of food (pages 107–108). As the zebras feed, they also trample the grass, which further stimulates its growth. Wildebeests prefer the shorter grass, and they are also more selective than zebra, taking more of the leaf and less of the stem from the grass plant. The leaf contains a higher proportion of protein than the stem, as well as

Box 5.4 Dung, dung-beetles and competition
The dung of mammalian herbivores consists mostly of partly digested plant material. Often, large fragments are visible, enabling the species of plants to be identified. But that which is waste to the herbivore is food for coprophages, which are able to digest a significant proportion of the dung. Many dung-beetles can consume dry dung, as can various termites and other insects. But some beetles also use dung as a food store and lay their eggs in it, so that on hatching their larvae are surrounded by food. For this purpose the adult beetles need to acquire fresh dung.

At times, dung-beetles are attracted to dung in large numbers. In Tsavo National Park in eastern Kenya, two scientists counted the beetles as they came to a half-litre of elephant dung one evening – within fifteen minutes, 4000 beetles arrived! In such situations, there are usually many species of beetles, which we can group according to their behaviour.

(a) **Burrowers** dig rapidly beneath the dung pat, and bury dung at the bottom of a passage which can be as much as a metre below the surface.

(b) **Rollers** make a ball from the fresh dung and roll it away for a considerable distance (see Figure), where they either eat it or bury it in a chamber some 20–30 mm across. As with the burrowers, this chamber can be as deep as a metre. Eggs are laid in the dung.

(c) **Endocoprids** are mostly small beetles that eat the dung where it is.

Fresh dung pats are places of great activity, especially in the wet season. The rollers shape and roll their balls as fast as they can, the burrowers are hurrying to bury dung, and both have to contend with the endocoprids busily eating dung. Fights between individuals of the same and of different species are frequent.

The scientists mentioned earlier – their names are B. Heinrich and C. A. Bartholomew – interpreted these contests as evidence of fierce competition, and sought to discover the competitive advantages possessed by the winners. Whenever two beetles fought over a ball, for example, the winner was found to be larger, as expected; but also to have a higher body temperature, as measured by a fine thermistor inserted into its thorax. Since metabolic rates depend upon temperature, the winner's higher temperature probably gave it greater strength.

There are about 2000 species of dung-beetles in Africa, ranging from ones weighing less than 20 mg to others a thousand times heavier. Beetles are attracted to dung by its scent, and may fly considerable distances to reach it. But before they can take off they have to warm up, which they do by shivering their wing muscles. Larger beetles of course take longer to reach the threshold temperature for flight. Hence small beetles have the advantage of arriving first at dung pats, but larger beetles can displace them once they arrive – provided there is any dung left. Thus each size has its advantages, which helps to explain the many species at one place. The species also vary in their preferences for dung of different age or origin, as well as in what they do with it.

The intense competition, involving beetles racing to arrive early, and then to eat, roll or bury dung at top speed, made the scientists wonder why all the beetles wait until nightfall before appearing. The answer, almost certainly, is that there are many prospective predators, especially birds such as hornbills and guineafowl, but these are only active by day.

One result of the competitive activity is the rapid disappearance of dung. This had led to the introduction of dung beetles into Australia, whose native insect fauna could not remove the quantities of dung pats produced by introduced cattle. The pats disappeared slowly, mainly consumed by larvae of flies which became a great nuisance to the local human inhabitants. The introduction of dung-beetles in some areas has successfully solved the problem.

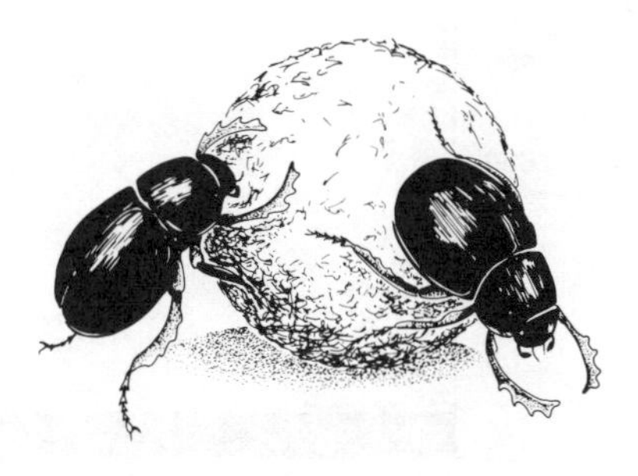

Two scarabaeid dung-beetles with a ball of dung which the one on the right is rolling whilst the beetle on the left is attempting to capture it. Competitive interactions of this type are frequent.
(From Ewer & Hall, 1978, Book 2 reference in chapter 1.)

Figure 5.25 Seasonal population variations of three grazing mammals in the Serengeti National Park, Tanzania, in relation to rainfall and grass height. There were year-to-year variations, but the monthly data for the years 1965–67 have been averaged. Note the different vertical scales for the three ungulates. The grass responded quickly to rainfall, so that the peaks for grass height and rainfall occurred at almost the same times. These were also the times of peak populations of the zebras, with wildebeests following 2–3 months later, and Thompson's gazelles another 2–3 months behind them. The timing varies with rainfall, which peaks in different months in different parts of the Park, hence causing the ungulates to migrate (page 170).
(Based on Bell, R. H. V., 1982 in *Ecology of Tropical Savannas* Huntley, B. J. & Walker B. H., (eds) Springer-Verlag: Berlin.)

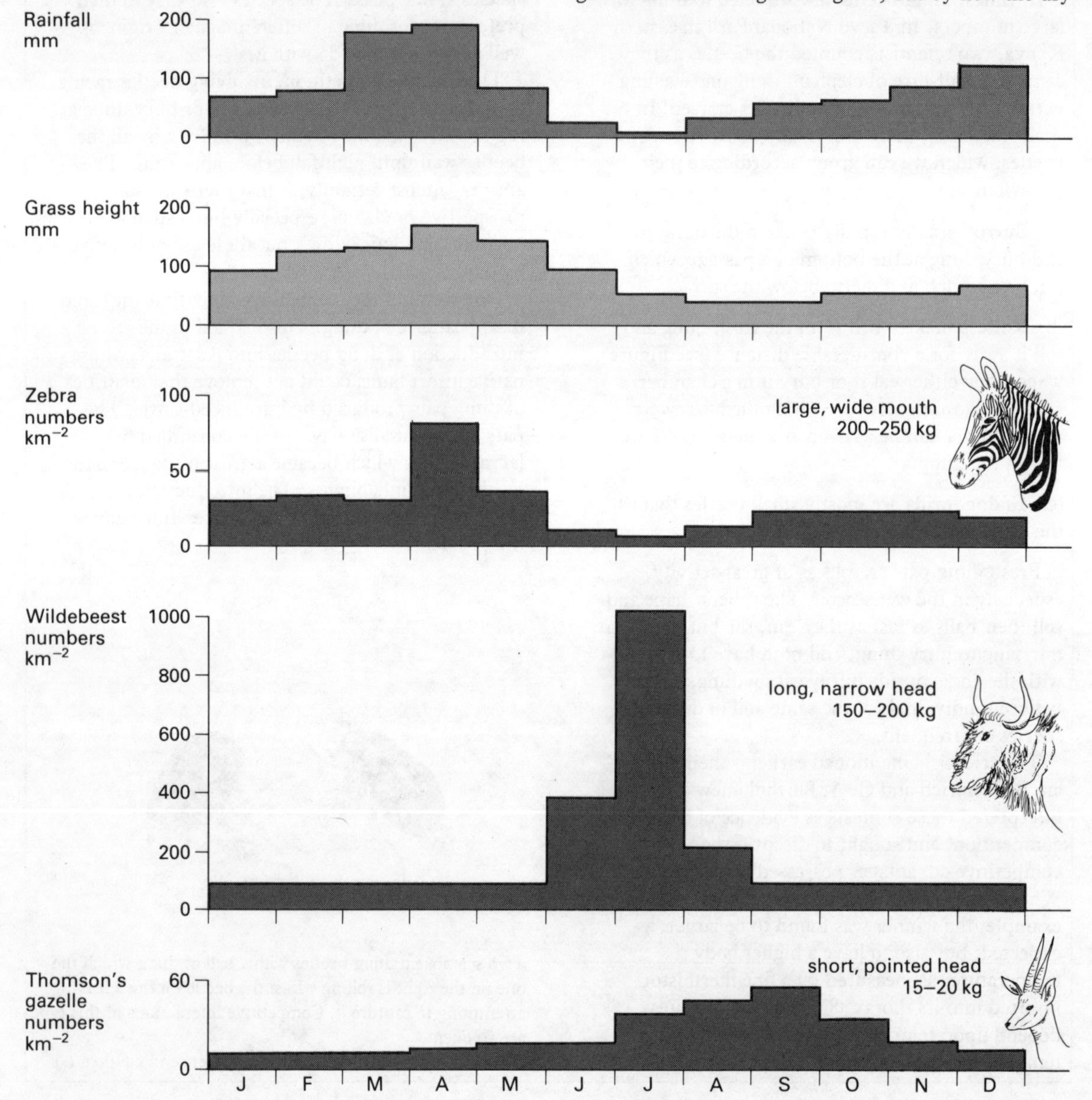

less fibre. Larger mammals require relatively less protein, and the zebra has particularly large molars which assist it in grinding the tough stems. The wildebeest has a long, narrow mouth, and this enables it to reach between the grass tussocks and to select the parts of the plant that it wants.

There are more than a million wildebeests in the Serengeti area, and they frequently aggregate into enormous herds which migrate extensively, following the zebras. The largest herds stay for less than a week in any one place (Figures 6.38, 6.39). As the wildebeest herds move on, the grass they leave behind has been trampled and chewed close to the ground. This again stimulates the grass to grow, but by now the dry season is approaching and growth is limited. The shorter grass also facilitates the growth of herbs. The resulting combination of herbs and short grass is ideal for the small Thomson's gazelle, a fine grazer which takes about 40 per cent of its diet as herbs in the dry season, and only about 20 per cent as grass stems. Competition between these three mammalian species is apparently avoided by their respective preferences for long or short grass, and for different parts of the plant; there is little or no selection for particular species of grass. The buffalo, however, does select particular grasses, preferring species with a relatively large amount of leaf, such as *Digitaria macroblephara*, and *Themeda triandra*, to those with less leaf and more stem, such as *Sporobolus pyramidalis* and *Pennisetum mezianum*. Cattle show similar preferences, and tussocks of *P. mezianum* frequently stand out, almost untouched, in heavily grazed semi-arid areas (Figure 5.26).

Figure 5.26 Clumps of unpalatable grass (here *Pennisetum mezianum*) remain ungrazed, even at the end of the dry season, when all other grasses have been grazed down to the ground. This is a clear instance of selective grazing. (D. E. Pomeroy.)

Selection of different parts of the grass plant was shown by a study of the food of sheep and two species of gazelles. Sheep select strongly for leaf (Figure 5.27), which may form more than 80 per cent of their intake in the dry season. This involves intense selection, since grass leaf is scarcest at that time. Thomson's gazelles take rather less leaf, whilst Grant's gazelles take stem, leaf and leaf-sheath in almost equal quantities. In the dry season, when food is scarce, the three species overlap only slightly in their diets, suggesting that they are not competing with each other then. In the wet season, food is no longer in short supply, competition does not occur even though the overlap in the food preferences of the three species is now much greater (Figure 5.27).

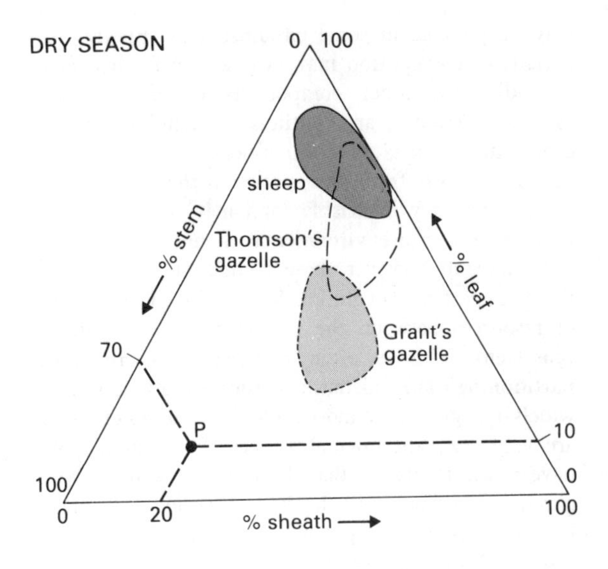

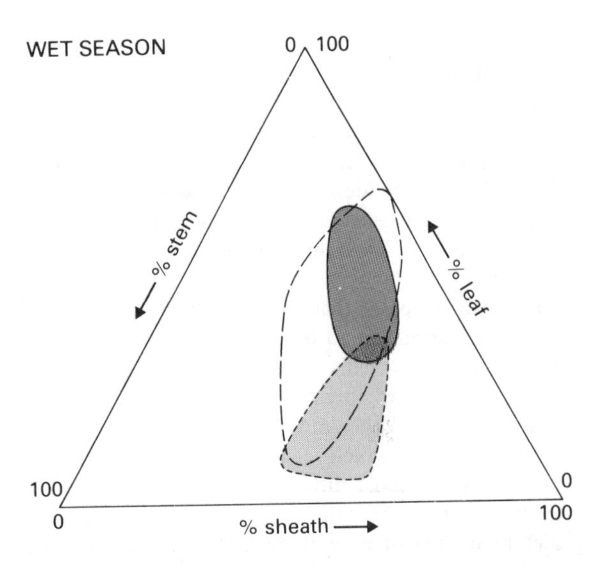

Figure 5.27 A diagram representing preferences of sheep and two species of gazelle for different parts of a grass plant in dry and wet seasons. The shapes enclose points made by plotting individual observations and they indicate the range of food taken. As an example, an observation at P would indicate that the animal has selected 10% leaf, 20% sheath and 70% stem in its food.
(Based on Pratt, D. J. & Gwynne, M. D. (eds.) 1977. *Rangeland Management and Ecology in East Africa*, Hodder & Stoughton: London.)

Conclusions concerning competition

Permanent coexistence between two or more species which share a common resource – and which therefore compete whenever the resource is in short supply –

may be possible in some circumstances. Firstly, the periods of competition may be few and far between. Secondly, the other components of environment, especially weather, are continually changing, so that even during times of competition, the competitive advantage shifts from one species to the other(s) and back, since they invariably (and inevitably) differ in their responses to environmental changes.

In general, however these studies and others like them suggest that evolution has resulted in a sharing of resources between the species that coexist in an ecosystem. This is often referred to as **resource partitioning**. The existence of such situations is now widely recognized by most ecologists, but so far there are very few experimental studies. For example, we have reason to suspect that sheep do not compete with gazelles. We should test this hypothesis experimentally by setting up several paddocks containing sheep, but differing from each other in the presence or absence of gazelles. We should then compare some measure of the secondary production of the sheep, such as growth rates. Such experiments have yet to be done, and until they have, we cannot really say whether or not the gazelles compete with the sheep.

In some cases, at least, it is likely that we should find the productivities of the two populations when together totalled less than the sums of each of them taken separately. Thus the presence of species A might depress the productivity of species B, and vice versa, but not to the point that either becomes extinct.

A final conclusion to be drawn from these investigations of interspecific competition is that animals' diets can be remarkably complex. Not only do species differ from each other in respect to what they eat, or how much of which foods, but there may be pronounced seasonal differences too. Nevertheless, it must be assumed that the net result is for each species' diet to be consistent with its nutritional needs, and there is evidence that that also affects its selection of food items (cf. Box 3.3).

Parasites

Parasitism is an association between two individual plants or animals in which one (the parasite) lives and feeds temporarily or permanently either in or on the body of the other (the host). Frequently the host is harmed in the process, i.e. its chances of survival are reduced. Parasitism is usually interspecific, the host being of a different species to the parasite, but this broad definition also covers intraspecific parasites such as males of certain. Crustacea and of angler fish (Order Lophiiformes) on their females. Logically, the dependence of the mammalian foetus on the mother is similar!

Ticks and lice which remain feeding on their hosts for many days or weeks are regarded as ectoparasites, and mosquitoes can similarly be regarded as ectoparasites, although they settle and feed on their hosts for only a few minutes. Most parasites, however, are endoparasites. They range in size from the protozoan malarial parasites of the blood (e.g. *Plasmodium falciparum*) to much larger animals such as tapeworms (*Taenia* spp.) which are found in the gut. Some parasites are non-pathogenic in their hosts; that is, they cause them no detectable harm. *Entamoeba coli*, for example, is ubiquitous in man's alimentary canal, but most people are apparently unaffected by it. But many – perhaps most – parasites exhibit some degree of pathogenicity and cause sickness in their hosts, or even death. The term **pathogen** is often loosely used to mean small parasites such as bacteria and protozoans, while the term parasite is confined to larger animals, but this is an unsatisfactory arrangement and it is simpler to call all such organisms parasites, or even micro- and macroparasites.

Many parasites complete their development in a single host, and infection occurs through close contact with the hosts or with their food. Other species have a life-cycle that is completed in more than one host. Where two or more hosts are involved, they may be of the same or different species, depending upon the parasite. Some parasites require a **vector** for their transmission. Most vectors are insects, mosquitoes being the best known. Virtually all species of animal are subject to infections by several different parasites. Hundreds of species are known to parasitize man: they range from viruses to nematodes and protozoans to lice. Many, especially the intestinal ones, are non-pathogenic, causing mild symptoms, if any. They absorb some of the food ingested by the host which will only show adverse effects at times of food shortage. In other instances the infected host may become sick, weakened and show atypical behaviour. Such sick animals are presumably more prone to predation as they are usually easier to catch (see the next section). Animals weakened by parasitic infections are less able to cope with extreme climatic conditions or a reduced food supply. They have less chance of surviving the rigours of migration, and are thus more likely to die in such stress situations.

Sometimes parasitic infections kill the host. This may be disadvantageous to the parasite too, because its supply of food suddenly stops, but some parasites may be unable to complete their development without killing the host. For example, certain parasitic nematode worms infect mosquito larvae, and invariably kill them when the worms bore through the cuticle to emerge into the water in order to complete their life-cycle. In this instance they no longer need their hosts, and so their own chances of survival are not reduced by killing the mosquito larvae.

A rather special form of parasitism has evolved in some insects, mainly members of the Orders Diptera and Hymenoptera. In these **parasitoids**, the adult females, which are free-flying, lay their eggs inside the living body of another species of arthropod (usually an insect), using a long and very sharp ovipositor. The life-cycle of the parasitoid is so closely synchronized with that of its host that it completes its feeding on the host's tissues at almost the same time that the host dies.

Some host–parasite relationships have become very complex, the parasites being incapable of completing their development except in one particular species of host. The distributions of these host-specific parasites are obviously restricted to those of their hosts. If the host becomes rare, their lack of choice increases their difficulty in infecting new hosts, always a major problem for parasites. Other parasites adopt a safer but less specialized life-strategy and can complete their development in a variety of different host species. Hence if one host becomes rare, or even extinct, they have alternatives.

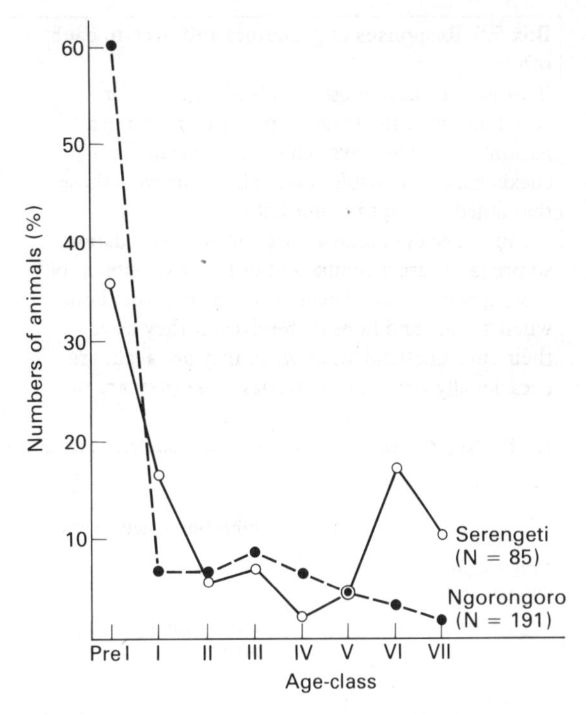

Figure 5.28 In northern Tanzania, spotted hyenas have developed the unusual behaviour of hunting in packs, and are important predators of wildebeests. The figure shows the percentage of wildebeests of different ages killed by hyaenas in two areas N is the number of skulls in the sample; skulls, which are found at most kills, can be aged on tooth wear and other criteria). A large majority of prey belonged to the youngest age-class; and in the Serengeti National Park predation increased again in older wildebeests.
(From Delany & Happold, 1979, reference in chapter 1.)

Predators

Predators occupy a similar trophic level to parasites. But whereas parasites are usually smaller than their hosts, predators are nearly always larger than their prey. Further, if the prey is caught, it nearly always dies. In this section we confine ourselves mainly to a qualitative approach, to predators as the components of an animal's environment. Predation takes very many forms; and whilst some predators are active hunters, others use poison, traps or deception to acquire their prey (Box 5.5). Looked at from the point of view of the prey, there are many possible strategies for avoiding predation, some of which are summarized in the Box.

Amongst the predators that chase their prey, one would suppose that individual prey which are weakened by parasites, or those suffering from loss of condition due to inadequate food supplies, are at greater risk. But there is surprisingly little evidence to show whether this is actually so, perhaps because of the difficulty of obtaining convincing facts. However, a number of studies have shown that predators catch mainly the youngest and oldest individuals of the prey population, the example in Figure 5.28 being fairly typical. In such cases, the relative importance of predators as components of the prey's environment varies with their age.

An important strategy of predation is to trap or poison the prey. Web-spinning spiders use methods of capture which rely upon the activities of the prey, and predation is almost random, in the sense that there is a considerable element of chance as to which individual is caught. Some prey species are more easily caught than others though, and some insects that are caught are distasteful to spiders, which remove them from their webs and release them.

Evolution is a consequence of natural selection, and predators are one of the most important selecting agents and important mortality factors. Evolution has resulted in a great variety of adaptations on the part of the prey to evade capture, and a range of adaptations by the predators to catch their prey. We enlarge on these topics when considering the effects that predators and prey may have on each other's populations, on pages 193–195.

Box 5.5 Responses of predators and prey to each other

Because they have mostly evolved together for a very long time, the various species comprising a natural ecosystem have achieved a level of coexistence that enables them all to survive – those that failed to adapt became extinct.

Most species of animals are subject to predation, so predators are a component of the environment of most species. Even elephants are at risk from lions when young, and lions themselves, if they leave their cubs unattended, or when they are aged, are occasionally attacked by hyenas. The prey are of course an essential component of the predator's environment – their food.

This table shows some of the many ways in which predators seek to catch their prey, and how, in many cases, the prey have evolved ways of escape. Examples of each category are mentioned; some are described in more detail in this chapter or the next, and most others will be familiar.

Just as predators and prey have coevolved, resulting in the survival of both, so have plants coevolved with the animals that feed on them – see pages 83–85.

A Predator strategies – some of the ways that prey are caught

Category	Method	Examples
Prey caught randomly 'wait and see' techniques	**Filtration** – nets, webs	Filtration of water, e.g. rotifers, flamingoes, blue whales Filtration of air, e.g. spiders' webs
	Deception – prey fails to see predator	**Hiding & stealth**, e.g, trap-door spiders, ant-lions, leopard **Camouflage**, e.g. chameleon, python, mantids
Prey hunted actively and individually	**Prey mostly conspicuous** and/or large, in comparison to the predator	**Poisons**, e.g. jellyfish, scorpions, and some ants, bees, wasps, spiders **Speed**, e.g. dragonflies, sharks, skinks, falcons, cheetahs **Strength or endurance**, e.g. cobras, starfish, lions, hunting dogs
	Prey small and/or inconspicuous	**Careful searching** (gleaning), e.g. some ants, some fish, many warblers, whiteyes and other insectivorous birds

B Prey responses – some strategies evolved by prey species to minimise predation

Escape – getting out of reach	**Running**, e.g. many insects, lizards, rodents, antelopes **Flying**, as in some insects, most birds and bats **Climbing**, e.g. many primates and cats **Swimming**, e.g. many fish and other aquatic animals
By behaviour or deception	**Camouflage** and cryptic behaviour, – e.g. mantids, insects living on bark, chameleons, rodents **Mimicry** being harmless but mimicking an aggressive or distasteful species, e.g. many butterflies, some flies and beetles
Defence against attack	**Aggregation** 'one of a crowd' or 'safety in numbers', e.g. most social insects and vertebrates **Aggression**, fighting, e.g. cornered snakes, buffaloes, rhinoceroses **Physical** protection, e.g. molluscan shells, crustacean carapace, tortoise shell, porcupine quills **Poisons** deter many predators and are produced by sea anemones, scorpions, snakes and some fish **Chemical** – being **distasteful**, as in many caterpillars, beetles and bugs

This section has been concerned with the kinds of interactions between species, and their significance as components of environment. A concluding example illustrates the extraordinary range of interactions which occur, in this instance among some of the fish which are found in great variety and abundance on African coral reefs: Box 5.6 has the details.

Man

We put man into a separate category because his effects are so various, and sometimes so catastrophic, that he often becomes the most important of all components of an animal's environment. Later, in chapter 8, we consider man from a human point of view; here we consider him from the point of view of the animals we are studying.

The most important ways in which man affects animals' chances of survival are these:

(a) *Habitat change* We noted earlier that ecosystems could be considered as natural, or as having been modified by man (introduction to chapter 3). The modifications can be of many kinds, from the selective removal of a few trees by logging, to complete destruction, as when a natural ecosystem is cleared and burnt

Box 5.6 Cleaner fish – an example of complexity in inter- and intraspecific relationships

The coral reefs of tropical oceans are remarkable for their productivity, and they support a biomass of animals which is about the highest of any ecosystem (Table 3.2). They are particularly remarkable for the great variety of fish, mainly small and brightly coloured species. Two of these are of special interest amongst the many on the coasts of eastern Africa: the cleaner fish *Labroides dimidiatus* and the blenny *Aspidontus tractus* (see Figure). Both are about 10 cm long.

The cleaner fish earned their name from their habit of removing ectoparasites from larger fish such as snappers and parrotfish. This symbiotic relationship has obvious benefits for both species. The cleaner fish, which are strongly territorial, have particular places where they await customers, and the customers soon learn where to come to have their parasites removed. Although some of the customers, such as parrotfish, are active predators, they allow the cleaners to move all round freely, and even to enter their mouths, without harming them.

The cleaner fish also have a remarkable intraspecific organization. Each territory apparently contains one male and several females, with the male being the dominant member of the group (cf. pages 137–138). If he dies, the most dominant of the females takes his place – and in doing so changes sex to become a functional male. (Such sex changes are not uncommon amongst reef fishes.) The hierarchy is a linear one, with all young cleaner fish starting life as females, and all the survivors moving up one position in the hierarchy whenever the male, or a higher placed female dies.

Apparently cleaner fish have evolved a very successful symbiotic relationship with their customers, because they have been mimicked by another unrelated species, the blenny. The two species are almost identical in appearance, except for the position of the mouth. But the blenny is not a cleaner; it waits for a fish to come along and then bites a mouthful of its flesh! Its true position is thus that of a predator.

(Figure redrawn from Wickler, 1968. – reference in chapter 6)

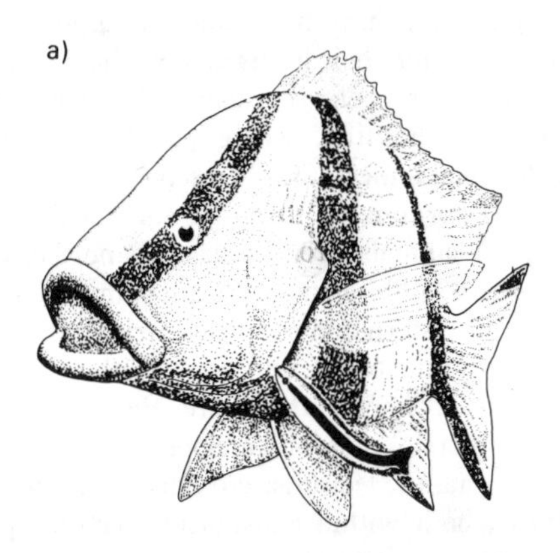

(a) A cleaner fish attending a red snapper fish *Lutianus sebae*. The blenny (c) bears a remarkable resemblance to the cleaner fish (b). The blenny's mouth is ventral, however, whilst that of the cleaner has claw-like lips used in prising ectoparasites from its customers.
(From Wickler, 1968, reference in chapter 6.)

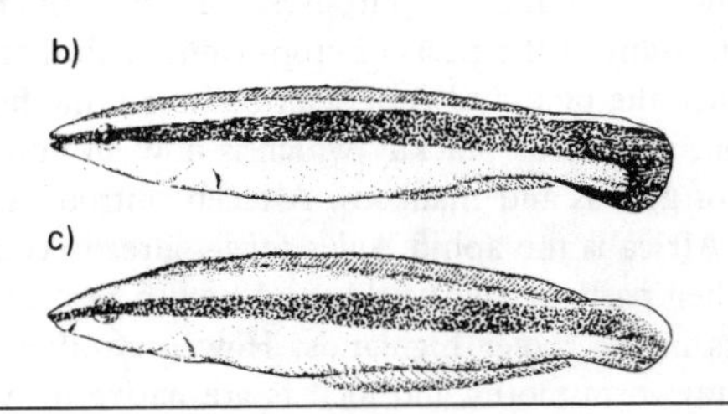

and replaced by crops or houses. In the process, new man-made ecosystems are created, but the animals inhabiting them are usually quite different from those that were there before. For example, most of the birds counted in an area of small scale farming near Kibwezi, Kenya, belonged to species that were uncommon or completely absent from an adjacent area of natural vegetation. Conversely, few of the birds characteristic of the natural area were seen in the cultivations. The total numbers of birds in the two areas were remarkably similar, however (cf. Figure 3.26).

Often the effects of man on his environment are indirect, as when lakes are polluted by industrial wastes, or eutrophication is induced by high levels of nutrients derived from fertilizers.

(b) *Hunting, fishing and poaching* These activities affect only a small number of species, but for them the results can be dramatic. In tropical Africa, both species of rhinoceros are faced with imminent extinction in the wild, as are the three Asian species, and in many countries the elephant will soon be gone (Figure 5.2). Both have been killed in large numbers by man, rhinos for their horns and elephants for their tusks. Killing for profit is usually termed **poaching**, in contrast to **hunting**, where people kill animals mainly to obtain meat for themselves. Hunting has practically exterminated many of the medium- and large-sized mammals from much of northern and western Africa. Wild animals are a less important part of the diet in eastern and southern Africa, where they have survived in greater numbers. Within 20 km of Nairobi, for example, there are three species of monkeys, several kinds of antelopes, zebras, hippopotamuses, giraffes, leopards, crocodiles and numerous large birds, both within and outside national parks, some in significant numbers.

(c) *Introductions* Many plants and animals which thrive in Africa today were brought by man. The list is long, and includes almost all of the major crop plants, all species of domestic animals including dogs and chickens, and many of the trees which are planted for firewood or building purposes, or for ornamental value. Many of the pests of crops came with them, for instance the pink and American bollworm moths with cotton, and citrus blackfly which is now an important pest of guavas and mangoes. A recent introduction to East Africa is the aphid *Aphis fabae*, already common in other parts of the world, and now a pest of bean plants in the cooler highlands. However, others such as armyworm moths and locusts are native to Africa.

(d) *Conservation* Although most human activities have an adverse effect on the native African plants and animals, there is an increasing awareness of their value. As a result, almost all governments have set aside areas where nature is to be conserved. Some of these areas are extensive, and well managed, but not all. Meanwhile many major ecosystems, especially forests, are still disappearing at a rate considered alarming by such responsible bodies as the United Nations Environment Programme (UNEP).

Responses to unfavourable environments

We have now considered the major components of animals' environments, and have seen how most of these have optimal values for particular species. (For most other species, the optimum numbers of humans is zero!) The probability of all components of a species' environment being favourable simultaneously is extremely small, although it might be achieved experimentally. Under ideal conditions in the laboratory, animals survive and reproduce at a maximum rate and very rapid increases in population are possible (pages 176–182).

With less favourable conditions, the population increases more slowly, and if conditions continue to worsen, the rate of increase may fall to zero or become negative. The species then decreases and in extreme cases can become extinct. Where some components of environment, such as rainfall and temperature, are subject to marked seasonal cycles, there is a strong selective advantage for species to synchronize their life-histories to the seasons. As an example, the young of most animals are produced at times of greatest food abundance. Different species adopt different strategies to survive unfavourable seasons. One solution is to migrate to a more favourable place (pages 165–172), but this is common only in birds, although of course it is also well known and also in some mammals and insects.

Becoming inactive

Animals that cannot escape unfavourable conditions by migrating may do so by going into an inactive state, thus minimising their requirements. Animals in a prolonged state of inactivity are said to be **dormant**. Dormancy can take several forms, but all involve reduction of the basic metabolic rate to low levels or, in exceptional cases, to zero. Once an animal has become dormant, it cannot move, so it usually seeks a hiding place beforehand, and may also be cryptically coloured.

In higher latitudes, many animals **hibernate** for several months during the cold winters. The habit is widespread amongst exotherms, but some endotherms such as bears and squirrels hibernate too. Dormancy in the tropics is confined to exotherms, and is usually a response to dryness, especially when accompanied by high temperatures. This form of dormancy is known as **aestivation**. Both hibernation and aestivation are usually preceded by an accumulation of food reserves, mainly lipids, although some gastropods store carbohydrates as their major reserve.

The West African lung-fish, *Protopterus annectens*, frequents slow-moving inland waters. At the onset of the dry season, it buries itself in the mud of a swamp or river bed, and enters a state of aestivation, which may last for as long as seven months. With the return of the rains the fish emerges from its aestivation chamber and this is soon followed by breeding. The East African species, *P. aethiopicus*, also aestivates during dry periods. Its horizontal aestivation tunnels are mainly formed in river banks and often become partially filled with water.

Aquatic snails of the genera *Bulinus* and *Biomphalaria* are important intermediate hosts of schistosomiasis. They became buried in mud, especially at bases of marginal vegetation and grass, at the beginning of the dry season when ponds, pools and ditches are drying out. These snails have no protective operculum to seal the opening of their shells but they secrete a mucus substance which affords considerable protection against desiccation. They can aestivate for about eight months in dry mud. In some snails, infections of the parasitic stages of schistosomiasis persist. The snails thus provide a reservoir of infection during dry periods, for transmission when their habitats are reflooded.

Terrestrial snails of the family Achatinidae also aestivate successfully during dry periods. For example, the giant land snail *Achatina fulica* withdraws into its shell and secretes a protective coating over its opening. This epiphragm keeps out pathogens, but probably does little to prevent desiccation. In savanna regions *Achatina* may aestivate for several months, whereas in wetter forested areas aestivation may be necessary for only a few days or weeks. When the rains arrive, the epiphragm is broken and the snail resumes activity (Figure 5.29). Some species of Achatinidae bury themselves in the ground before aestivation, others remain on the surface or under a few leaves and plant debris. Other species inhabiting arid regions can aestivate for several years when forced to do so.

In some insects, development of the eggs, larvae or adults may cease, although external conditions may appear favourable. In this instance the insect is said to have entered a state of **diapause**. Diapause can be obligatory in every generation as a part of the life-history strategy of the insect, or only occur in response to the onset of unfavourable conditions. In many cases, but not all, changes in daylength both induce and terminate diapause and hibernation: aestivation is usually broken by a return to moist conditions.

Figure 5.29 The African giant snail is only active in wet weather
(D. E. Pomeroy).

Eggs of many *Aedes* mosquitoes such as *Ae. aegypti*, the yellow fever mosquito, regularly enter diapause in order to survive the dry season when their larval habitats dry out. Even when these are flooded, the eggs may not hatch unless there is a reduction in oxygen content of the water. In West Africa, larvae of the millet grain midge pass the dry season as diapausing larvae in millet heads. When sufficient rain has fallen to wet any unharvested millet ears remaining in the field, larval diapause is broken and pupation occurs. This allows the appearance of the first adults of the new season to coincide with the flowering of the millet. During the wet season there is no larval diapause, and four to five overlapping generations of the midge can occur. Another African pest of cereals, the white rice-stem borer, which is a small moth, often enters larval diapause. In Sierra Leone larvae may remain in diapause in rice stubble for as long as about five months.

Another mechanism for overcoming severe adverse environmental conditions is **cryptobiosis**: this is a term used to describe the peculiar process by which an animal shows no obvious signs of life and where metabolic activities have ceased. The only insect definitely known to exhibit cryptobiosis is found in Nigeria. It is a chironomid fly called *Polypedilum vanderplankei* that occurs, together with the mosquito

Aedes vittatus, in small water-filled rock pools exposed to the sun. These pools dry out completely from about October to April, and whereas *Ae. vittatus* survives in the dry bottom mud as drought-resistant diapausing eggs, larvae of *P. vanderplankei* become dehydrated and lose as much as 92 per cent of their normal water content. In this condition, they are able to withstand both severe desiccation and high or low temperatures. Larvae can withstand total dehydration for more than three years; in the laboratory they can even survive for several days at −190 °C. When their pools are reflooded by rain water, the larvae hydrate and development resumes.

Losing 'condition'

At times when an animal's food supply is inadequate, if it remains active it will inevitably lose 'condition' (pages 107, 125 and 130). An animal's 'condition' is its overall fitness and health; when in good condition it will also have reserves of stored energy. During favourable times, when an animal has enough to eat, and is not growing, the following equation holds true:

$$\text{Energy intake} + \text{Energy reserves} = \text{Heat production} + \text{Work output} + \text{Energy storage} + \text{Energy losses in urine and faeces}$$

The energy intake is mainly from food; direct solar energy sometimes contributes a little. If the food supply declines in quality, quantity, or both, the total energy intake may cease to be adequate to maintain the animal in good condition. This will be noticed as a loss of weight, due mainly to mobilizing part of the energy store, which is usually in the form of fat. After prolonged hardship, muscles and other organs may atrophy; they are metabolised to provide energy.

To maintain their condition, tsetse flies need to take blood meals from mammals or reptiles about once every three days, although the speed of digestion is affected by temperature, so that the interval decreases in hot weather and increases when it is cold. Vertebrate blood contains about 19 per cent of proteins by weight, but only about 0.1 per cent carbohydrates and 0.6 per cent lipids. In newly emerged tsetse flies of both sexes, most of the ingested nutrients are used for synthesis of muscle proteins, whereas in mature females which are pregnant large amounts of protein are needed to nourish the developing single larva which the fly produces.

A pregnant fly normally requires four to five large blood-meals for the successful maturation of the larva, which develops to an advanced stage inside the female. When hosts are scarce, the number of larvipositions decreases and the larvae produced are small. This results in small flies in the next generation. Insufficient food also results in abortions, that is the birth of dead larvae which are not fully formed. The success and health of a tsetse population can be judged from the weight of its puparia. Energy is stored in the tsetse larva and puparium as fat and this has to keep the puparium alive for some three to five weeks, until the adult emerges from it.

Mosquitoes are also affected by the quantity and quality of their blood-meal. Species that normally feed on birds produce few or no eggs if they are forced to feed on mammalian blood. In the absence of blood-meals, adults can survive for many weeks, but in most cases females are unable to mature their ovaries and thus cannot lay any eggs. In a few species, however, sufficient nutrients are carried over from larval development to allow the females to lay at least one batch of eggs without a blood-meal. This is known as **autogenous** development, and it has particular survival value for species inhabiting areas where suitable hosts are rare, such as in arctic and desert regions. It has also been observed that when the population densities of larvae are high, they often produce large numbers of small adults, which in turn take smaller blood-meals and thus lay fewer eggs; whereas when larvae are uncrowded, and take more food, fewer adults are produced but they are bigger and lay more eggs. This is a good example of a density-dependent effect (pages 192–206).

Suggested reading

The general approach to animal ecology followed in this chapter is based upon Andrewartha, 1970 which in turn was derived from Andrewartha & Birch, 1954. Both books discuss components of environment in some detail, with many examples. Glasgow's 1963 study of tsetse flies is an example of the use of this approach in practice. Although mainly concerned with insects, Varley *et al.*, 1973, also cover many of the topics discussed in this chapter.

More general texts, giving African examples relevant to this chapter, include Ewer & Hall, Book 2, 1978, Cloudsley-Thompson, 1969, and Owen, 1976.

The effects of climate on animals are described in Bligh *et al.*, 1976, whilst Hardy, 1972, deals specifically with temperature. A number of the chapters in Townsend & Calow, 1981, are concerned with both the ecology and physiology of resources, especially food.

Competition is discussed in most ecological textbooks, for example there are good accounts in chapter 3 of Colinvaux, 1973, chapter 12 of Krebs, 1978, and chapter 2 of Whittaker, 1975. Hassell, 1976, provides a good summary of modern ideas on both competition and predation, whilst Curio, 1976, writing solely on predation, has many African examples. Ecological parasitology is reviewed by Kennedy, 1975, whilst Nnochiri, 1975, concentrates on parasites of medical importance in the tropics. Several recent books, such as Rosenthal & Janzen, 1979, and Crawley, 1983, deal thoroughly with the subject of herbivory: two particularly good ones about herbivorous insects are Hodkinson & Hughes, 1982, and Strong *et al.*, 1984.

The distributions of animals in Africa are described in various books. Beadle, 1981, deals with inland waters and their fauna, and there are maps showing the distributions of mammals in Dorst & Dandelot, 1970, which deals with the larger species over Africa as a whole, and Kingdon, 1971–82. Kingdon's series of books is concerned primarily with East Africa, and is remarkable for the detailed accounts of contributions, and the factors affecting them. Hill, 1975, has good maps of insect pests in the tropics, whilst Brown *et al.*, 1982, have distribution maps for African birds.

Elton, 1958, in a classic book, reviews the effects resulting from introductions of animals and plants by man to various parts of the world.

Essays and problems

1 Review the importance of water to terrestrial invertebrates.

2 Write an essay on temperature and animal life. Starting points are Bligh *et al.*, 1976, or Hardy, 1972.

3 Assess the importance of parasites as components of animals' environments.

4 Describe the range of aquatic ecosystems to be found in your country.

5 Choose any species of animal that interests you – it could be a common local lizard or bird, or a pest such as a leaf miner or cockroach. Describe the components of its environment so far as you can, using the scheme in this chapter. Use your own knowledge of the species, as well as published information about it. Include a map of its distribution in Africa. What remains to be discovered about the ecology of this species?

6 'From the plant's viewpoint, most of the social bees are a detrimental component of the ecosystem. They are primarily flower visitors rather than pollinators' (Janzen, 1975, pp. 19–20: full reference in chapter 2). Discuss this and other interactions between bees and plants.

7 'An animal's life is mostly about eating – or avoiding being eaten'. Is this an oversimplification?

8 Using examples, preferably from your own country, describe the importance and role of diapause and aestivation in animals' life-history strategies.

9 Contrast the different adverse conditions facing animals living in semi-arid areas and subalpine areas. How have they overcome the harsh environmental conditions of living in such extreme climates?

10 Discuss the proposition that wind can be an asset or hindrance to animals.

References to suggested reading

Andrewartha, H. G. 1970. *Introduction to the Study of Animal Populations*, (2nd edn). Methuen: London.

Andrewartha, H. G. & Birch, L. C. 1954. *The Distribution and Abundance of Animals*, University of Chicago Press: Chicago.

Beadle, L. C. 1981. *The Inland Waters of Tropical Africa*, (2nd edn). Longman: London.

Bligh, J., Cloudsley-Thompson, J. L. & Macdonald, A. G. 1976. *Environmental Physiology of Animals*, (2nd edn). Blackwell: Oxford.

Brown, L. H., Urban, E. K. & Newman, K. 1982. *The Birds of Africa*, Vol. 1, Academic Press: London.

Cloudsley-Thompson, J. L. 1969. *The Zoology of Tropical Africa*, Weidenfeld & Nicholson: London.

Colinvaux, P. A. 1973. *Introduction to Ecology*, Wiley: New York.

Crawley, M. J. 1983. *Herbivory*, Blackwell: Oxford.

Curio, E. 1976. *The Ecology of Predation*, Springer-Verlag: Berlin.

Dorst, J. & Dandelot, P. 1970. *A Field Guide to the Larger Mammals of Africa*, Collins: London.

Elton, C. 1958. *The Ecology of Invasions by Animals and Plants*, Methuen: London.

Ewer, D. W. & Hall, J. B. (eds). 1978. *Ecological Biology* 2 Longman: London

Glasgow, J. P. 1963. *The Distribution and Abundance of Tsetse*, Pergamon: Oxford.

Hardy, R. N. 1972. *Temperature and Animal Life*, Edward Arnold: London.

Hassell, M. P. 1976. *The Dynamics of Competition and Predation*, Edward Arnold: London.

Hill, D. 1975. *Agricultural Insect Pests of the Tropics and their Control*, Cambridge University Press: Cambridge.

Hodkinson, I. D. & Hughes, M. K. 1982. *Insect Herbivory*, Chapman & Hall: London

Kennedy, C. R. 1975. *Ecological Animal Parasitology*, Blackwell: Oxford.

Kingdon, J. 1971–1982. *East African Mammals*, (7 volumes), Academic Press: London.

Krebs, C. J. 1978. *Ecology*, (2nd edn). Harper & Row: New York.

Nnochiri, E. 1975. *Medical Parasitology in the Tropics*, Oxford University Press: London.

Owen, D. F. 1976. *Animal Ecology in Tropical Africa*, (2nd edn). Longman: London.

Rosenthal, G. A. & Janzen, D. H. (eds). 1979. *Herbivores; their Interaction with Secondary Plant Metabolites*, Academic Press: London.

Strong, D. R., Lawton, J. H. & Southwood, T. R. E. 1984. *Insects on Plants*, Blackwell: Oxford.

Townsend, C. R. & Calow, D. 1981. *Physiological Ecology, an Evolutionary Approach to Resource Use*, Blackwell: Oxford.

Varley, G. C., Gradwell, G. R. & Hassell, M. P. 1973. *Insect Population Ecology*, Blackwell: Oxford.

Whittaker, R. H. 1975. *Communities and Ecosystems*, (2nd edn). Macmillan: New York.

6 Behavioural ecology

An animal's ability to survive and reproduce – its fitness – is affected profoundly by its behaviour, which is consequently of interest to us as ecologists.

The behaviour of animals is reflected in their dispersion – a word which describes their spacing. Related to dispersion is sociality. Living in groups can be advantageous, giving greater alertness to predators and, especially in some of the highly social insects, division of labour. Sociality may simply involve one individual helping another, as with some birds when nesting, or be highly specialized and with various castes, as in soldier and worker termites and ants. Many social species have evolved complex mating systems. But we also find that living in a group can sometimes be disadvantageous, as groups attract predators, facilitate disease transmission and may intensify competition.

Animals communicate in many ways with their own and other species. The songs of birds and frogs attract mates and proclaim territories; odours including sex pheromones also attract mates, while obnoxious smells repel attackers. Coloration imparts information too. Brilliant displays of colours can attract mates or alarm predators, in contrast to the camouflage of cryptic colouration. The evolutionary advantages of warning coloration in deterring predators have led to mimicry, a strategy extensively exploited by insects.

Feeding behaviour usually involves both foraging for food and consuming it. Studies of animals' foraging strategies have shown that they maximize their food intake in relation to the costs involved.

Most animals are capable of moving about, but often their movements are restricted to a limited area, the 'home range'. Some animals defend an area – their territory – from other individuals, which are usually, but not always, of their own species.

We conclude by considering activities which are seasonal. For most species a clear example is in the timing of breeding which occurs at the time of year when resources are most abundant and can support an influx of young into the population. After breeding, individuals of many species disperse or migrate. Migration typically implies a long-distance movement of a predictable nature, the animals later returning to the place where they started. Migration of birds and whales can involve travelling thousands of kilometres a year.

An animal's behaviour is part of its response to its environment, and is, therefore, highly relevant to a study of its ecology. However, the traditions of studying ecology and behaviour have developed separately; the study of behaviour, especially in the field, becoming known as **ethology**. This division was unfortunate, since to consider one without the other results in a narrow view of how animals are adapted to their environments. In the last few years there has been a welcome trend amongst biologists to integrate these two aspects of an animal's biology, and there have been rapid developments in the field of **behavioural ecology**, in which the central theme is to seek a fuller understanding of the adaptiveness of behaviour.

One has only to spend a short time studying animals in the field to become aware of the variety of behaviours to be seen. Sometimes the reasons for an observed piece of behaviour will be obvious, but often, at least to start with, they will not. Nevertheless biologists generally accept the premise that almost all behaviour has survival value; that is, it is adaptive in the sense that it enhances the animal's fitness to its environment. The concept of **fitness** is a very important one: it refers to the contribution and individual makes, through its descendants, to the next generation. Hence increasing fitness means having more offspring which themselves survive and thus promote the spread of their parents' genes through the population.

We have already discussed some aspects of animals' behaviour, such as how prey react to predators, and vice versa (pages 125–127). We also noted some behavioural responses by animals when their environments became unfavourable (pages 128–130). In this chapter we explore further aspects of behaviour which have a direct bearing on an animal's fitness.

Behaviour towards other individuals of the same species

Dispersion

Just as with plants (chapter 4), individual animals can be spaced in various ways relative to each other: for example, some animals are social and live in groups, whilst others are not. Individuals and groups can be arranged in different ways within a habitat. These arrangements are best described by the term **dispersion**, although 'distribution' is sometimes used in this sense. Notice, however, that the word distribution is more correctly used to mean the total geographical area where the species occurs (pages 93 and 94), whilst 'dispersion' is confusable with 'dispersal' (pages 164 and 165).

Ecologists recognize three main types of dispersion: regular, random and aggregated (Figure 6.1). These three categories arise from different kinds of behaviour.

In a **regular dispersion** the spacing between individual animals is similar (Figure 6.1(a)). This is a situation typical of species which are territorial. The process of defending their territories from neighbours leads to even spacing – provided that the habitat is uniform. This is encountered, for example, in bird

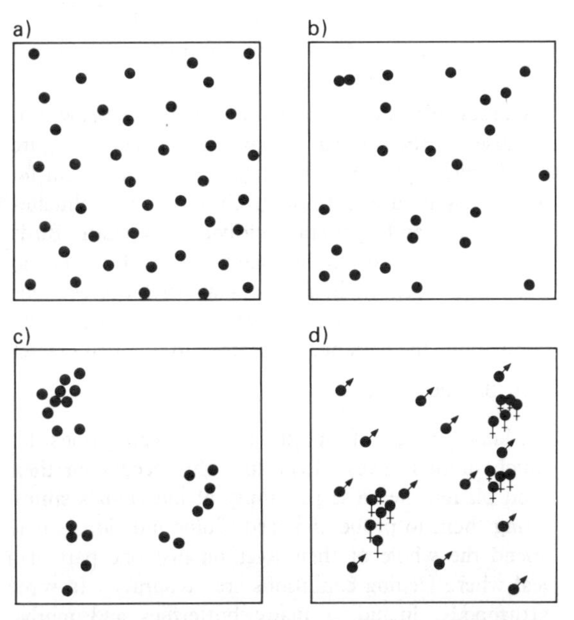

Figure 6.1 Types of dispersion: in diagrams (a), (b) and (c) each dot represents an individual animal. When the individuals are spaced at approximately equal distances (a) the dispersion is **regular**; when in groups they are said to be **aggregated** or clumped (c); whilst between these two it is possible to have a **random** dispersion (b).
Actual patterns of dispersion are usually more complex. For example, males of some antelopes are territorial and in a uniform habitat can be fairly evenly spaced (d, ♂). The females, however, live in groups which occupy home ranges (♀); (cf. pages 138–143) within these ranges their movements have an element of randomness. (See also Figure 6.9.)

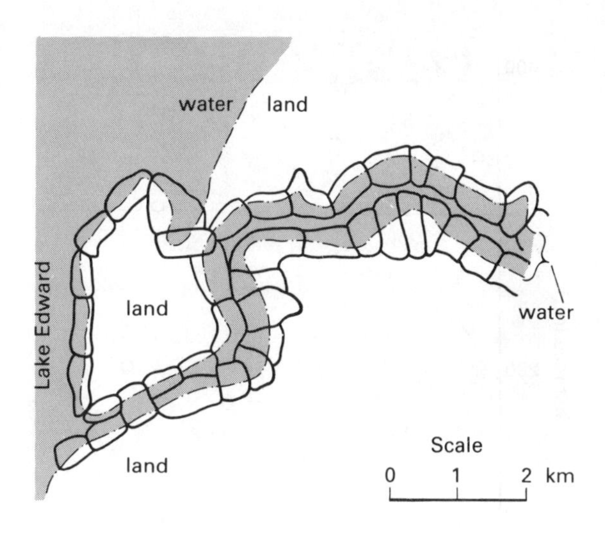

Figure 6.2 Even spacing of fish eagles in part of the Queen Elizabeth National Park, western Uganda. The approximate territorial boundaries are shown: each territory was occupied by a male and female. Every suitable site was occupied. (Redrawn from S. J. A. Sumba, 1983. Unpublished Ph.D. thesis, University of Nairobi, Kenya.)

populations in which breeding pairs or groups defend their territories from intruders (Figure 6.2). Regular dispersions are less common amongst invertebrates, although one can sometimes see whole colonies of termites showing a regular spacing (Figure 6.3). But although dispersions of this kind, tending towards regularity, are rather uncommon, they are significant in that they provide evidence of territoriality which, as with termites, might otherwise be difficult to demonstrate. An important characteristic of a regularly dispersed population is that when sampled, the numbers of animals found in successive samples show rather little variability. The variance (s^2) of the samples in these populations is low, and always less than the mean ($\bar{x}$), i.e. $\frac{s^2}{\bar{x}} < 1.0$; in a perfectly regular dispersion, there is no variability at all, so $\frac{s^2}{\bar{x}} = 0$.

In a **randomly dispersed** population (Figure 6.1b)), there is an equal likelihood of any individual occupying any point within the area, and the presence of one individual neither attracts nor repels others. There is no organised arrangement, just a random scattering of individuals. Very few animals are randomly dispersed, because few habitats are uniform, and furthermore individuals of most species interact, either negatively, tending to produce regularity, or positively, leading to aggregations. If a perfectly randomly dispersed population existed, then samples taken from it would have a variance equal to their

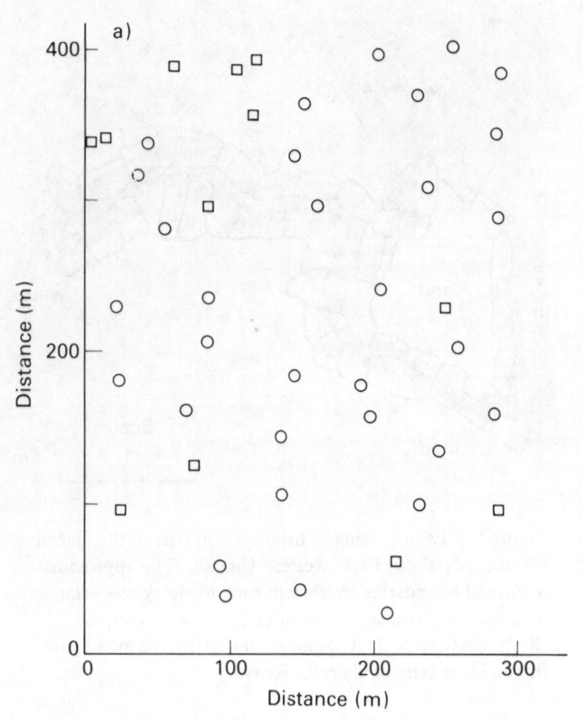

Figure 6.3 (a) Dispersion of mounds built by two species of termites at Emali, Kenya: *Macrotermes subhyalinus* (○) and *M. michaelseni* (□). The site was a flat, uniform grassland. Termites are believed to be territorial and in this case it seems that the two species are also mutually exclusive, since their mounds are well spaced regardless of species. By eye, a fairly regular pattern is apparent on the map, but there is another factor. One would expect larger colonies to possess larger mounds and to command larger territories.
(b) Evidence for this is seen in the graph which shows the distance from each closed mound to its nearest neighbour. Generally, large mounds were further from their neighbours than small ones. (Only mounds containing live colonies are included.)

mean (i.e. $\frac{s^2}{\bar{x}} = 1.0$), but exact equality is exceptional in the extreme. Something approaching a random dispersion is however seen in aphids, when the winged forms first settle onto a crop. But it does not persist, since they soon begin to aggregate in the more favourable feeding sites.

Aggregated dispersions (Figure 6.1c) are by far the commonest kind; they are also known as contagious, or clumped. As most habitats are not uniform, it is hardly surprising that individual animals prefer some parts of their habitat to others. There are numerous examples of both vertebrates and invertebrates showing markedly aggregated dispersions, which are usually explicable in terms of the dispersion of their resources, particularly food and shelter (see below); or because of their social behaviour, or both (Figure 6.1(d); page 133). However, aggregations can occur for other reasons. Bees form large clusters in cool climates and consequently generate individual warmth. Birds and mammals also huddle together in cold weather to share their warmth. In a dry climate such aggregations may also minimize rates of water loss. In aggregated populations the variance is greater, often much greater than the mean (i.e. $\frac{s^2}{\bar{x}} > 1.0$).

Many species of aphids live in aggregations for much of their lives. Typically, this occurs on their food plants, the close proximity of individuals stimulating them to probe and feed. Some individuals may spend the whole of their lives on just one part of a leaf where feeding conditions are favourable. In some arthropods, including many butterflies and moths, each new generation is at first highly aggregated. The young caterpillars are typically clumped together because eggs are laid in batches, but the degree of aggregation decreases with time, as the growing caterpillars become more dispersed. Conversely, insects colonizing a new habitat often show a random dispersion at first, becoming more aggregated later.

A striking example of animals aggregating at a food source is vultures coming to a carcase (Figure 6.4). To a lesser extent, the same effect can be seen at rubbish tips, which as well as vultures, attract kites, crows and

Figure 6.4 A 'vulture party'. The carcase of a buffalo has attracted many vultures, which are of several species, African white-backed and Rüppell's being the commonest. At the carcase are two black-backed jackals. (F. Hartmann.)

flies such as greenbottles and blowflies; elsewhere these species are relatively uncommon.

Sometimes a more detailed investigation is needed to demonstrate an association. In Uganda, a population of the land-snail *Limicolaria martensiana* was sampled using quadrats each of one square metre. The numbers of snails per quadrat ranged from 0–202, and the highest numbers were in quadrats which also contained an abundance of their food plants.

Before leaving the subject of dispersion we should repeat an important point: that it is affected by evenness of the habitat as well as by the behaviour of the individuals. Thus, the finding of an aggregated population may imply that the individuals behave socially, or that their favoured habitat is patchy, or both.

Kinds of sociality

Relationships between the individuals that constitute a species can be of many kinds. Those species that are found in groups of two or more are said to show **sociality**. The ethologist Aubrey Manning defines an **animal society** as a 'stable group whose members intercommunicate extensively, and bear some semi-permanent relationship to one another'. Clear examples are the social insects. Manning contrasts these animal societies with aggregations of animals, such as the vultures at a carcase (Figure 6.4), which come together only because of the food; they exist as a group so long as the food lasts, then disperse individually. The interactions between the members of a 'vulture party' are mainly negative; whereas social insects exhibit many positive interactions between the individuals – for example some are fed by others. As we shall see, membership of a society can confer mutual benefits.

Animal societies are not usually so complex as those of the social insects: there is in fact a whole range of complexity. Amongst the simplest are the rather loose gatherings seen in some bird flocks, or locust swarms, where the individuals may stay together for some time, but show little active co-operation (Figure 6.5). More specialized are many mammalian societies, where long-term bonds between individuals are frequent (pages 139, 154 and 155). Members of these societies may have differing roles, apart from those directly related to sex (Figure 6.6).

Amongst insects, true societies are confined to two orders, the Isoptera (termites) and Hymenoptera (ants, bees, wasps and their relatives). Being gregarious may have been a first step towards the evolution of more complex social behaviour. Amongst the bees and wasps, a series of stages can be seen, from species which are solitary, to those which form colonies with only a few individuals and no division of labour, and finally to the most specialised such as honey bees.

Social behaviour can confer various advantages. There are more ears, eyes and noses, and hence predators have difficulty in approaching undetected. Once detected, predators are likely to be further deterred by being 'mobbed' or even attacked by groups of individuals. To be stung by one bee may be tolerable, but to be stung by many of them is another matter. Predators that chase their prey tend to become confused by a fleeing group, chasing first one individual and then another, and ultimately catching none.

In insects particularly, the evolution of complex societies has made possible the construction of

Figure 6.5 Flamingoes are gregarious, forming large flocks which may stay together for long periods. But the individuals do not constitute an organized 'society'; for instance there is no dominance hierarchy or division of labour. (D. E. Pomeroy.)

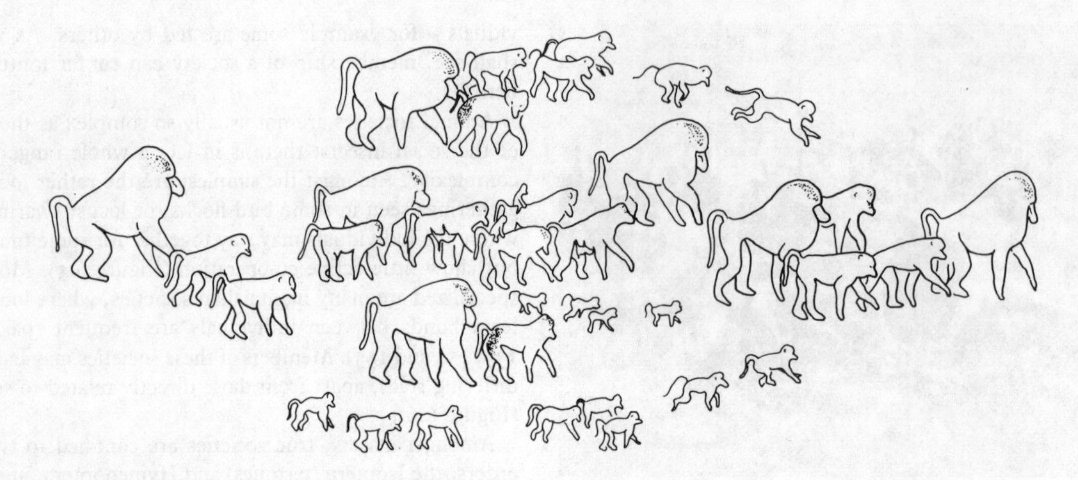

Figure 6.6 When a troop of baboons has to cross open ground, they tend to arrange themselves in a characteristic way, as shown here. At the centre are females with young, clinging to their undersides or riding on their backs. They are accompanied by adult males. The two females with dark rumps are in oestrus and each is closely attended by a consort, one of the more dominant males. Juveniles and subordinate males are around the outside. Should danger threaten, the juveniles would retreat to the centre, whilst the males attempt to defend the troop.
(From Delany and Happold, 1979, after Hall, K. R. L. & de Vore, I. 1965, pp. 53–110 in I de Vore (ed.) *Primate Behaviour*, Holt, Rinehart & Winston: New York.)

complex hives and nests. The internal structure of the termite *Macrotermes bellicosus* nests in parts of West Africa is probably the most remarkable animal construction in the world (Figure 6.7), and implies highly organized and co-ordinated behaviour, largely mediated by pheromones (pages 155–156).

There is evidence, too, that communal behaviour can promote exchanges of information. In birds such as the queleas, this apparently occurs at the overnight roosts. Flocks of queleas which have had a poor day foraging tend to follow other flocks the next morning, and may in this way be led to better feeding grounds.

Relatively few predators are social, at least whilst hunting, since they would tend to scare each others' prey. Spiders, mantids, eagles and leopards are all solitary. There are exceptions, however. At times, lions and hyaenas co-operate in hunting, as hunting dogs invariably do, and insectivorous birds commonly forage in parties which involve several different species. Safari ants have highly organized feeding columns. White pelicans swim in a line which, when they locate a shoal of fish, encircles the prey and surrounds them. Suddenly, all the pelicans dip their heads towards the centre and scoop up the fish.

Social behaviour seems to favour parental care, through which the chances of individual survival are increased. The protection afforded to the youngest individuals may be extended to older ones and even to adults. This behaviour is well seen in social species of mammals, such as most of the larger antelopes. Female elephants live in groups that persist throughout their lives (page 140) and in many primates semi-permanent groups of both sexes are found. Baboons provide an excellent example of this type of sociality (Figure 6.6). In general, the extent of parental care is correlated with the degree of sociality.

In the more complex social groupings, there is commonly a division of activities between individuals. In captivity, at least, individuals of primates, or social fish, can survive by themselves, although in nature they would probably not. However, in the most advanced cases of sociality, notably the termites and some ants, there are pronounced morphological differences between the **castes** – that is, the workers, soldiers and reproductives. Here the individuals are totally dependent upon each other.

Amongst social birds and mammals, group size is normally larger in species inhabiting open places, such as lakes or grassy plains, than in forest species. We shall return to this point later when discussing antelopes (Box 6.1), but it also applies to other species. In birds, very few forest species form flocks (although some join mixed-species foraging parties). But in fields and open plains, flocks of weavers, starlings, doves and other birds are a common sight. Many of the birds seen on lakes also form flocks – particularly ducks, coots and cormorants. These observations reinforce the argument that flocking (or herding) is primarily a reaction to predators. In circumstances where both predator and prey can see each other easily, the prey

(a)

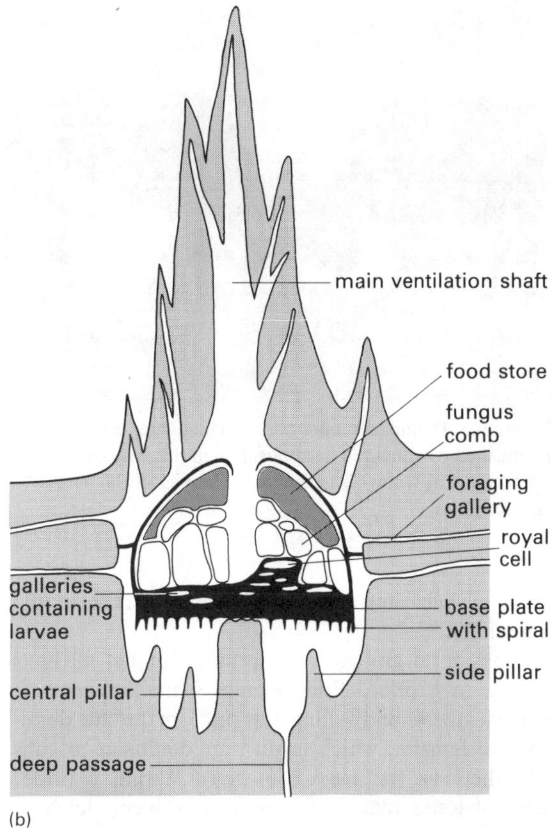

(b)

Figure 6.7 (a) *Macrotermes bellicosus* is widely distributed in tropical Africa where its large mounds are sometimes very numerous. Those in Nigeria seem to be particularly complex, with a large plate beneath the hive on whose underside is a spiral ridge (b) The whole of the hive (the 'living area', consisting of food store, fungus combs, royal cell etc.) is supported on a central pillar; thus access is restricted to a few points. This presumably enables the termites' defenders – the soldiers – to concentrate their activities if the colony is attacked by predators such as safari ants, *Dorylus*. (Photo: D. E. Pomeroy; drawing based on Collins, N. M., 1979, *Insectes Sociaux*, **26**, 240–46.)

tend to respond by forming flocks, but there is less advantage in doing so in a forest, where the thick vegetation makes visual detection of prey difficult.

Dominance hierarchies are found in many social species, especially vertebrates, but also in bees and other social insects. The simplest type of hierarchy is that in which one individual from a social group dominates all of the others; the second individual in the hierarchy dominates all except the first; and so on to the bottom in a linear sequence. This is the system amongst groups of the domestic fowl, and in various wild birds and other animals, for example cleaner fish (Box 5.6).

Dominance is established by contest. In buffaloes and many species of antelope it takes the form of head-to-head pushing (Figure 6.8), but fighting rarely involves serious injury, and deaths are exceptional. Often it suffices for one animal to adopt an aggressive stance, or for a bird to sing loudly, to establish the winner. The acquisition of a territory is commonly related to strength, which in turn depends largely on size. Males of many species with dominance hierarchies continue to grow throughout life, though at a decreasing rate. Examples include most antelopes, elephants, baboons, crocodiles and some fish.

Dominance hierarchies result in a sharing of resources, usually food, shelter or mates. The division is unequal, with the most dominant individuals obtaining disproportionately more than the least dominant, which in extreme cases get nothing. Such an arrangement is only advantageous to the individuals at the upper end of the hierarchy and is therefore most likely to persist in species which experience shortages of resources quite frequently. Since the more dominant individuals usually have a better chance of surviving and reproducing, the genes maintaining this

Figure 6.8 Dominance amongst social ungulates is usually established by ritualized head-pushing, in this case by buffaloes. The winner is generally the heavier of the two. (D. K. Jones.)

form of behaviour are favoured and persist in the population.

Lions form groups called prides, but not all lions belong to a pride. Pride members are dominant to non-members, and within the pride males are dominant to females, which in turn are dominant to cubs of either sex (cf. pages 145–147). Within a pride, lions of lower rank in the dominance hierarchy have to wait for those above to finish feeding before they can do so. In plains antelopes such as impala, there are separate hierarchies in males and females. The same applies to various primate species and also to hunting dogs.

A study of baboons near Gilgil, Kenya, revealed a remarkably complex system. There were several hierarchies within one troop, but the most consistent was that amongst the adult females. Each adult female had her place in this hierarchy. As the troop of baboons moved, each of the adult females was usually accompanied by her kin group; this consisted of her offspring of several previous years, and they each had the same rank as she did. Thus even her smallest offspring ranked higher than the leading female of the group below. There was a separate hierarchy amongst adult males, but it was less rigid than that of the females, with individual males quite frequently leaving, joining or changing rank. On reaching adolescence, young males usually left their troop to join another, often some distance away, whereas young females remained in their kin groups, and eventually one would take over when the group leader died. Adult females usually had particular 'consort' males, but these were not necessarily the most dominant males. Dominance in both males and females was established by contest, usually by just threatening behaviour but sometimes actual fights occurred. However, although the most dominant males were necessarily the most aggressive, the dominant females tended to dislike aggressive males. Aggressiveness in adult males is nevertheless important for the troop, since it is they who protect the rest from predators.

Other studies of baboons in different parts of Africa have revealed similar social organizations, but in some cases there were important differences. Thus one study found that the largest males moved from troop to troop, becoming the most dominant within a short time of joining a new one, and mating with any of the females that were in oestrus.

Evidently behaviour is variable and presumably adaptive; a system suited to one set of circumstances may be disadvantageous elsewhere. Students of behaviour (like those of ecology) soon learn to be wary of making generalizations until they have obtained confirmation for their observations from other localities.

We have seen several advantages in social behaviour, not least in facilitating behaviour which is more complex and presumably more efficient also. But only a minority of animal species show any marked degree of sociality, and fewer still have evolved true societies. For whilst sociality facilitates detection of predators, large groups are often conspicuous and may therefore attract predators. Moreover, diseases spread more easily between members of a group than between solitary individuals.

Territories and home ranges

The term **territory** refers to a place which an animal defends against other individuals, usually of its own species. In contrast, a **home range** is not defended, but is simply the area an animal is familiar with and does not leave voluntarily. These ideas sound simple enough, but in practice the two terms are used to cover many different situations. The examples which follow illustrate their variety.

Dik-diks

These are amongst the smallest antelopes, inhabiting drier areas of Africa. They have clearly defined territories, each occupied by an adult male and female, although only the male is active in defence. The territorial boundaries are conspicuously marked by their owners' droppings: they defaecate and urinate at 'latrines' which occur at intervals of a few metres all round the perimeter. In favourable areas the territories are quite small, typically between two and five hectares, and they adjoin each other, so that every boundary is common to two adjacent pairs. Each pair stays together throughout the year, usually keeping within sight of each other, and never leaving their territory. They can survive without drinking, and they find sufficient food and cover within the territory. Their young also stay with them; usually there is only one. When the young reach puberty, they leave the territory, sometimes being chased out by their parents, and have to seek a place of their own. This they are likely to find difficult, for most of the favourable places are already occupied, and vacancies only occur when an older animal dies. Many young fail to secure territories and are taken by a predator, such as a leopard, or a martial eagle.

Single-pair territories like those of the dik-dik are found in some other mammals, and many species of birds, e.g. eagles, plovers and boubous (such as the bell-shrike). Typically, these species have a strong pair-bond, which in many cases is probably for life.

Figure 6.9 Schematic representation of the territories of an antelope – such as kob or impala – where the dominant males are territorial but other individuals have home ranges. Solid lines represent boundaries of territorial males. The territories vary in size, reflecting their quality and perhaps the prowess of the owner. The waterhole is used by all members of the population and could not be defended by any male. Territorial males tend to be near the centres of their territories, ready to move to any boundary if necessary (cf. central place foraging behaviour, Figure 6.25). Shading indicates the home ranges of two female herds; the darker part being the 'core' area which is the most frequently used. Notice that the home ranges overlap; and their boundaries are less fixed than those of territories. 'Bachelor herds' of younger and subordinate males also have home ranges, but these are not shown.

Impalas, gazelles, waterbucks

These and various other medium-sized antelopes of grasslands and open woodlands have a more complex system. The details vary with species and to some extent geographically, but the basic pattern is similar in each case. Only dominant males hold territories, which they defend vigorously against other males. The females, unlike the males, form small herds which also include their young. They move within quite a large home range, extending over several males' territories (Figure 6.9). During the time that a female herd is within a particular male's territory, he will mate with any female in oestrus. However, for most of the time, there are no females within his territory, although there may be young males. The male offspring leave the female herds as they reach puberty, and form their own bachelor herds. These also move freely through the territories of the territorial males, where they are tolerated so long as they do not attempt to mate with any females that are present. As they become older, males from the bachelor herds begin to challenge territory-holders and eventually may succeed in displacing an old male and taking over his territory.

The dominant males' territories are primarily for mating in these antelopes. Typically the males obtain most of their food from their territories, but may leave occasionally to drink.

A rather different and unusual case concerns the kob, a species that is widely distributed in grasslands from Senegambia to Ethiopia. In one part of its range, western Uganda, male kobs defend territories so small that they provide only a fraction of their food requirements. These 'mini-territories', each only 15–30 m in

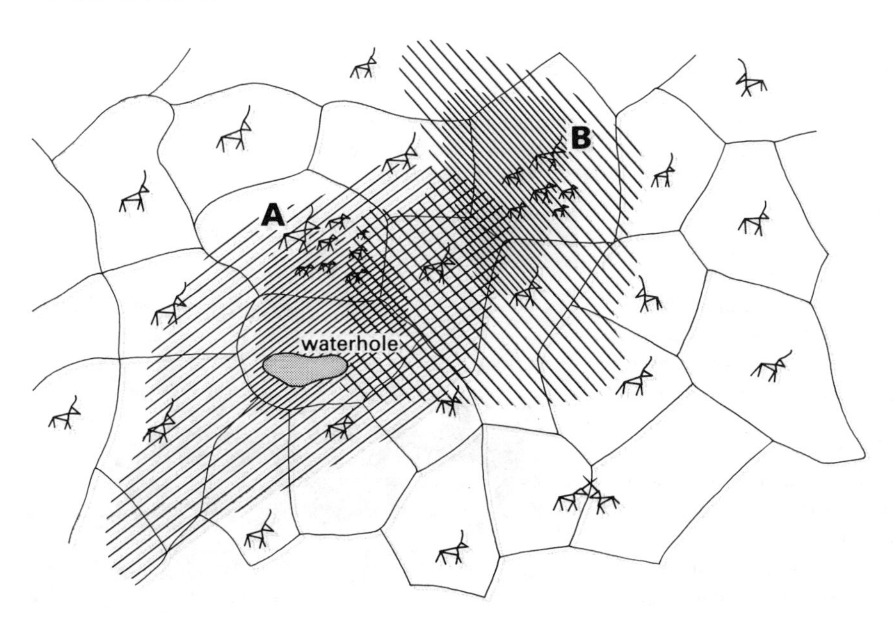

diameter, occur in groups of 10–15, termed **leks**. Female herds are strongly attracted to the leks, and most mating takes place there. But because of the minute size of his territory, a male can stay for only a few days before having to leave in search of food. After a period of rest, usually with a bachelor herd, the male is likely to attempt to regain one of the lek territories by displacing a hungry male.

Wildebeests

Wildebeests are amongst the largest species of antelopes, and they sometimes occur in huge herds – those of the Serengeti Plains in Tanzania being the most famous. The herds are compact and rapidly consume all the grass wherever they are, and consequently need to move frequently. Their movements in the Serengeti take the form of regular migrations (pages 170–172). It is therefore impracticable for the males to defend a particular piece of ground, so instead they attempt to defend a particular group of females. As the herd moves on, so do their territories, with the males frequently circling 'their' females. Fortunately for the males they do not have to keep this up for long, because the females are only receptive for a few weeks each year.

Antelopes

There are about seventy species of the family Bovidae in Africa, all but a few of them being antelopes. Most species have been studied in some detail, and hence their social organizations can be compared. One scheme groups antelopes into five classes, based on a range of related adaptations (Box 6.1). The dik-diks and other small species (class A in the Box) are selective feeders in forests and woodlands, usually living in pairs in permanent territories. At the other extreme, the largest species (class E) are inhabitants of open grassy plains; they form herds and never hold territories. They are also much less selective in their diets.

The small forest-dwelling species largely avoid detection by predators through being cryptic (see pages 149 and 150). If they suspect that a predator is near they 'freeze'. But herds of animals in open grasslands cannot avoid being seen by predators. On the other hand, the many pairs of ears and eyes in a herd ensure that approaching predators are detected. Medium-sized antelopes react to the approach of a predator by running away. However, larger species are rarely attacked. A herd of buffaloes can afford to stand their ground, but they remain alert whilst predators are near, the adults encircling their young.

The scheme set out in Box 6.1, and the explanations just given, form a set of hypotheses. They seem eminently reasonable, but the possibility exists that with further observations or experiments, some may prove to be wrong, or at least to need modification.

Thus kob leks, as we have seen, are exceptional in that the territories are so small that males can hold them for only a few days at a time, whereas kobs away from leks, as well as other species of antelopes in group C, hold territories for lengthy periods. A hypothesis to explain the evolution of leks might be that males on leks achieve more matings in the short time that they are there than do non-lek males holding permanent territories, because the leks are much more attractive to females. This hypothesis has yet to be tested.

Elephants

The African elephant is not territorial but does have a home range. In the drier parts of Africa a herd will occupy one area during the wet season and another during the dry season, moving between them along traditional migration routes. The herds are very loose gatherings, but within them are female groups whose composition is almost constant. Typically, each group consists of an old female (the matriarch) and two or three adult females who are almost invariably her daughters, together with their young (Figure 6.10). The members of these female groups usually remain together throughout their whole lives, which may be 50–70 years, but the males leave when they reach the age of puberty. Adult males are sometimes solitary, but they also form temporary bachelor groups, within which the largest member is dominant.

Figure 6.10 Female elephants live in groups, whose individuals sometimes remain together throughout their lives. The adult female on the right has just given birth: note her enlarged mammary gland.
(Photo: W. Leuthold, from Delany & Happold, 1979, reference in chapter 1.)

Box 6.1 An ecological scheme for classifying African members of the family Bovidae (mainly antelopes). This version of a scheme originally proposed by Jarman is modified from Delany & Happold, 1979.

Class	Group size	Adult body weight (kg)	Main habitats	Food and feeding behaviour	Seasonal movements	Territoriality	Response to predators†	Examples*
A	Usually pairs, sometimes with offspring	Mostly 3–20, a few to 60	Forest and woodland	Selective browsers	None	Permanent territories, both sexes	Freeze, remain hidden	duikers dikdiks (klipspringer)
B	Pairs or occasionally small groups	20–80	Open woodland, grassland	Selective browsers and grazers (varies seasonally)	Local movements in relation to food availability	Some males territorial	Inconspicuous, but flee on close approach	reedbucks gerenuks oribi
C	Usually between 6 and 60	20–250	Open woodland, grassland	Graze in wet season – may browse in dry; not very selective of food plant species		Males territorial, at least in breeding season	Watch predator, run when it approaches	gazelles impala waterbucks (kob)
D	From 6 to 200 or more	90–270	Grassland	Graze; selection of plants by size and palatability	Migrate if suitable food locations vary seasonally	Males territorial only in limited breeding season	Move away slowly but run if attacked	hartebeests wildebeests (roan, oryx)
E	Herds of up to several hundreds	300–900	Grassland	Coarse grazers – some dry season browsing	Local movements	None	Very little – buffalo may attack predator	elands buffalos

* those in brackets are atypical in some respect, e.g. klipspringers inhabit rocky outcrops
† members of classes C, D and E depend to a large extent on the 'safety' in numbers' principle

Agamid lizards

These well known reptiles have territories which contain a dominant male, with several females and subdominant males. The territory is sufficiently large for them to find most of their food within it. A dominant male of *Agama agama* is very conspicuous, with bright orange head and blue tail, in contrast to the subdominant males which resemble the brown females. Only the dominant male defends the territory, which he does primarily by displaying his bright colours to any potential intruder. The territory usually contains a prominent rock or post on which the male sits for much of the day. He is easily seen by other males looking for a suitable territory, and that is sufficient in most cases to prevent them from entering. Displays by male lizards, like those of many birds, are an integral part of their breeding behaviour (page 144).

Village weavers

Many species of weaverbirds, including village weavers, nest in colonies. The nests are built by the males, which display noisily to the females by hanging beneath their nests and flapping their wings vigorously (Figure 6.11). Each male builds several nests, and if he succeeds in attracting more than one female, each uses a separate nest in which to lay her eggs. During this time, the male defends a small volume of air around his group of nests, sometimes including nearby twigs, and chases away any other male who attempts to intrude. Nevertheless all the birds from a colony mix freely whilst feeding.

Figure 6.11 A small male weaver at his nest. He will display vigorously each time a female approaches. (D. K. Jones.)

Other examples

Territorial behaviour is not exclusive to land vertebrates, although it has been studied most in them. Some fish defend territories, such as many of those inhabiting coral reefs (e.g. cleaner fish, Box 5.6).

Territorial aggression can often be observed between male dragonflies. In Kenya males of *Orthetrum chrysostigma* arrive before the females at potential breeding places and will defend small territories against other males by aggressive displays backed up by actual physical encounters. Females, however, are not attacked, and on entering a territory they are mated by the male.

Praying mantids may also defend territories against other members of their species, but their aggressive displays are ritualistic. When mantids strike at prey they close their spined tibia against the spined femur to kill the prey, but when defending territories the tibia is not closed against the femur, thus the 'display strike' does not injure intruding mantids.

In the social insects, such as ants and termites, a territory may be defended by a whole colony. Many of the numerous ant species found in West African cocoa plantations are dispersed in well-defined mosaics (Figure 6.12). Small numbers of ants are scattered throughout the plantations but generally a single species predominates in a particular small area, defending a territory vigorously against members of its own species from other nests (intraspecific aggression) as well as against other ant species (interspecific aggression: cf. *Macrotermes* termites, Figure 6.3). The red tree ant, *Oecophylla longinoda*, and a smaller black ant, *Macromischoides aculeatus*, establish mutually exclusive territories. The former prefers to nest amongst young unshaded trees with their nests exposed to sunshine, while *Macromischoides* favours older and more shaded cocoa trees, building its nests on undersides of leaves. Some species of ants, however, share their territories, as happens with

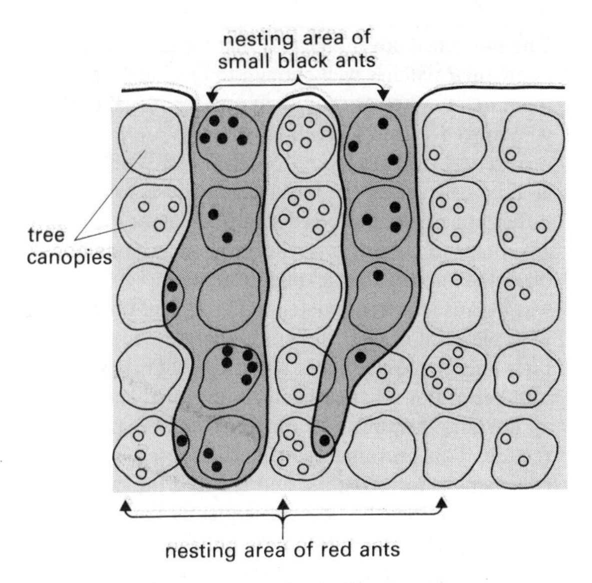

Figure 6.12 Part of a plantation in Ghana, with 30 cocoa trees, showing the distribution of colonies of the small black ant *Macromischoides aculeatus* (black circles) and the red tree ant *Oecophylla longinoda* (open circles). Note how the distributions of the two species are mutually exclusive. (Modified from Majer, J. D. 1972. *Bulletin of Entomological Research*, **62**; 151–60.)

Crematogaster castanea and *O. longinoda* and with *C. clariventris* and *M. aculeatus*. Territoriality is maintained by competition for food and nesting sites, heterogeneity of the environment and by aggressiveness of the ants.

An interesting and unusual defence of a resource – and therefore a form of territoriality – is shown by some parasitic wasps, which lay their eggs inside other insects. When laying her egg, the female 'marks' the host, physically and chemically, which has the effect of stopping other females from laying further eggs in the same host. This ensures that each larva will have sufficient food to complete its development.

Many hymenopterous insects such as sphecid wasps as well as most species of ants vigorously defend a territory occupied by the whole colony.

Territory-holders know their territories well, including the best places to feed, roost or hide from predators. Knowledge of a particular area is also a likely explanation for having a home range. Home ranges differ from territories in not being defended, and usually lack definite boundaries. However, there is often a central 'core' area, where the animal spends most of its time. Unlike territories, where the territory-holders benefit and non-territory holders suffer, home ranges benefit all the individuals involved.

Why do some animals defend territories whilst other do not? Studies of sunbirds help to provide at least one possible explanation (Box 6.2). These nectar-feeding birds have been shown to defend territories only so long as it was economical to do so, in terms of their energy budgets. This happened when the supply of nectar was abundant in the flowers where they fed. When the supply of nectar decreased below a certain level, they abandoned their territories, apparently because they could no longer afford the energy necessary for chasing intruders. This particular case is also interesting in that it provides a further example of interspecific territories – others being seen in ants (Figure 6.12) and probably in termites (Figure 6.3). Interspecific territoriality is apparently commoner in the tropics than in temperate zones, where it is generally rare.

The expenditure of energy for territorial defence can be considered disadvantagous, in that it reduces the energy available for other purposes. Moreover, territory-holders usually advertize their presence by conspicuous displays, which can serve as both a warning to potential intruders and as a means of attracting a mate. But in making themselves conspicuous to prospective mates or rivals, the territorial males expose themselves to increased risks of predation. Thus in hartebeests, males stand on termite mounds where they are easily visible, and in many birds, including sunbirds, the males are brightly coloured and also announce their presence by songs. To make territoriality worthwhile, the extra risks of predation must be more than balanced by the advantages of possession.

The existence of territorial behaviour in a species implies two things: that there is something within the territory that is worth defending, and that it is defendable. Sunbirds abandoned their territories when the flowers became sparse and territorial antelopes in drier regions do not attempt to defend waterholes – they simply could not (Figure 6.9).

A territory might be considered as a resource, and its defence as a form of intraspecific competition (pages 114 and 115): when favoured sites are in short supply, then territory-holders have to defend them against those without a territory. This is a form of 'contest' competition (pages 143 and 200).

Mating systems

Natural selection favours those individuals which leave most offspring, and the process of choosing a mate is itself subject to strong selection. However, there are many kinds of mating systems in animals, each presumably having adaptive advantages in particular circumstances. Of special interest to evolutionists and

Box 6.2 Territorial behaviour in sunbirds

Sunbirds are primarily nectar-feeders, and many plants that are adapted for pollination by these birds produce copious quantities of nectar. The golden-winged sunbird is a common and attractive highland species, ranging from eastern Zaire to Kenya and Tanzania. It feeds from a variety of flowers, including sweet potatoes, but its favourite source of nectar is the conspicuous orange-flowered labiate *Leonotis*, a species which is locally abundant, particularly along roadsides.

Frank Gill and Larry Wolf made detailed studies of golden-winged sunbirds outside their breeding season. They found that some birds held territories, but not all. Territorial birds could be of either sex (only rarely did a pair share a territory), and there were four other species of sunbird that also held territories containing *Leonotis*. Individual birds defended their territories by chasing off intruding sunbirds of any species. Territories were held for relatively short periods – from one to seventeen days – and not all the birds in the area held territories.

Why did some sunbirds hold territories whilst others did not? Gill and Wolf found that during their main study period, all territories contained similar numbers of *Leonotis* flowers: between 1000 and 2500; but the territories ranged in size by a factor of about 100 times, depending mainly on how close together the flowering stems were (see Figure). This led the researchers to suggest that the quantity and quality of the birds' food supply was the key to understanding their territories. They noticed that flowers varied in their nectar contents, and they were also able to show that if the average contents in a territory fell below about 2 μl per flower, then the bird would probably abandon its territory.

What then were the advantages to a territory-holder of spending energy on its defence? One presumed advantage was that birds came to know their patches well enough to be able to search them more systematically than non-territorial birds, who moved within a much larger area. Another researcher, Graham Pyke, took Gill and Wolf's data and, by calculating the birds' energy costs for different activities, showed that the biggest advantage in holding a territory was that it enabled birds to *minimize* their expenditure of energy. This was because they only defended groups of flowers with plentiful nectar, and could therefore meet their energy requirements by visiting a fairly small number of flowers. Occasionally they had to evict an intruder, but much of the day could be spent resting. This presumably made them less conspicuous to predators, and reduced the 'wear and tear' on their bodies; consequently their chances of survival to the next breeding season were increased. But when flowers were far apart, or their nectar contents low, the costs of defending territories exceeded the benefits, and the birds abandoned them.

(Based on Gill, F. B. & Wolf, L. L., 1975, *Ecology*, **56**, 333–48, and Pyke, G. H., 1979, *American Naturalist*, **114**, 131–45.)

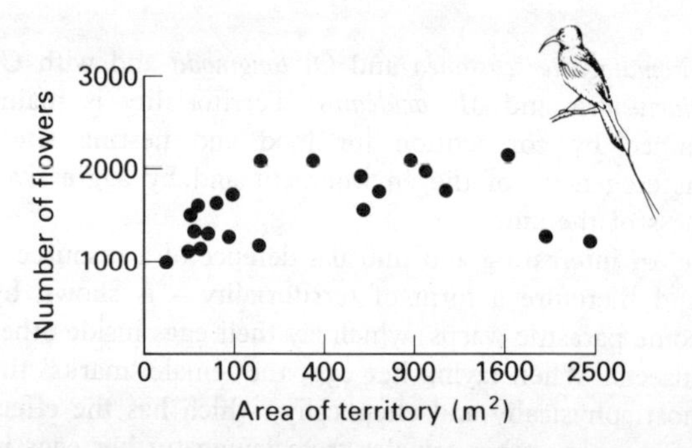

The numbers of flowers (which occur on compound flower-heads) in territories of various sizes. Notice that the horizontal scale is exponential.

geneticists, as well as to ecologists, is the question how do an animal's strategies and mating behaviour affect its reproductive potential?

In many species, the events preceding fertilization are straightforward, such as the synchronized release of male and female gametes by many marine forms. The process of mating is necessarily more complex in terrestrial animals, but in these the extent of courtship varies from none to the extremely complex.

There are many species whose males are more brightly coloured than their females. This is true of birds as varied as ostriches (Figure 6.13) and weavers, as well as agamid lizards, cichlid fishes and numerous insects. Furthermore, there can be a size difference between the two sexes. In terrestrial vertebrates, the males are usually larger than the females, especially in species where males engage in contests to defend a territory or a rank in a dominance hierarchy, where success depends largely upon size. Male rhinoceros beetles with their massive horns are often twice as long as females, while the male of the parasitic worm causing schistosomiasis in man is much fatter than the

Figure 6.13 A clear case of sexual dimorphism – ostriches. The black-and-white males sit on the eggs at night, whilst the more cryptic females incubate during the day.
(D. K. Jones.)

female, which lies in a groove along the male's body.

In many other species it is the female which is larger, sometimes by a factor of many times. In spiders, females are usually much bigger than their mates, who are frequently eaten as soon as mating is completed! Males of a few species evade this fate by temporarily binding the female with silken threads during copulation. Extreme sexual dimorphism is exhibited by spiders of the genus *Nephila*. For example, in West Africa the female of *N. turneri* has a body length of 36 mm, compared to 5 mm in the male.

In general monogamous species do not show sexual dimorphism. In these animals, if there is parental care, it involves both sexes. In polygamous species, the male invests more effort in adornment and little or none in parental care. Where the males are larger, or more highly adorned, they take the lead in displays, but it is the females who do the choosing. Features such as bright colours, long tails or a good voice almost certainly enhance a male's chances of being selected (Box 6.3). In this way the adornments increase his fitness, as measured by the number of progeny he is likely to leave. On the other hand, conspicuousness must also increase his chances of being predated. This presumably sets a limit to the evolutionary trend for brighter colours of longer tails, although in species such as the long-tailed widowbird (Box 6.3), or peacock, it is remarkable how far the limit is extended.

In social species, there is an inverse correlation between the degree of sociality and the proportion of individuals involved in reproduction. In colonies of termites, which are perhaps the most highly social of all animals, there are only two individuals which are reproductively active, even in colonies numbering several millions.

Amongst social species of birds and mammals, mating is also restricted. Amongst lions, any male who does not belong to a pride has no access to the adult lionesses. The lionesses are the only permanent members of the pride and they are usually closely related. This kinship favours co-operation in both hunting and the rearing of young (cubs). Although there is considerable variation, a typical pride might contain three to five adult females and their cubs, plus two or three mature males. However, prides vary in size from two to twenty or more, and individuals periodically leave or join. Young lions reach the age of puberty at about three years old, when the males are invariably chased from the pride, and sometimes the females too (Figure 6.14). The inexperience of the young leads to heavy mortality, especially amongst young males. They may have to wait until they are six or seven years old before being able to re-join a pride, whilst young females usually find another pride within a short time.

A study made in the Serengeti Plains of Tanzania showed that male lions normally formed groups of two or three, who again are quite likely to be relatives, often brothers. By co-operating, male groups improve

Box 6.3 Sexual selection in widowbirds
Widowbirds are relatives of weavers; their name refers to the male's breeding plumage, which is predominantly black. The long-tailed widowbird is found in Angola, Kenya and South Africa. The male has an exceptionally long tail during the breeding season: it is three or four times as long as the rest of him. Males take up territories early in the breeding season, and periodically each male flies in a characteristic slow way over the long grass in his territory. Females, which are plain brown with short tails, build their nests in the grass.

Malte Andersson, who studied these birds near Kiñangop, Kenya, set out to discover whether the length of a male's tail was important for his mating success. Andersson selected 36 territory-holding males, and divided them into nine groups of four. The birds in each group were treated as follows:
bird 1 had its 50 cm tail cut to about 15 cm
bird 2 had its tail lengthened, by gluing cut tail feathers from bird 1 on to its own tail, which was thus extended to about 75 cm.
bird 3 had its tail left intact, as control (I); and
bird 4 (control II) had its tail cut and then glued back on again, as a check that the cutting itself was not harmful.

These tail treatments were performed half-way through the breeding season. Previously, the number of nests in each territory had been recorded (see Figure (a)); the average was about 1.5 in each case. After tail treatment, the number of new nests in each male's territory was again recorded. There were considered differences (Figure (b)), and these were statistically significant ($P < 0.05$). So Andersson concluded that the male's adornment – his long tail – did indeed attract the females, and the longer the tail, the more attractive the male.

Since males with longer tails attract more females, one may ask why the tail has not evolved to become even longer than at present? It seems probable that possession of a long tail is a handicap for a male trying to escape predators, such as falcons (although so far there is no direct evidence for this). During evolution, the tails probably continued to lengthen until a point was reached where the mating advantage of any extra length was balanced by a comparable increase in the risks of predation. (Based on Andersson, M., 1982. *Nature*, London **299**, 818–20.)

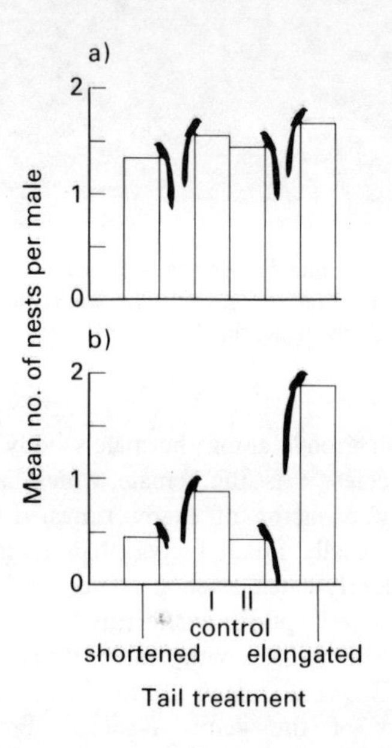

The mean numbers of nests in territories of male long-tailed widowbirds, before (a) and after (b) tail treatment.

their chances of taking over a pride, and then of retaining it. Nevertheless they will probably be displaced in their turn after two or three years (Figure 6.14). After that, their chances of survival as elderly males will be poor, since they have to find their own food, whereas within the pride most of the kills are made by lionesses. This is why old male lions tend to become scavengers. The lionesses live longer, and ages of fifteen and even twenty have been recorded in the wild.

Many African species of birds have evolved a kind of sociality known as **co-operative breeding**. The behaviour is so widespread that it must have evolved independently many times. For instance, speckled mousebirds form parties consisting of a breeding pair and their offspring which help the parents to feed the young of successive broods. Other well-known species exhibiting similar behaviour are the ostrich and various bee-eaters, starlings, hoopoes and shrikes.

The phenomenon of co-operative breeding in birds has considerable theoretical interest. If natural selection favours those who contribute most offspring to the next generation, why should some individuals help to raise the young of others? There appear to be two reasons. Firstly, in most of the cases that have been studied in detail, the 'helpers' are closely related to the actual parents and hence to the young they jointly raise. Because of this close relationship, they share by

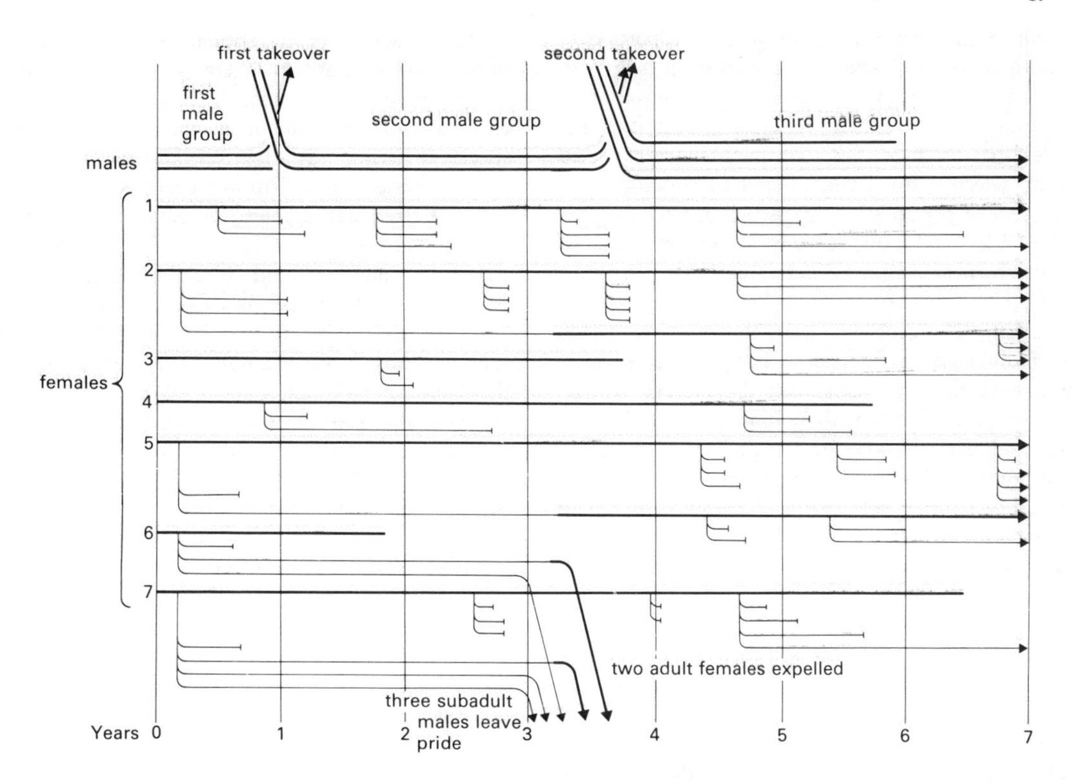

Figure 6.14 The history of one pride of lions during a seven-year period. Where lines end (except in arrows) the animals concerned died. Twice in seven years the male lions were challenged by outside groups and displaced. Each time the new males killed some of the cubs – particularly noticeable after the second takeover, which was followed by all seven of the pride's lionesses giving birth within a period of three to four months. Notice that only two lionesses joined the pride, both daughters of existing pride members. Four died and two were expelled. At the end of the study the pride was relatively large: 2 lions, 5 lionesses and 11 cubs, a total of 18.
(Bertram, B. L., 1975. *Scientific American*, **232**, 54–65,)

descent a high proportion of the same genes (e.g. 50 per cent in the case of brothers or sisters, 25 per cent in the case of cousins). Secondly, the helpers are often young birds, and by assisting the parents they gain experience which probably improves their own chances of successful breeding when they are older. Such arrangements are said to be the result of **kin selection**, a process favouring behaviour by an individual that increases the fitness of its close relatives, with whom it shares many genes.

In a few species of birds, such as wood-hoopoes, the helpers may include individuals that are not close relatives. Here, the benefit to the helpers is presumably the experience that they gain; although unrelated they are improving their own chances for successful reproduction in the future. Thus helping is a form of investment, whose benefits are most likely to be recovered in long-lived species, typically those that are 'K-selected' (pages 199–200).

There are thus several things which can contribute to an individual's **fitness**, that is how widely its genetic constitution will be represented in the next generation. In summary, these are:

(a) how well it is adapted to its environment, i.e. its own individual fitness, resulting from processes of natural selection. This applies to all species.

(b) whether it can increase the genetic contribution of its close relatives to the next generation, through helping them; such behaviour, if heritable, comes through kin selection.

(c) whether, by co-operating with other individuals, though not necessarily relatives, it can obtain benefits that will contribute to its own future reproductive success, through gaining experience. The second and third of these ways apply only to some species.

Together, all of these processes determine an individual's overall or **inclusive fitness**.

In most animal species, the numbers of males and females in a population are similar, at least at the time of birth. Sometimes the sex ratio (the proportion of males to females) changes with increasing age, with females usually outnumbering males. Since one male can inseminate a number of females, and it is the females that produce eggs, one might have expected natural selection to favour populations with more females. The reason why this is not usually the case can be understood if one remembers that natural selection acts by favouring those individuals that leave most offspring – i.e. pass on more of their genes to the next generation. If a genotype were to appear in which most individuals were females, then the average male would have more partners than the average female, and would therefore contribute more genes to the next generation than the average female. So, if a mutant favouring maleness then appeared in the population, it would be strongly selected for. This, in turn, would lead to a rapid increase of males in the population, and eventually to an equilibrium with equal numbers of both sexes.

But although the numbers of adult males and females are similar in most species, the females are often much bigger (or live longer) and hence the biomass of females may considerably exceed that of males.

There are some species in which unequal sex ratios are normal. Often these are a result of mortality rates being higher in one sex than the other. A more complex case is that of *Romanomermis culicivorax*, an aquatic nematode of North America whose immature stages parasitize and kill mosquito larvae. If three or more immature nematodes infect a single mosquito larva then they are all likely to develop into male nematodes, and the corresponding shortage of females will lead to a reduction in the nematodes' population size. This is a nice example of density-dependent population regulation (pages 192 and 193) based on sex. It would clearly be disadvantageous to the populations' survival if their numbers became so large that all mosquito larvae were parasitized and killed. Notice that multiparasitism does not affect the survival of individuals, but their fitness is reduced if they leave fewer progeny.

In certain reptiles and invertebrates, the sex ratio of the offspring is determined by some component of environment, such as temperature. For example, the eggs of marine turtles on Aldabra (part of the Seychelles archipeligo) can develop into hatchlings of either sex. If laid in cool, shady places, most eggs develop into males, whilst eggs laid in warmer places produce mostly females. What the advantage of such a system might be is something of a mystery, but it could have important ecological consequences, which are still being studied.

Mate-selection is of special interest during the process of speciation. There are many groups of animals in which speciation is apparent from the existence of several (sometimes many) closely related species differing only slightly from each other. These are sometimes referred to as 'species swarms'. Examples are frequent amongst insects: termites of the genus *Macrotermes*, malaria vectors of the genus *Anopheles*, and simuliid blackflies such as those belonging to the *Simulium damnosum* complex. This last comprises some twenty six very closely related species, whose larvae can only be separated reliably by the banding on their giant polytene chromosomes; some adults can be distinguished by biochemical methods. Although such species are so closely related as to be morphologically indistinguishable (or nearly so) they often differ significantly in their ecological requirements (Figure 8.23).

Closely related species can often be separated by their mating behaviour. Interbreeding would result in hybrids adapted to the environment of neither species, and hence there is a strong selective advantage for each species to evolve a system which prevents interbreeding. A good example of how this may happen is seen in a series of seven closely related termites of the genus *Microtermes* from the Guinea savanna region of Nigeria. They all produce their alates (the winged reproductives) at the beginning of the rains, normally in April. But each species has a characteristic time of day for its alates to fly, and the species vary also in the number of days that elapse between the first rains and their flights. In combination, these two variables effectively separate the different species.

But whilst mating between closely related species is disadvantageous, so is mating with close relatives. There is evidence that many species, especially social species, are able to distinguish their close relatives from others. Often this is done by smell, although in other cases it may include sight and vocal signs as well. Whatever the mechanisms it seems that natural selection tends to favour outbreeding.

Communication

Animals communicate with each other in many ways, but not always intentionally. For there are situations where it is disadvantageous for an animal to signal its presence, such as to predators. We shall take 'communicate' to imply any action or feature of one individual that produces a response in another individual, whether of the same or another species. Animal communication is principally through sight, sound and olfaction (taste and smell). For communication to be efficient, the emission and reception of signals has to be clear and unambiguous.

Communication by sight

Specific recognition

Animal coloration can serve various purposes. It frequently concerns intraspecific communication, especially the attracting of mates (pages 144 and 145). Patterns of colours can also be effective in warning other animals, particularly predators, or they may confer protection, as in camouflage.

Clearly for a signal to be effective there has also to be a means of receiving it. Thus we would expect that brightly coloured animals are also able to see colours. In general this is true: many species of insects and birds are brightly coloured and for those that have been tested experimentally, there is evidence that they do indeed possess colour vision. In mammals, most of which are black and white, or shades of brown, colour vision is lacking except in the primates. And the primates are almost the only mammals where bright colours occur, notably in the perianal region – bright blue in vervet monkeys, for example. Other primates have bright colours on their faces.

Animals in which the sense of vision is well developed, including those that cannot see colours, often have distinctive patterns, which are doubtless important to them as means of identification. However, experiments with birds have shown that a single key feature, such as a conspicuous red throat, may be all that is required for an individual to recognize another of its own species. Similarly, when a large bird with a long tail flies overhead, it produces alarm in most small birds, although not all large birds with long tails are predatory.

In social species, the members of a group recognize each other as individuals and this too may depend upon sight. Baboons probably look as different to each other as our friends do to us.

Protective coloration

Coloration often features in animals' defence strategies; it can serve to make them inconspicuous to predators, or conversely make them alarmingly conspicuous so that predators are deterred from attacking. These are referred to as **cryptic** and **aposematic** colouring respectively.

Many animals are cryptic because of countershading, that is, their upper surfaces are darker whilst below they are paler. This is readily observed in fishes (Figure 6.15(a)) and in many caterpillars. Countershading works because in bright light it reduces the contrast between the strongly lit upper surface and shaded lower surface. Amphibia often have much

a)

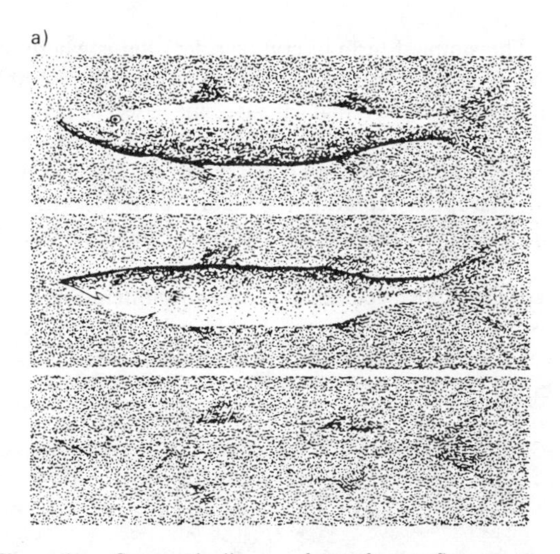

Figure 6.15 Countershading as a form of camouflage or cryptic coloration. (a) diagrammatic explanation of countershading. The top fish is uniformly coloured but appears darker underneath because it is lit from above by the sun. The middle figure shows a countershaded fish which is darker above, and which appears darker above when viewed from the side, but when seen from above (bottom illustration) it is well camouflaged (Cott, 1975)
(b) Thomson's gazelles are darker above than below, a characteristic of many species of antelopes. In nature, this contrast is greatly reduced by the lower part being in shadow. The result is that the gazelles merge easily into their background.
(M. W. Service.)

(b)

paler skins ventrally than the brownish or green coloration of the back. Antelopes are also frequently paler underneath than along their sides and backs (Fig. 6.15(b)). In contrast the wings of some butterflies are darker on the under-surfaces. This is because when resting with closed wings on trees or amongst vegetation the wings present a sombre appearance which merges into the background.

The simplest form of cryptic coloration is where the colour and pattern of the animal so closely resemble the background that it is, in effect, invisible (Figure 6.16(a), (b), (c)). Zebras and giraffes are good examples of another type of cryptic coloration – that is, **disruptive coloration** (Figure 6.16(d)). In the open, these animals are quite conspicuous but in wooded habitats they can be surprisingly difficult to see. Disruptive coloration is also encountered in small fishes living amongst the highly coloured mosaic of coral reefs, in tree frogs and in many insects and birds.

(a)

(b)

(c)

(d)

Figure 6.16 Cryptic coloration. (a) This large, flattened spider has been disturbed and is easily seen. But when at rest amongst lichens, as at the bottom of the photograph, its pale and dark markings are hardly distinguishable from those of the background. (D. E. Pomeroy). (b) This looks like a picture of stones and a few grass plants. Look carefully at the base of the right-hand grass stem and you will see the head of a nightjar, its body stretching low to the left. These birds sit very tight on their nests, but when you walk through the grass one may fly up from your feet, showing a flash of white in wings or tail. (F. Hartmann.). (c) Bustards are large birds, weighing several kilograms – but they are difficult to see when they move through long grass. (D. E. Pomeroy.). (d) The coat patterns of zebras are highly distinctive, but not always as conspicuous as might be expected. In situations like the one shown, they may have a disruptive function. (F. Hartmann.)

Another defence strategy is to resemble an object which a predator is likely to ignore. Caterpillars of some swallowtail butterflies are black with a white stripe on the back, and thus resemble bird droppings. However, when they become too large for this trick to work, they adopt the tactics of warning coloration. Stick insects and some mantids look remarkably like living or dead twigs or leaves, while some Hemiptera resemble thorns of *Acacia* trees and there are spiders which look like flowers. Perhaps the best examples of camouflage are the chameleons, which can adjust their coloration to match that of their background. Young of the side-striped chameleon are brown and the ability to change their colour is limited to shades of brown; adult females are either predominantly green or brown, but males are all of one colour type – greenish-blue with a reddish crest, yellow patches and a reddish purple lateral line. Both sexes become darker at lower temperatures or in bright sunlight, and paler at higher temperatures and at night. Thus colour polymorphism is associated with age and sex, but both sexes can change colour to a limited extent and in so doing become less conspicuous.

Several species of butterflies and moths which are camouflaged at rest by exposing only the dark undersides of their wings, have another line of defence. If, despite their protective coloration, they are attacked, they open their wings rapidly, suddenly revealing eye spots, which resemble vertebrate eyes. This often succeeds in frightening away predators such as small birds. In other species, eye spots are not concealed but visible when the butterflies are resting, but, in such cases, they are often located towards the extremities of the wings. Predators mistake these spots for eyes on the head of prey and strike at them, which although causing the butterfly some minor damage, such as a torn wing, usually enables it to escape.

Cuttlefish use an unusual form of camouflage to escape predation. When attacked they discharge a cloud of dark pigment from an 'ink sac', which confuses the predator, and at the same time the cuttlefish turns very pale. This combination of setting up a 'smoke screen' and becoming less visible usually allows them to escape.

Instead of camouflage, some animals adopt an opposite strategy. They advertize their presence by being particularly conspicuous, that is, they show aposematic or **warning coloration**. Potential predators associate these colours with being distasteful or else possessing some unpleasant properties such as a sting, irritating hairs, or fluids that smell or taste bad.

Birds, and probably most other vertebrate predators, usually have to learn to associate warning coloration with distastefulness. This inevitably results in some of the prey being eaten. But if distasteful prey live in family groups, the relatives of those eaten are likely to survive; and relatives share many genes in common. For example, nymphs of the cotton-stainer bug are bright red and black in colour. They are distasteful, conspicuous and live in kin groups resulting from newly hatched individuals staying together. Interestingly, the adult bugs, though distasteful, are solitary in behaviour and they are also much less brightly coloured.

Most warning colours are conspicuously bright; red, orange and yellow being particularly common. These colours, frequently combined with black and white, and arranged to form bold, easily recognized patterns, are thus readily learnt by predators, which is advantageous to both predators and prey.

Grasshoppers are frequently preyed upon by birds such as hornbills and shrikes. The elegant grasshopper, a pest of cassava in West Africa, has a striking black and yellow pattern (Figure 6.17) and a distinctively unpleasant smell – as a consequence it is sometimes called the stink grasshopper. This combination effectively protects it from bird attacks. Many wasps are coloured black and red or yellow, and reinforce their warning with stings; most birds avoid them, although in this case the reaction is probably innate rather than learnt.

North American skunks, and in Africa, weasels and zorillas, are good examples of warning coloration – in these animals contrasting black and white markings (Figure 6.18) are linked with the ejection, when attacked, of a foul-smelling fluid. Similarly, the black and white banded quills of the crested porcupine advertize their unsuitability as prey.

A few predators have evolved the means of coping with defended prey. Thus some species of cuckoos eat

Figure 6.17 The elegant grasshopper is strikingly coloured – and distasteful.
(D. E. Pomeroy.)

Figure 6.18 This white-naped weasel is not uncommon in Central and southern Africa, but it is nocturnal and not often seen: the same applies to its close relative the zorilla. Both show striking black and white patterns which serve as warning colorations. The zorilla has been described as the fiercest animal for its size in Africa. The weasel also has scent glands which produce pungent odours. The net result is that larger predators avoid those animals.
(Ewer & Hall, 1978, reference in chapter 1)

caterpillars whose irritant hairs protect them from most predators, whilst bee-eaters do as their name suggests. These brightly coloured birds, having caught a bee in flight, carry it to a branch, and knock it to discharge the sting before swallowing it whole.

Mimicry

Since distasteful animals generally have distinctive warning colours which help to protect them against predators, it is not surprising that some palatable species have evolved coloration that effectively copies them and thus affords protection to the palatable species too. The naturalist H.W. Bates, working on butterflies in South America, was responsible in 1862 for drawing attention to this form of **mimicry**, or **false warning coloration**, which is now referred to as **Batesian mimicry**.

Because the warning coloration of a distasteful animal (called the model) is real, and that of the mimic is false, several conditions are necessary for it to be effective. Thus (a) the model and mimic must occur together; (b) the mimic must resemble the model closely enough that predators usually mistake it for the model; and (c) the model must be considerably commoner than the mimic, since it is from the models that the predators learn to associate warning colorations with unpalatibility, thus allowing the mimic also to escape predation. There are numerous cases of Batesian mimicry amongst African insects. They include syrphid flies and some hawk moths which though harmless resemble either bees or wasps, and first instar nymphs of certain long-horned grasshoppers which are mimics of stinging ants.

In tropical Africa, unpalatability is common in butterflies of two large families, the Acraeidae and Danaidae; they in turn are mimicked by other butterflies, especially members of the families Papilionidae and Nymphalidae. The situation is complicated by polymorphism of the mimics, and several different colour forms of the same species may coexist in the same area, each form closely resembling a different model, the models being distasteful butterfly species.

Frequently only females are mimics and exhibit polymorphism, males not showing any mimicry. For example, males of *Papilio dardanus* are yellow and black and resemble typical swallowtail butterflies, whereas females are completely dissimilar to the males and exist in several polymorphic forms, each of which mimics a distasteful model species belonging to the families Danaidae or Acraeidae (Figure 6.19). A possible explanation for this sexual dimorphism is that mate selection, as in many insect species, is made by the female. Consequently a male having an atypical pattern because it was mimicking another butterfly species – the model – might be rejected by females of its own species. Presumably the advantage of being selected as a mate outweighs the advantage conferred by mimicry.

In 1879 the biologist F. Müller, also working in South America, discovered another form of mimicry in which several species, all of which are distasteful, share the same coloration. He argued that young predators have to learn that certain colours and patterns were associated with distastefulness and during this process of learning a number of warningly coloured individuals will be killed. The total numbers of individuals of any one species killed will be reduced during this learning period if two or more prey species share the same warning coloration. Strictly this is not mimicry, because there is no imitating of a distasteful species by a palatable one. Nevertheless the evolution of a 'common warning coloration', has become known as **Müllerian mimicry** (Figure 6.20). Characteristics of Müllerian mimicry, in addition to all species being distasteful and warningly coloured, are that all species can be equally common, and resemblance between forms need not be very close, because neither species deceives a predator but instead reminds it that it is distasteful.

Now, edibility of prey in part depends upon the

Figure 6.19 Butterfly models and their mimics, an example of Batesian mimicry. On the right are three polymorphic forms of females of the African butterfly *Papilio dardanus*, each of which is a mimic of one of the three quite separate butterfly species on the left (the upper two are species of *Amauris* and the lower one is a species of *Bematistes*). The models on the left are distasteful to predators. The warning colours are contrasting orange (grey), yellow (plain) and black.
(Owen, 1976, reference in chapter 1.)

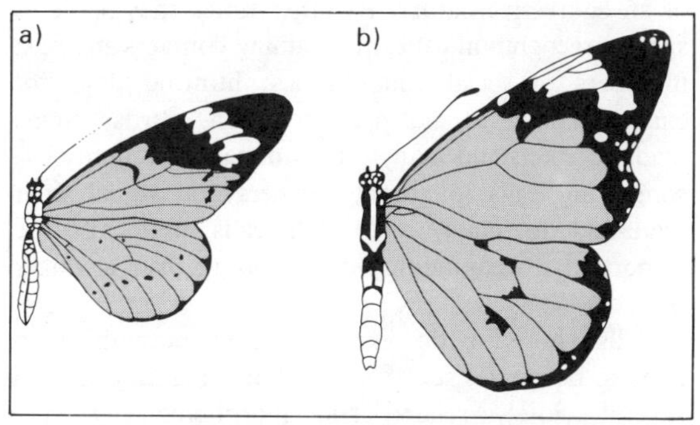

Figure 6.20 An example of Müllerian mimicry. These two African butterflies are not only different species (the very common and widespread *Acraea encedon* on the left, and the equally common *Danaus chrysippus* on the right), but belong to different families. Both are distasteful to predators and have similar patterns of warning colours orange, (grey) white and black, though they differ in size.
(Ewer & Hall, 1978, reference in chapter 1; after Owen, D. F. 1971. *Tropical Butterflies. The Ecology and Behaviour of Butterflies in the Tropics with special reference to Africa*, Clarendon Press: Oxford.)

level of hunger of the predator, and consequently it becomes hard to define edibility accurately. Hence it is often difficult to determine whether Batesian or Müllerian mimicry is occurring between two species having similar warning colours. Indeed, in butterflies, examples of both types of mimicry sometimes occur in the same group.

Mimicry is most common in insects and spiders, but is also found in a variety of vertebrates such as fish, snakes, birds and mammals. The black flycatcher, a bird of African savannas, is often found in company with the similar-sized drongo, which is also black and is recognized by predators as being distasteful; the flycatcher presumably benefits from this association.

Communication by sound

As with visual signals, animals communicate a wide variety of different kinds of information by sounds. Many species of both invertebrate and vertebrate animals produce sounds, which may be either vocal or non-vocal. Vocal sounds are produced by most terrestrial vertebrates, including the croaking of frogs and toads, singing of birds, trumpeting of elephants, screeching of primates, and howls of dogs. Non-vocal sounds are produced as stridulation by male crickets, by buzzing of wings in some bees and wasps, head-rattling in termites, quill-rattling by porcupines, foot- and tail-thumping by various mammals, drumming by woodpeckers and chest-beating by gorillas.

In many cases sounds are produced to warn off intruders or help defend territories, that is, they are of an aggressive nature. At other times they serve as species recognition calls, maintaining contacts amongst members of social communities; hunting dogs for example howl to aid pack cohesion. Birds, frogs, toads, crickets and many other animals commonly use songs and calls to attract partners and signal their readiness to mate, and such calls may also be important in maintaining pair-bonding over a longer period.

Calls whose function is intraspecific communication need to be species specific. A marshy area may contain several closely related and morphologically very similar frogs or toads but their mating calls are quite distinct so that attempts at interspecific courtship are rare.

Not all sounds are species specific, however. Many kinds of small birds produce high-pitched alarm calls if a predator, such as a hawk, is seen above; all species nearby respond to this warning by taking cover. High-pitched calls are hard to locate, so the bird giving the alarm is unlikely to reveal its own position. In response to a tree snake, such as a boomslang, a rattling alarm call is given. This low-pitched call is more easily located, and all birds nearby are thus warned of the snake's position and hence avoid it.

Sounds produced to attract mates or maintain species aggregations or communities can be dangerous if they attract the attention of predators. It is a common observation that as one approaches the apparent source of a male cricket's song it suddenly ceases chirping, and then one is distracted by the chirping of another nearby cricket in a different patch of vegetation. Such behaviour will reduce the likelihood of predators finding crickets by their calls.

Most species of primates produce a wide range of sounds, each with a distinct meaning. As an example, the grey-cheeked mangabey, a monkey of East African forests, produces two distinct sounds of equal volume. One is a low-pitched 'whoop-gobble' which carries a long way and warns members of other troops to keep out of its territory; the other is a higher-pitched scream which is used in aggressive encounters within its own troop, and, because it does not need to carry so far, it is less likely to attract predators.

When caught, the death's head hawk-moth emits a hissing sound by passing air through its epipharynx and pharynx, and this is often effective against predatory birds or geckoes, causing them to drop the moth. The buzzing of bees and wasps probably warns predators not to attack.

The fruit-bats of tropical Africa have good eyesight which they use during their nocturnal feeding flights. But the smaller insectivorous bats have a remarkable system of echo-location, detecting obstacles as well as their prey by responding to reflected sound waves. They emit ultrasonic sounds of 30 to 60 kHz (the upper limit for humans is around 20 kHz). Their large, rotatable ears pick up the reflected sounds (Figure 6.21).

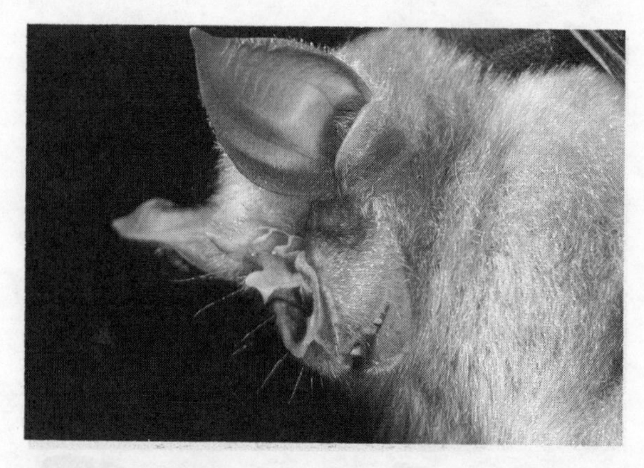

Figure 6.21 In flight, insectivorous bats emit bursts of high-pitched sounds. The time taken for the reflected echo to return to the bat (a fraction of a second) is proportional to the distance of the reflecting object from the bat; and changes in frequency of the sound, on being reflected, give information about the type of object – a leaf, or tree-trunk, or a moth, for example. This horseshoe bat's large external ears undoubtedly assist its hearing.
(Photo by Stephen Dalton/NPHA.)

Whales and dolphins also emit sounds which, as in bats, are used in echo-location to detect the distance, direction and nature of objects in the sea, including each other and their prey. The echoes enable them to create a topographical picture of their surroundings; whales can respond to sounds of 100 kHz and above. Many fish produce non-vocal sounds involving the vertebrae, operculum, swim-bladder, gills or teeth. The sounds are used to keep shoals together and also to attract mates; surprisingly the sounds emitted by some fish are of a low enough frequency to be audible to man.

Bird songs are the most complex animal sounds produced other than by man (and perhaps whales), but despite world-wide studies much remains to be discovered and explained about the subtleties of bird songs. In general, birds living in forests have songs rich in pure tones and low pitched calls and whistles, whereas songs of savanna birds are of a wider frequency range and have more trills and rapidly repeated notes. The reasons for these difference are not certain but may be due to the behaviour of sound in different types of habitat. After a bird has produced a song the noise is altered by attenuation (loss) and by degradation and distortion. Now, a forest bird's song is affected by echoes from trees and leaves, but in more open habitats wind is likely to be the major factor affecting sound quality. Reverberations leading to distortions, as will occur in forests, are more pronounced with higher than lower frequencies, and echoes from trees would seriously alter trills and their recognition. In contrast in open areas where sound is just temporarily masked by gusts of wind there is a need for short repetitive bursts and trills to maintain species contact.

Amongst tropical birds it is common for a pair, once mated, to remain together throughout their lives. This has been shown (by marking individuals) to occur in species as diverse as starlings, shrikes and fish-eagles. Often in those species, both male and female sing, and sometimes their songs are developed into highly synchronized duets, one calling and the other replying immediately. In black-headed gonoleks and Vieillot's barbets, for example, the duet consists of a call made by the male followed by one from the female. Their synchronization is accurate to within 0.01 of a second, and indeed it is hard to believe that two separate birds are involved. The frequency of duets varies seasonally but they occur throughout the year. A probable explanation is that the duet not only helps to maintain the pair-bond but also enables the pair to harmonize their gonadal cycles and thus to be ready for breeding at the same time.

Olfactory communication

The sense of smell, like those of sound and sight, enables animals to communicate information of various kinds: recognition of the species, sex, physiological condition of an individual, identification of fellow members of the same group or colony; as well as for demarcation of territories, sexual attraction and stimulation, trail-laying and detection of food including prey, and in defence. Whereas sound emitted by a single species can vary greatly in frequency, loudness and harmonic qualities, odour is less variable, and has the advantages of requiring much less energy to produce and in many instances persisting for a long time.

The term **pheromone** is used for a substance secreted by an animal to affect the behaviour of others of the *same* species. Although this term is now widely applied to insects, it is equally applicable to some of the odours produced by vertebrates. There are many different types of pheromones. Some, such as those produced by cockroaches and stink bugs, stimulate aggregation and are hence called aggregation pheromones. In contrast are the spacing pheromones, produced for example by flour beetles. These odours are tolerated by other individuals so long as the population density is low, but if it increases above a certain level, the accompanying increase in odour becomes repellent, causing the beetles to disperse.

Alarm pheromones are released when honey bees sting and result in other aggressive bees joining the attack. They are also produced by ants defending a colony from attack, and in this case they attract hoards of other ants to aid in defending the colony. (In higher termites, however, the soldiers are alerted by a characteristic head-rattling sound.)

Recognition pheromones play an important part in enabling ants and termites to distinguish members of their own colony. Non-members, whether of their own or other species, are attacked. But if an intruder ant is placed in an ant colony and protected from attack (e.g. by keeping it in a fine mesh cage) it soon acquires the colony's odour and when released is accepted by other ants. Alarm pheromones are also secreted by aphids and bedbugs and cause them to scatter and disperse (Figure 6.22).

Worker honey bees are attracted to the queen bee and attend her due to her secretion of hormones. Bees and other social insects are usually very fastidious in disposing of dead individuals from their nests; this results from the production of so-called funeral pheromones by the dead insects. Probably the best known pheromones are the sexual ones that attract mates. These are commonly found in moths, especially

Figure 6.22 An example of the effects of alarm pheromones on aphids. A small triangle of filter paper, which contains the alarm pheromone of the peach–potato aphid, is placed amongst feeding aphids. In less than five minutes, they have scattered and a few have dropped from the leaf. Such pheromones are produced when danger threatens, e.g. approach of a predator.
(Based on Mathews, R. W. & Mathews, J. R., 1978. *Insect Behavior*, Wiley: New York.)

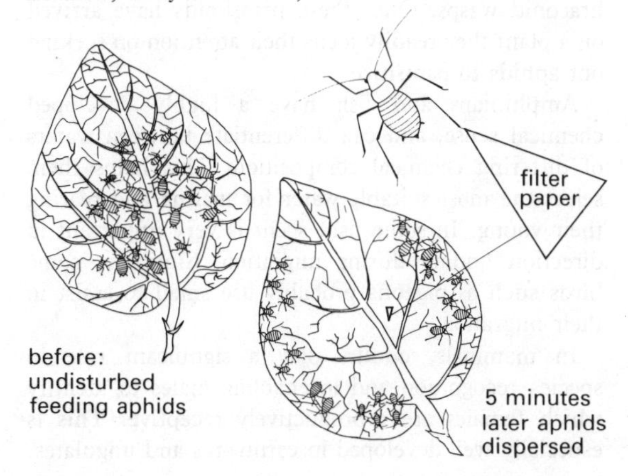

females, and mates can be attracted from far away. Sometimes after such attraction copulation will not occur until chemical aphrodisiacs, produced mainly by males, make the opposite sex receptive. Contact pheromones occur in some insects, such as tsetse flies, and stimulate mating when the flies are in contact or very near each other. Natural or synthetic contact pheromones soaked into a boot lace can induce male tsetse flies to attempt to mate with it! This principle is used in the design of some traps.

Many ant species are able to produce trail hormones. When a worker ant of a species which feeds on immobile prey has discovered a good food source it secretes from special abdominal glands minute quantities of scent onto the ground while returning to its nest. Other workers can then locate the food by following the scent trail. The odour trail volatilizes within a few minutes unless it is enhanced by other workers going to and from the food. This allows fresh scent trails to be made to other food sources when the original one has been used up, thus preventing confusion with old trails leading to a depleted food source. Other ants, such as those feeding on vegetation and consequently having a more permanent food source, lay longer lasting scent trails. Termites do the same.

There are two other groups of chemical messengers which affect individuals or populations of different species. One group termed the **allomones** are more beneficial to the sender than to the receiver; an example is the smell produced by skunks to deter attackers. The other group, termed the **kairomones**, are of adaptive value to the animal receiving the olfactory stimulus. Carbon dioxide in exhaled breath, and sweat and other body odours produced by man, attract blood-feeding arthropods such as mosquitoes and ticks. Likewise, many phytophagous insects find their host plants by their odours. Mustard oils in *Brassica* crops (e.g. cabbages) attract some aphid species, and even more interestingly they also attract certain braconid wasps. Once these parasitoids have arrived on a plant they readily focus their attention on seeking out aphids to parasitize.

Amphibians and fish have a highly developed chemical sense, and can differentiate between waters of differing chemical composition. This helps them select the most suitable water for mating and rearing their young. In some fish, scent is very important in direction finding during migration. Similarly, some birds such as pigeons probably use smell to assist in their migration.

In mammals, odours play a significant role in species recognition and in enabling males to identify which females are reproductively receptive. This is especially well developed in carnivores and ungulates. The role of odours can be readily observed in encounters between dogs, where the male nearly always sniffs the perianal region of the bitch to determine whether or not she is 'on heat'. In a number of mammals, including most antelopes, the males actually stimulate the females to urinate, and then use their tongues to take a sample of the urine into their mouths. In doing this, a male pulls back his upper lips to create a characteristic facial expression known as 'flehmen'. A sensitive organ inside his mouth enables the male to assess the hormone levels in the female, and thus how close she may be to the time of ovulation.

Some territorial mammals have the habit of urinating frequently at points along their territorial boundary, thus marking it. This behaviour can be readily observed in domestic dogs; fence posts, tree trunks and other objects are thus marked. Territorial antelopes, such as dik-diks and impalas, use their faeces to mark the extent of their territories. They have a series of 'latrines' at intervals around the boundary and these sometimes accumulate considerable quantities of droppings. Similar behaviour is seen in various mammalian carnivores. To be effective, the dung of each individual must have a unique combination of odours; these are secreted by glands just inside the anus. Members of the cat family also have well-developed scent glands in the skin, especially the cheek and neck, and commonly rub these against trees. When domestic cats rub their bodies against their owners' legs and furniture they are not merely deriving pleasure but placing a mark of ownership.

Feeding strategies

We stressed in chapter 3 that food is always an important resource, usually the most important. So it is hardly surprising that animals devote much time to locating and consuming food (Figure 6.23). And since the satisfaction of their food requirements is crucial to animals' survival, we can safely assume that there will have been strong selective pressures for maximizing food intake.

So how does an animal maximize its food intake? Behavioural ecologists approach this problem in terms of **costs and benefits**. In this context, anything which increases an animal's fitness – its ability to survive and reproduce – is a benefit; anything reducing the ability is a cost. The benefits are measured as carbohydrates, or mineral nutrients, or joules, of the food consumed. However, it is more difficult to measure the costs. They include the energy required to locate, handle and digest the food, all of which could be expressed in joules. But an animal seeking food is also more

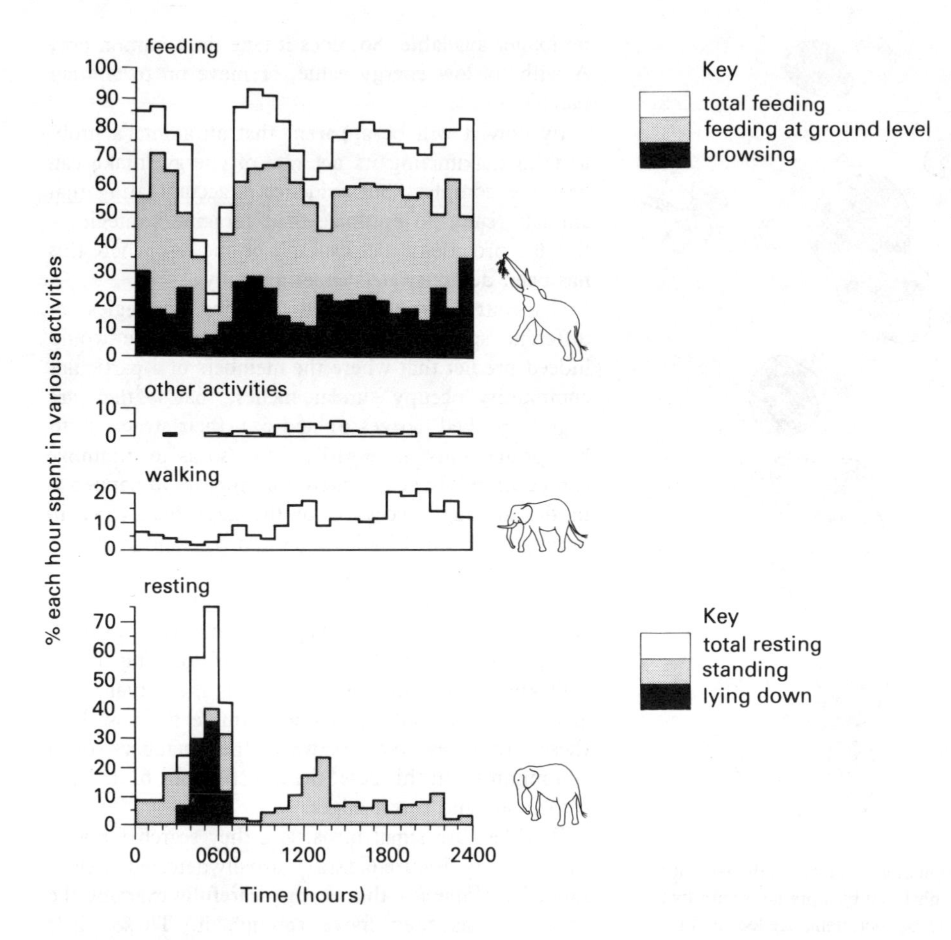

Figure 6.23 The African elephant spends up to three-quarters of its time, day and night, in feeding, according to the observations shown here, which were made in western Uganda. The 'time budget' represents the average activity pattern of 18 females. Feeding is time consuming for many animals, but the elephant probably represents an extreme case.
(From Eltringham, S. K. 1982. *Elephants*, Blandford Press: Poole.)

exposed to predators than one resting in its shelter, so that its chances of survival are thereby reduced, although usually by much less than if it went without food.

Since food itself is a benefit, whilst obtaining food is a cost, the overall gain to an animal from a spell of feeding is the benefit minus the cost, which we call the net gain. We can now refine our question to ask how does an animal maximize its net rate of gaining energy? To answer this question, we need to consider how animals obtain their food. Almost always they have to search for it – a process known as **foraging**. There are three main stages in this process, at each of which the animal may have to make a decision. First it has to decide where to go to look for food; secondly, how long to spend feeding at a particular place once food has been found; and finally, which type of food to consume when more than one type is available.

Some examples will help to illustrate these different aspects of feeding behaviour.

(a) Selection of feeding sites. It is usually the case that food is not randomly dispersed (pages 133 and 134), but aggregated in certain places, referred to as *patches* (Figure 6.24). These food patches often correspond with distinct features of the habitat, which animals recognize instinctively or learn to associate with favourable feeding sites.

(b) How long to stay? Optimal foraging theory suggests that an animal should stay in a particular place only so long as its net rate of energy gain is above a certain level: that level being equal to the average for the whole foraging area. It is not easy to obtain data to test such hypotheses but there is some supporting evidence. For instance a study of ant-lions

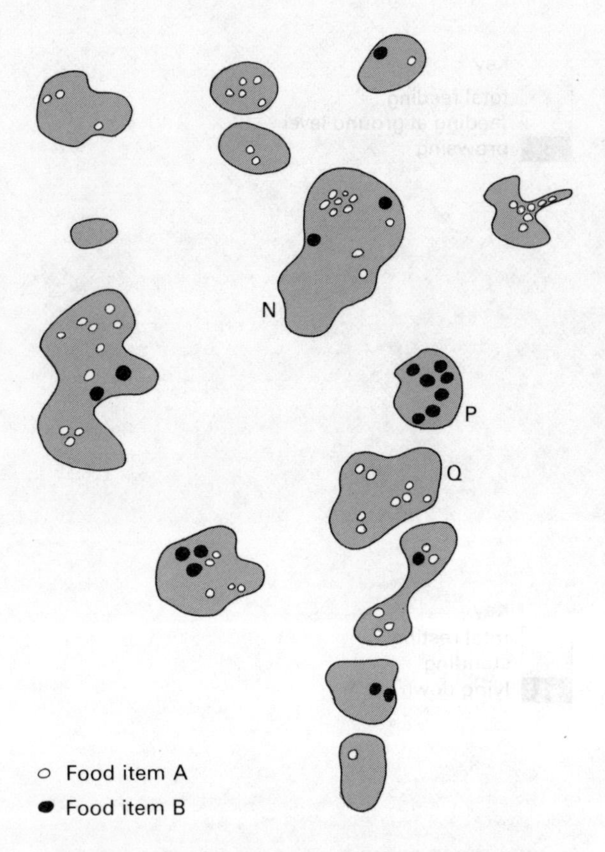

Figure 6.24 A representation of an area where the habitat is not uniform. The place is inhabited by a predator with two species of prey, A and B. These food items are located in restricted parts of the habitat, termed **patches**. The patches could be thickets in a grassland, particular tree species in a forest, termite mounds in a woodland, or anything similar. They could also be macrophytes in a pond. The patches N, P and Q are referred to in the text.

in Sierra Leone showed that the larvae moved their pits more frequently at times of hunger than when they were well fed.

(c) What to eat? Potential food items have various attributes, such as abundance, size, energy content and ease of capture. The sort of choice that an animal may have to make is whether to concentrate on small, common prey that yield only a little energy each, or large, rather scarce prey that – once found – provide much energy. Thus in Figure 6.24, prey A may be like the first mentioned, and B the large scarce prey. Even the scarce prey could be locally abundant, so the optimal strategy for a predator in patch P would be to concentrate on the more profitable prey B, but when it has consumed all of B that it can catch, and moves to say patch Q, it should soon notice that B is no longer available. So, does it take the common prey A with its low energy value, or move on to another patch?

By now it will be apparent that an animal's problems in maximizing its net rate of energy intake can be quite complex, but evidence is accumulating that animals really do optimize their foraging strategies – Box 6.4 provides a nice example of one case where this has been demonstrated experimentally.

Of course, the optimal foraging strategies of different species will also be different. We would indeed predict that where the members of a particular community occupy similar niches, that is they are 'tightly packed' (pages 61 and 62), their strategies to be optimal must necessarily differ so as to minimize competition. One case where this appears to happen is in the grazing succession of the large herbivores in Serengeti National Park (pages 120, 122 and 123).

However there are, as yet, few studies on feeding behaviour in African animals with sufficient data to show whether or not they are following optimal foraging strategies. But one aspect of foraging that is well known in a number of cases is the actual strategies. We shall take just one example to show how these strategies vary between the members of a community – in this case, the insectivorous birds from woodlands and forest edges.

Warblers are small birds, and they search for very small insects (Figure 6.25(a)), usually detected at close range by *gleaning* – that is, they carefully examine the vegetation as they move through it. These birds mostly weigh less than 12 g, and have small, slender bills with which they remove their prey from the plant, swallowing them whole.

Larger insects form the prey of birds like bulbuls (Figure 6.25(b)) and drongos. Their method of feeding can be described as **snatching**. The prey are usually spotted at some distance, whereupon the bird flies to the prey, snatches it, returns to a perch and swallows the prey whole. Birds that feed by snatching tend to eat larger prey than gleaners, but because they fly to the prey they also consume more energy.

Various species of kingfishers (Figure 6.25(c)) and rollers, most of them weighing more than 30 g and a few over 100 g, watch for prey from perches with a good view over the ground. They are attracted by larger insects such as grasshoppers and beetles, which they detect at distances of up to twenty or thirty metres. They fly to the prey, grab it with their beak, and return with it to a perch. Their prey are likely to need crushing before being swallowed, and they have powerful beaks for this purpose.

These birds form a series of increasing size. Several characteristics appear to be correlated with the birds'

Box 6.4 Foraging strategy in a crab

Shore crabs *Carcinus maenus* were studied by Elner and Hughes in Wales. The crabs were offered food in the form of bivalve molluscs *Mytilus edulis*, the edible mussel. All crabs used in the experiments described here were of medium size, but the prey varied in size.

In order to eat a mussel, a crab has first to break open its shell, using its chelae. The strength of a mussel's shell increases progressively with size, so that breaking larger shells becomes increasingly time consuming. After breaking the shell, the crab can scrape out and eat the soft body inside. The mussel's body size, and hence its food value to the crab, is proportional to the size of its shell.

Elner and Hughes recorded the time taken by crabs to crack shells of various sizes, and they determined the energy value of the molluscs' bodies. When they divided the energy value by the time taken to obtain the food, they obtained a measure of profitability (of mussels to crabs). In the graph (a), profitability (in joules per second) is seen to be greatest for mussels about 2 cm long.

In a second series of experiments, crabs were offered a choice of mussels of various sizes. The crabs showed a preference for prey whose size was similar to the most profitable size (histogram, (b)). Rather more smaller mussels were taken than might have been expected, perhaps because they were easier to break.

(Elner, R. W. & Hughes R. N., 1978. *Journal of Animal Ecology*, **47**, 103–16.)

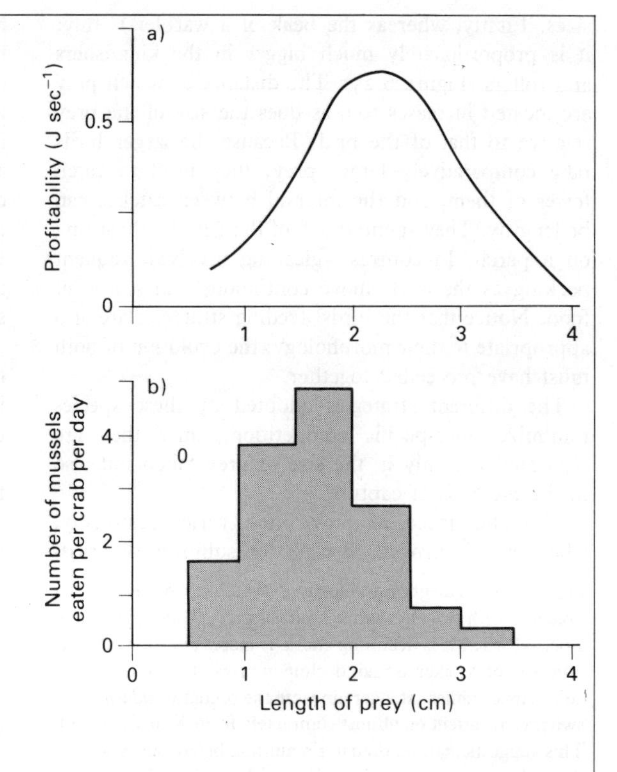

The profitability of mussels of various sizes (a) mesasured as the energy obtained per unit time. The actual sizes of prey most frequently selected by crabs (b) were similar (though rather smaller) to the most profitable size.

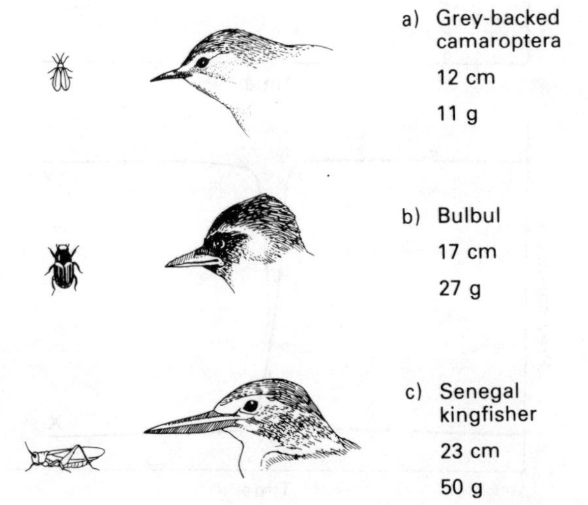

Figure 6.25 Three common birds of African woodlands and forest edges. The camaroptera (a) feeds by gleaning leaves, searching for tiny insects. Bulbuls (b) take a wide variety of foods, including insects which they spot from a short distance away, and capture by flying and snatching them. The kingfisher (c) mainly takes grasshoppers and other large insects from the ground. It detects them at a distance by sitting on a perch offering a good view of the ground. Notice the progressive increase in the relative size of the bills.

sizes. Firstly, whereas the beak of a warbler is tiny, it is proportionately much bigger in the kingfishers and rollers (Figure 6.25). The distance at which prey are located increases too, as does the size of the prey relative to that of the bird. Because the larger birds take comparatively larger prey, they need to catch fewer of them, and the interval between catches can be lengthy. They spend much of the day simply sitting on a perch. In contrast, gleaning involves frequent pecking as the birds move continuously in search of food. Notice that the birds' feeding strategies are also appropriate to their morphology: the evolution of both must have proceeded together.

The different strategies adopted by these species minimize interspecific competition, since they are separated not only in the size of prey taken but also in the methods of capture.

The abundance of prey often varies seasonally, whether it is *Leonotis* flowers for sunbirds or grass-

Figure 6.26 A diagram to illustrate 'switching' behaviour of a predator with two alternative kinds of prey, X and Y, during a period when X is declining and Y is increasing (above). The numbers of X taken do not decline in proportion to their abundance: rather, at a certain time the predator *suddenly* switches its attention almost completely from X to Y (below). This suggests that the predator's hunting behaviour is dependent on one major 'search image' at a time.

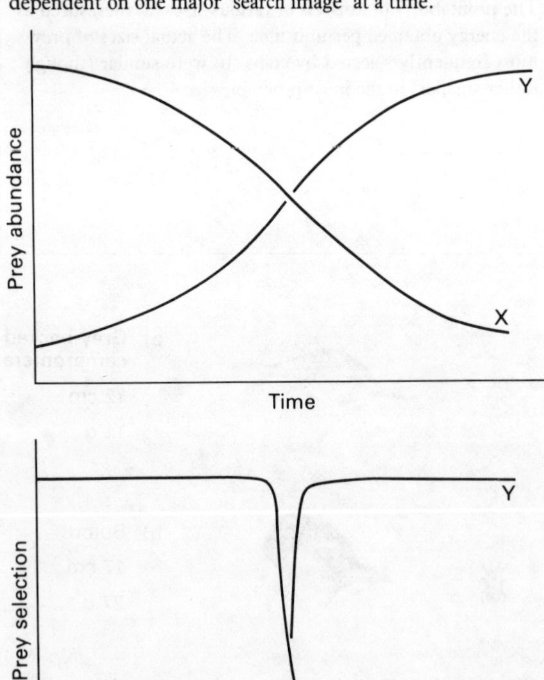

hoppers for kingfishers. Thus foraging strategies may need to change with time. Predators apparently develop specific 'search images' for a prey item that is temporarily abundant – they learn how best to find it, and take it in preference to other food items that are present. If, later, this food item declines in abundance, the predator may suddenly 'switch' to a different food item, often one that has been increasing in abundance (Figure 6.26). In so doing, the predator is exerting a density-dependent control on the prey species (cf. pages 193–194).

We have argued that optimal foraging strategies maximize the net *rate* of energy intake, and from this it follows that time is important (cf. Box 6.4). Time related to feeding activities can include several different components. First, time may be needed for foraging and then for actually catching prey once it has been spotted. Sometimes the prey has to be 'handled' before it can be consumed – for instance, it may have to be killed and dismembered (if it is animal prey), or removed from a plant.

Within their territory or home range, animals commonly have a 'central place' where they rest, and from which they go out to seek food. The central place may be one which is safe from predators – a refuge in which to hide – or it may be a nest of some kind, as in social insects or breeding birds. The expenditure of energy by an animal going to forage for food will be minimized if the 'central place' is in fact at the centre of the patches where its food occurs. Thus if we were to suppose that Figure 6.24 represents the territory of a predator feeding on prey items A and B, then patch N might be a suitable 'central place'.

There is considerable scope for research on African animals to discover to what extent the ideas we have just discussed – at present rather theoretical – do in fact help in understanding species' fitness.

Seasonality

Animals' activities often vary with the time of year. This **seasonality** may reflect variations in resources, especially food and water, or the fact that activity is only possible at certain times, e.g. when the soil is moist. Seasonality in behaviour patterns is clearly seen in the breeding cycles of most species, and in the timing of migrations and other kinds of movements.

Breeding cycles and life-histories

There are seasonal changes in the supply of resources for most animals, and for terrestrial and many aquatic species, rainfall is normally the biggest single cause of these changes. Furthermore, animals' requirements

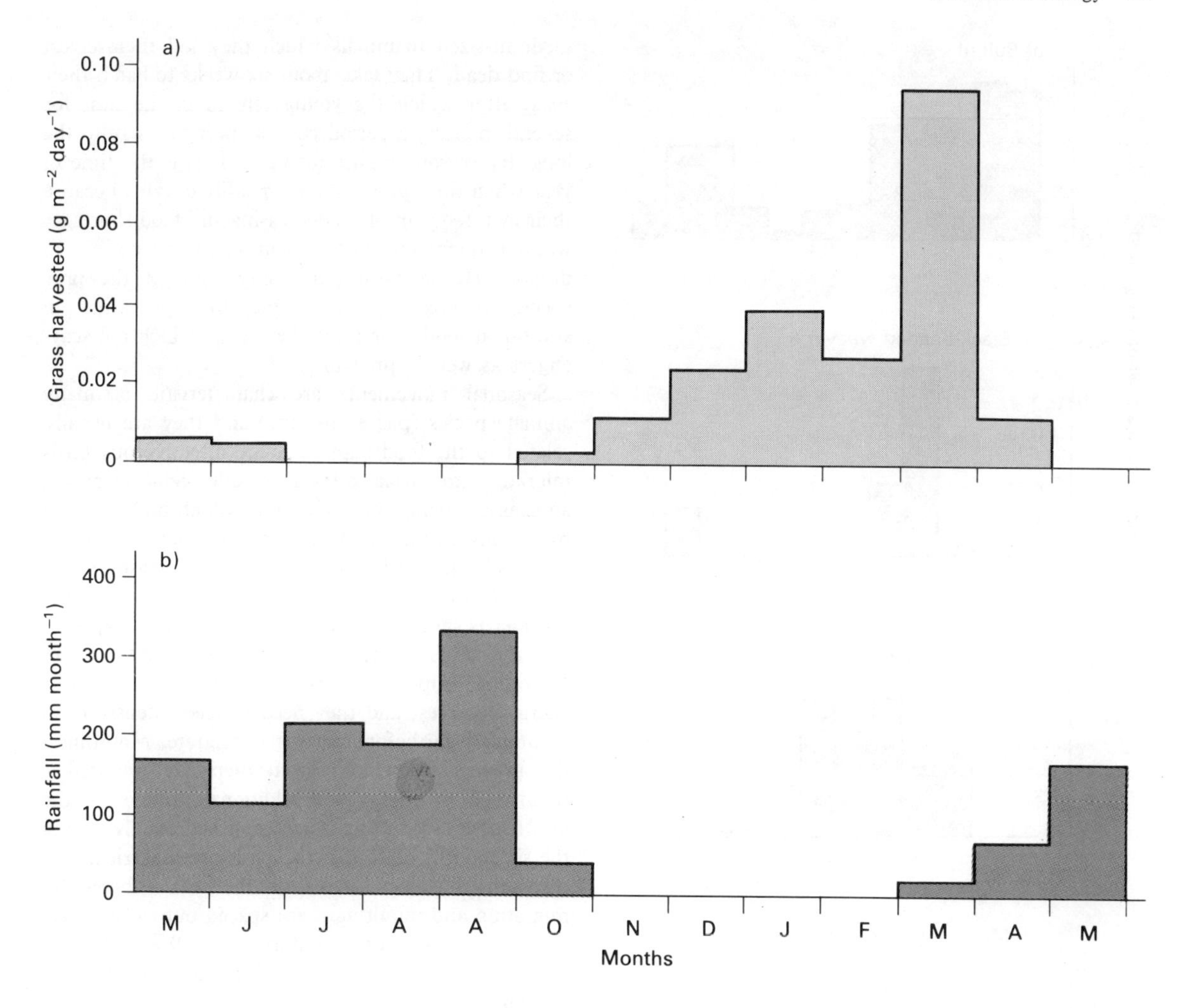

Figure 6.27 Seasonal feeding activity by a termite in the Guinea savanna zone of Nigeria. Rates of grass collection by the harvester termite *Trinervitermes geminatus* (a) compared to the monthly rainfall (b). The daily collection rates in (a) are averages for four plots, two of which were grazed by cattle. The rate of grass collection increased as the dry season progressed, but dropped rapidly during the rainy season. This was attributed to the termites' preference for dead grass, and to their dislike of heavy rain.
(Based on Ohiagu, C. E., 1979. *Oecologia*, **40**, 179–88.)

vary during the course of their life. Food is usually the most important resource, and the demand for food by a population is greatly increased following reproduction. Consequently, we should expect that animals' life-histories will be related to the seasons in such a way that their young are produced at a time of year when the food supply is increasing, or with species that breed throughout the year, that the reproductive rate increases during the more favourable seasons. The seasonality of reproductive behaviour is strikingly illustrated by termites: in most species, the emergence of winged alates follows closely the start of the rainy season. Feeding activity in termites is also highly seasonal, as is clearly seen in the case of a grass-eating species of West Africa (Figure 6.27).

The seasonality of breeding has been studied extensively in birds. Figure 6.28 shows three examples. Bulbuls are largely insectivorous, although they also feed on flowers and fruits. This broad diet enables them to find food throughout the year, and they have been recorded nesting in every month (Figure 6.28(a)), although with marked peaks in the rainy seasons. The black-headed or village weaver, like the quelea (page 170), is primarily a seed-eater. Its breeding season is so timed that the young leave their nests during the second half of the rainy season, when grass seeds are abundant. (There is an interval of about three weeks from the egg laying dates shown in Figure 6.28(b), to the time when the young leave the nest.) Tawny eagles feed largely on birds and small to

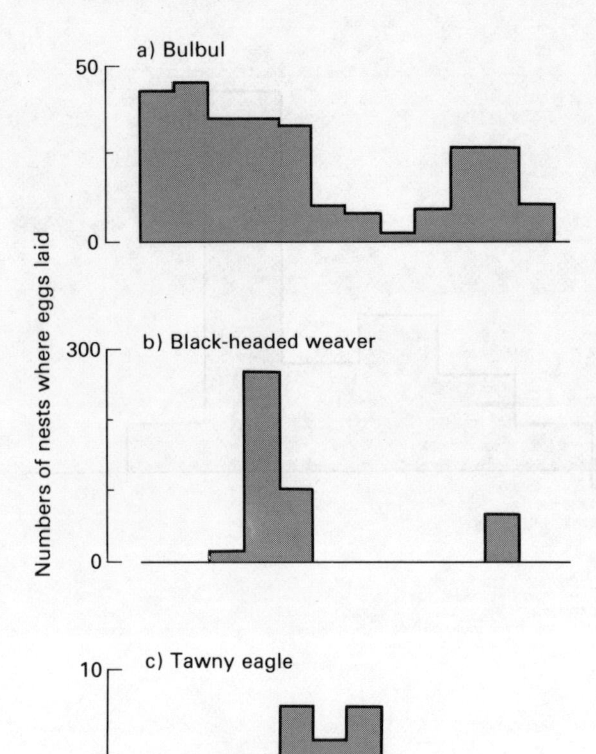

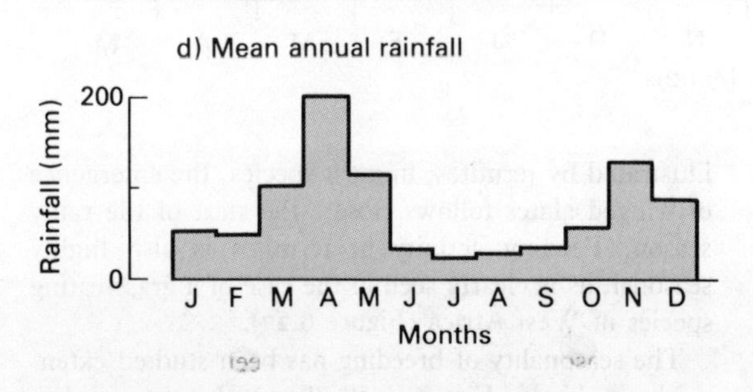

Figure 6.28 Seasonality of birds breeding in the Nairobi–Dar es Salaam region of East Africa. Breeding seasons are here taken to be the months in which eggs are laid, and the records refer to the numbers of nests found with eggs in each month. The bulbul (a) has a varied diet, in which insects play an important part. The weaver (b) is mainly a grain-eater, although it feeds its nestlings on insects. The tawny eagle (c) feeds on vertebrates, mainly mammals, which it kills itself or finds dead. Thus the different seasonalities of these birds depend on seasonality and availability of their food.
(Data from Brown, L. H. & Britton, P. L., 1980. *The Breeding Seasons of East African Birds*, East African Natural History Society, Nairobi.)

medium-sized mammals which they kill themselves, or find dead. They take about six weeks to hatch their eggs, after which the young remain in the nest for several months, a period corresponding closely to the long dry season (Figure 6.28(c)). This is the time of year when their prey are most readily caught, because their own food supply is decreasing and food shortages weaken them and make them more susceptible to disease. Hence natural mortality amongst the prey species is greater at this time, thus providing an additional food source for the eagles, which are scavengers as well as predators.

Seasonal movements are characteristic of many animal species (pages 165–172) and they are usually related to the food supply. Some insectivorous birds migrate from Africa to the Palearctic region to breed, an example being the ruff, a bird which finds its food by wading in mud and marshes, probing the ground with its beak. During the ruff's breeding season in the northern summer, insect life is abundant, but the summer is short and as it ends most insects disappear, and the birds return south to Africa (Figure 6.29). Migration imposes heavy demands on the birds' energy reserves, and they need to feed intensively to accumulate fat before starting to migrate. Moulting – the process by which the feathers are renewed – requires both energy and additional protein for the new feathers. Moulting is accomplished mainly during the six months or so that the birds are in Africa. The three main energy drains on the ruff – breeding, migrating and moulting – are spaced out so that they only do one at a time (Figure 6.29). We shall have more to say about the timing of migration in the next section.

Larvae of different species of chironomid midges coexisting in the same habitat can show markedly different seasonalities (Figure 6.30). *Cladotanytarsus pseudomancus*, *Cryptochironomus stilifer* and *Polypedlium deletum* occur together in West African reservoirs and lakes, but their abundance differs seasonally, especially that of *P. deletum*. This species is commonest in June–July during the rains, whereas *C. stilifer* is most abundant during the hot dry months of November–January. The reasons for these differences are not certain, but all species feed primarily on detritus and the heavy rains bring an accumulation of silt into lakes which may inhibit feeding of species such as *C. pseudomancus*. Now, detritus as well as silt is carried into reservoirs and lakes during the rains, and it appears that some species, possibly including *P. deletum*, can utilize fresh detritus as food, whereas other chironomids, possibly including *C. pseudomancus*, require detritus to have decayed before it can be successfully assimilated.

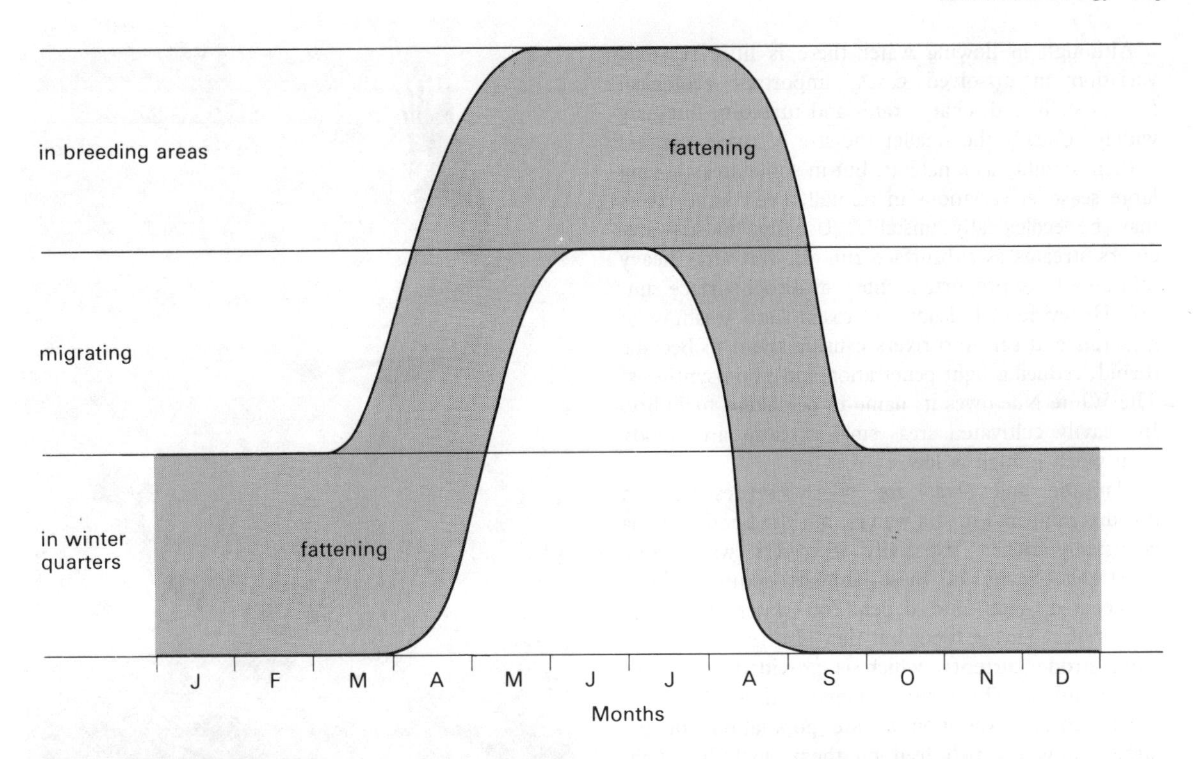

Figure 6.29 Seasonal movements of the ruff, a small wading bird which winters in Africa, where it is usually found on the edges of lakes and ponds. In northern Nigeria it is so abundant as to be a pest of ricefields, damaging young plants. Before migrating from Africa to breeding grounds in northern Europe and Asia, birds feed intensively and increase their weight by up to 60 per cent, mainly as fat. They do the same before migrating south again, about five months later. The main period of feather moult, which like migration is a drain on the birds' energy resources, takes place whilst the birds are in Africa.
(Based on Pearson, D. J., 1981. *Ibis*, **123**, 158–82.)

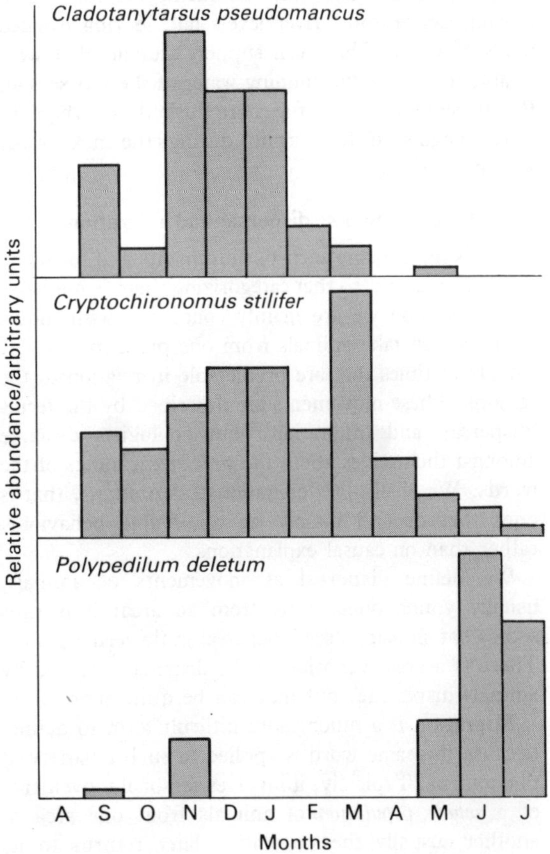

Cladotanytarsus pseudomancus

Cryptochironomus stilifer

Polypedilum deletum

Relative abundance/arbitrary units

A S O N D J F M A M J J

Months

Figure 6.30 Monthly variations in the relative numbers of chironomid larvae of three different species collected from a reservoir near Accra, Ghana. The reasons for these marked differences in abundance are probably related to their feeding habits and to seasonal variations in availability of food.
(Ewer & Hall, 1978, reference in chapter 1.)

Although in flowing water there is little seasonal variation in dissolved gases, important ecological factors such as discharge rates and turbidity fluctuate widely. Usually the smaller the size of rivers the less is their stability as a habitat, but in some areas having large seasonal variations in rainfall, even large rivers may be ecologically unstable. Usually, most water enters streams as subsurface run-off, but after heavy rainfall a large proportion enters as direct surface run-off. Heavy rainfall leads to erosion and washing of minerals and soil into rivers causing them to become turbid, reducing light penetration and photosynthesis. The White Nile owes its name to persistent turbidity. In heavily cultivated areas small streams are cloudy even when rainfall is low.

Plankton and algae are poorly represented in flowing compared to still waters, but the benthic fauna is usually richer, especially in rivers with stony substrates. Many of these animals require highly oxygenated water and depend on water flow as a means of acquiring food, but they also require shelter from strong currents, which is provided by stones and boulders. The benthic fauna is thus subject to much seasonal variation as are populations of the larger animals which feed on them, including fish. During dry seasons, river levels fall, leaving isolated pools of water. These can support animals that were unable to survive the running water of the wet season. But these temporary refuges are flushed out when the water begins to flow again during the next rainy season.

Figure 6.31 Migrating white storks. These birds breed in Europe, sometimes building their massive nests on a house or a church. In autumn they migrate to the savanna regions of Africa, often joining kites and other birds at grass fires, searching for animals killed by the flames.
(F. Hartmann.)

Periodic movements: dispersal and migration

Animals make many sorts of movements and for many different reasons, so that categorizing them is not easy. In this section we are mainly concerned with movements which take animals from one place to another, usually at times that are predictable in relation to the seasons. These movements are described by the terms 'dispersal' and 'migration', but zoologists disagree amongst themselves about the precise meanings of the words. We shall give 'operational' definitions, that is ones that depend mainly on observable behaviour, rather than on causal explanations.

We define **dispersal** as movements of animals, usually young ones, away from an area; it is thus somewhat similar to seed dispersal in flowering plants. There is a great variation in the distances covered by animals dispersing, but they can be quite short.

Migration is a much more difficult term to define, because the same word is applied to such a variety of phenomena. Typically, it involves seasonal movements of a *whole population* of animals from one area to another; usually the population later returns to its original place. Sometimes the individuals which began the migration are the ones that return, but in other cases it is their progeny that comes back. Migrations can, like dispersals, be over very short distances, as when plankton move up and down in water on a 24-hour cycle (Figure 5.5); but at the other extreme are birds which migrate half-way round the world and back, every year. Indeed, migrations are most marked in animals with the best powers of locomotion, notably insects, fish and birds (Figure 6.31). We shall now amplify our definitions of dispersal and migration by discussing them further, and by describing some examples in more detail.

Dispersal (*cf. Dispersion, page 132*)

All species have some method of dispersal. If all offspring were to remain with their parents, then in long-lived species at least, the population would become increasingly aggregated and the extent of intraspecific competition would increase too. Thus, some species disperse in response to overcrowding. In many territorial species the young are evicted by their parents as soon as they can fend for themselves. Aquatic animals, especially those inhabiting streams and rivers, are dispersed by water currents. Similarly winged adults of terrestrial insects are dispersed by wind.

Dispersal is of particular importance to animals that live in temporary or unstable habitats – the classic 'r-selected' species (pages 81–83 and 199 and 200). Places like pools of rain-water, or habitats in early successional stages such as abandoned gardens, are colonized remarkably quickly by both plants and animals.

The survival of many species – perhaps all – depends to some extent upon their ability to colonize or recolonize suitable habitats whenever these appear. For example, one of the greatest natural disasters to occur in Africa was the rinderpest epizootic of 1896, which killed vast numbers of wild ungulates in Tanzania, Zambia, Zimbabwe, Mozambique and the Transvaal, thus eliminating the natural hosts of tsetse flies such as *Glossina morsitans*. As a result, this insect disappeared from large areas of these countries. In due course, the afflicted areas were recolonized by wild ungulates, and soon the tsetse reappeared too.

In other instances tsetse flies have been exterminated from areas by mans' action, only to reinvade them later. In Nigeria *G. morsitans* was eradicated from savanna areas by the application of persistent insecticides such as DDT and dieldrin, but when spraying stopped, the tsetse flies returned, advancing at an average rate of about 1.6 km a month. In 1914 vigorous anti-tsetse measures eradicated *G. palpalis* from the small and remote island of Principe in the Gulf of Guinea, but by 1956 there was once again a large population of this species on the island. How this recolonization occurred is a mystery, although the island of Bioko, some 200 km away, is the most likely source (cf. Figure 3.29).

Another example of dispersion is found amongst blackflies of the *Simulium damnosum* complex, which are vectors of the disease known as river blindness, or onchocerciasis. Local breeding populations of the flies have been almost eradicated from much of the Volta River basin in West Africa by weekly dosing the streams and rivers inhabited by larval blackflies with the insecticide temephos. However, in most years the area is repopulated at the beginning of the rains by immigrant flies which apparently arrive from the south–west, most probably carried on winds associated with the Inter-Tropical Convergence Zone (ITCZ) (page 6).

Migration

Typically, migration is a response by animals to changes in their environment, and is almost always related to the seasons, especially in places where these are well marked. Although some migrations are daily, as with the plankton mentioned earlier, it is more usual for the migratory movements to have an annual pattern as in Figures 6.37 and 6.39.

The degree of regularity in migratory movements reflects the predictability of the environment. Eurasian swallows are small birds that feed on aerial insects and they commonly nest on buildings. In England, individual swallows have been found returning to the same locality to nest in successive years, despite having travelled many thousands of kilometres to Central and southern Africa during the intervening winter. They migrate at night, but whilst they are passing through tropical Africa, they can sometimes be seen resting by day in large numbers on overhead wires (Figure 6.32).

Figure 6.32 Eurasian swallows migrating through tropical Africa mainly fly at night. By day they feed, and rest, sometimes collecting in large numbers. They are fond of perching on overhead wires.
(D. E. Pomeroy.)

In contrast to these regular movements are those of birds and some other animals inhabiting the drier regions of Africa, notable for the unpredictability of their rainfall. Many of these species migrate in response to rainfall (or the lack of it) and their presence in any particular place therefore varies from year to year. An extreme example, which we shall return to shortly, is the desert locust.

The ability of some animals to navigate, sometimes over considerable distances, and to return to the same place after a long interval, has been established in many cases, as in the swallows. How they do so is not yet fully understood, but it is known that some species use the sun or stars as a guide and there is now convincing evidence that a variety of animals are able to respond to the direction of the earth's magnetic field. These include a number of species of birds such as pigeons, as well as honey bees, mice and some moths. But much remains to be explained, and so far most explanations as to how animals actually use magnetic information for navigation are far from complete. And in homing pigeons, at least, the birds apparently only use their magnetic sense when they are unable to see the sun.

Long-distance migration is a subject of great biological interest, and furthermore some migrants are serious pests. We now illustrate these points by examples.

Some examples of African migrations

The desert locust

In many ways the desert locust is an unusual and remarkable insect, as well as being a very important one to man (Box 6.5). This is the locust that was responsible for one of the biblical plagues of Egypt, some 3000 years ago. It breeds in more or less permanent 'outbreak' areas (Figure 6.34). In years when conditions are particularly favourable for them, the numbers of locusts in these areas exceed the carrying capacity, and the population forms swarms which emigrate into 'invasion areas'. In the past, some swarms were of incredible size (Figure 6.33), commonly containing up to 1000 million insects. The largest ever recorded was an estimated 40 000 million, in Ethiopia in 1958; it covered about 1000 km^2, causing widespread devastation and extensive food shortages. Gradually, the development of monitoring systems and more efficient control methods, have reduced the frequency and scale of locust swarms: thus the last major swarms of the desert locust in Kenya were in 1954; there has been none in West Africa since 1961. Since 1963 the desert locust has been declining, although small swarms of the gregarious form are still reported every year from one or more African countries. So long as some of the outbreak areas of the desert locust are subject to political instability, international monitoring and control will be difficult. In 1977–78, exceptionally heavy rainfall in semi-desert coastal areas along the Red Sea and Gulf of Aden allowed an explosive increase in desert locust populations there. Swarming behaviour (of the *gregaria* phase, see Box) was first

Box 6.5 Locust phases and migration

Locusts are not a distinct taxonomic group – the word locust is applied to any of about twenty species of grasshoppers (Family Acrididae) that under certain conditions form swarms. All locust species produce two or more forms or phases which differ morphologically from one another. The solitary phase is termed **solitaria**, and in the desert locust it is an inconspicuous green, grey or reddish colour, selective in its food and with individuals behaving as ordinary grasshoppers – avoiding contact with their neighbours.

From January to June, the solitaria phase of the desert locust lives and breeds in the 'outbreak areas' of North Africa and the Middle East (Figure 6.34), but in some years, when their environment is particularly favourable, large populations of gregarious 'hoppers' are produced; this phase is called **gregaria**. These hoppers are smaller than those of the solitaria phase, they are longer-winged, more brightly coloured (having contrasting black and yellow stripes), have a higher metabolic rate, and eat more. Gregaria adults lay fewer, but larger eggs than solitaria, resulting in bigger nymphs, which because of their larger food reserves are better able to survive periods of starvation.

After some time, the gregaria locusts stop feeding and become very active. They aggregate and finally in late June or early July adopt swarming behaviour. They suddenly take off and fly upwards towards the sun, reaching heights of 1000 m or more. In so doing many are swept southwards by the prevailing winds.

Moisture level appears to be one of the most important factors regulating both the survival of locusts and their population density, and consequently their phase formation, swarming and mass dispersal. High rainfall and humidity stimulate plant growth, increasing the abundance of the locust's food. High but not excessive moisture in the soil is needed for egg survival, although if the weather is too wet this causes increased mortality, mainly of the immature stages. Evidently a critical balance of moisture conditions is required for maximum population growth. But despite much work on the physiology, behaviour and ecology of various locust species, the exact relationships between the factors leading to migration are still not fully understood.

Figure 6.33 Part of a swarm of the desert locust in 1958 which spread over 1040 km^2 in Ethiopia, and which invaded the cultivated areas around Gigiga.
(Tropical Development and Research Institute, London.)

apparent in 1978, and swarms reached as far as India, Pakistan and Ethiopia. By 1979, Kenya was also threatened, but fortunately control measures were then able to destroy the remaining swarms, thus halting the migration.

There was also an upsurge in numbers in West and North-west Africa in 1980–81, but fortunately that too was soon controlled.

Migration of locusts occurs at considerable heights where wind speeds are in excess of the locust's flight speed of about 15 km h^{-1}, and hence their direction of dispersal is largely determined by the prevailing winds. Winds in the north of Africa are usually northerly, whereas south of the Libyan–Sahara deserts they blow from south to north. It is along the belt where the two meet (the ITCZ, page 6) that locust swarms are most likely to congregate (Figure 6.34).

Swarms of desert locusts often migrate from North Africa into dry northern savanna areas of West Africa (Figure 6.34(a)), but south of Niger and Mali, they have rarely been serious pests and they seldom breed south of 12° north.

In eastern Africa, during April to October, a high speed jetstream sweeps up from Mauritius through Madagascar, Kenya, Ethiopia, Somalia and across the Indian Ocean to the west coast of India, and beyond. It is highly likely that this air movement, which is associated with the ITCZ over the Arabian Sea and western India, and is unusually low (1200–2000 m), influences the disperal of various species of insects and birds in Africa from the south to the north and north-east.

The African armyworm moth

Armyworms are so-called because the caterpillars swarm like a marching army across fields in search of food, mainly grasses and also maize, sorghum and rice. They are widely distributed in Africa and a major pest of cereals. Young caterpillars are greenish, but older ones occur in two colour forms. At low population densities they are green with longitudinal brown

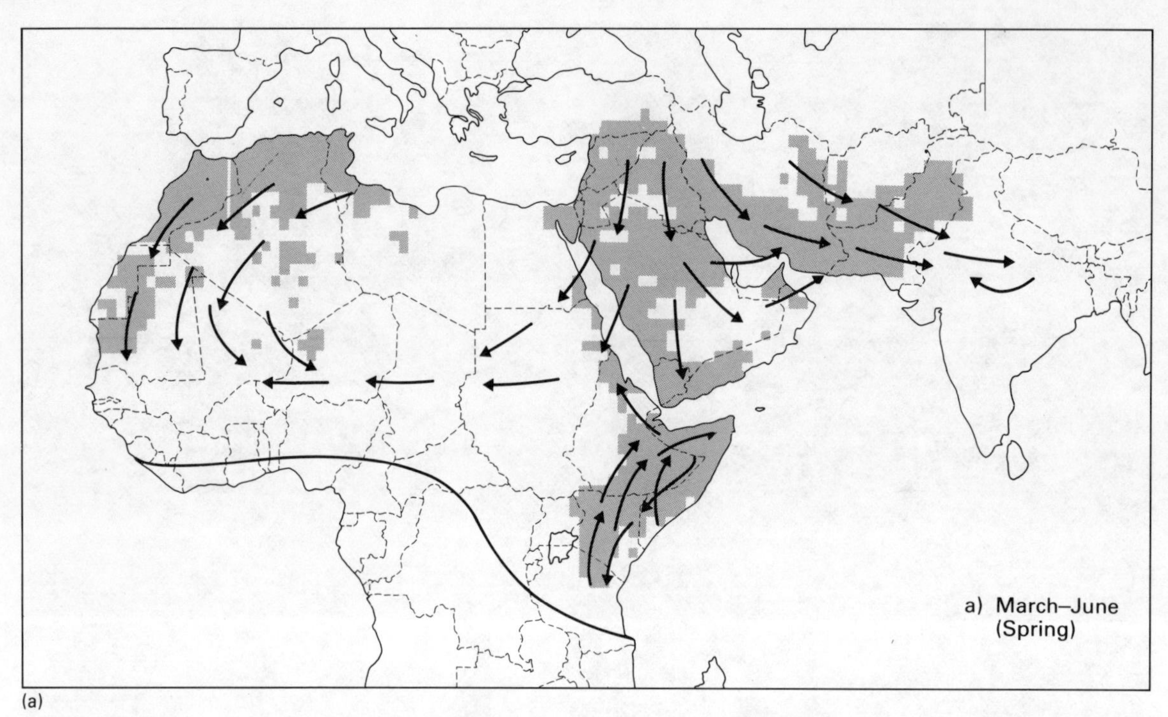

(a)

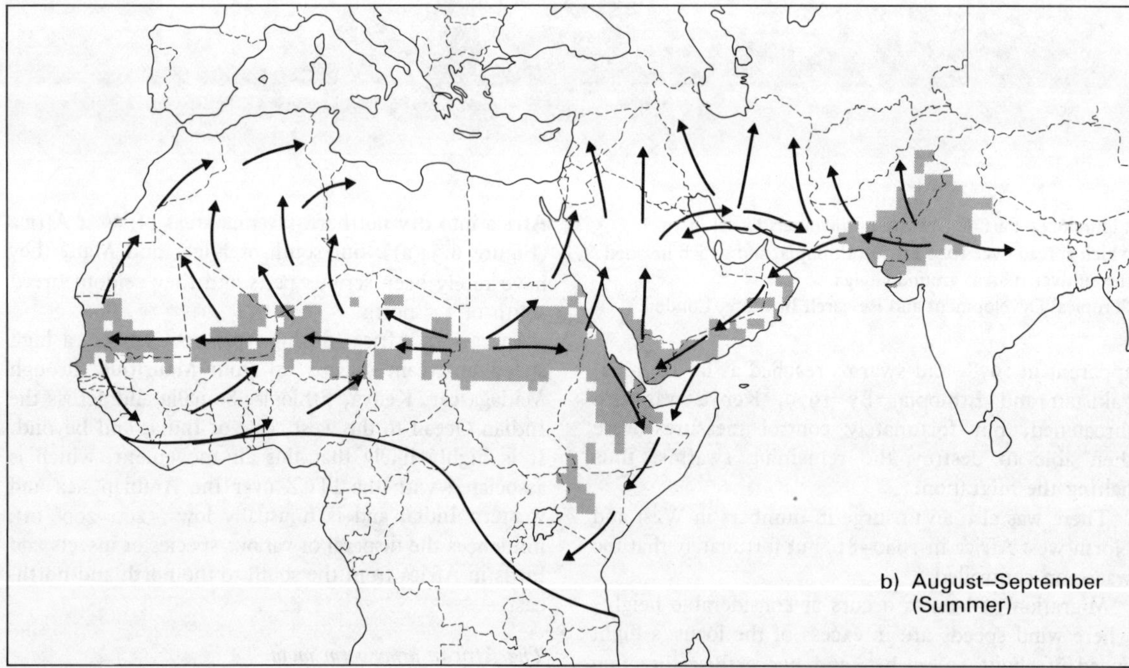

(b)

Figure 6.34 Seasonal breeding areas and migration routes of the desert locust during invasion years. Shaded areas show the principal breeding areas, thick lines demarcating the southern limits of breeding areas, while arrows show the main directions of swarms. In the spring (a) swarms move southwards across the Sahara towards the ITCZ where they become concentrated and summer breeding sites (b) are established. During the summer (b) winds generally blow from south to north and locusts migrate northwards, while in the cooler winter months (c) breeding sites are greatly reduced, and mainly confined to the horn of Africa, and smaller numbers migrate into the Arabian peninsula. (Modified from maps supplied by the Tropical Development and Research Institute, London.)

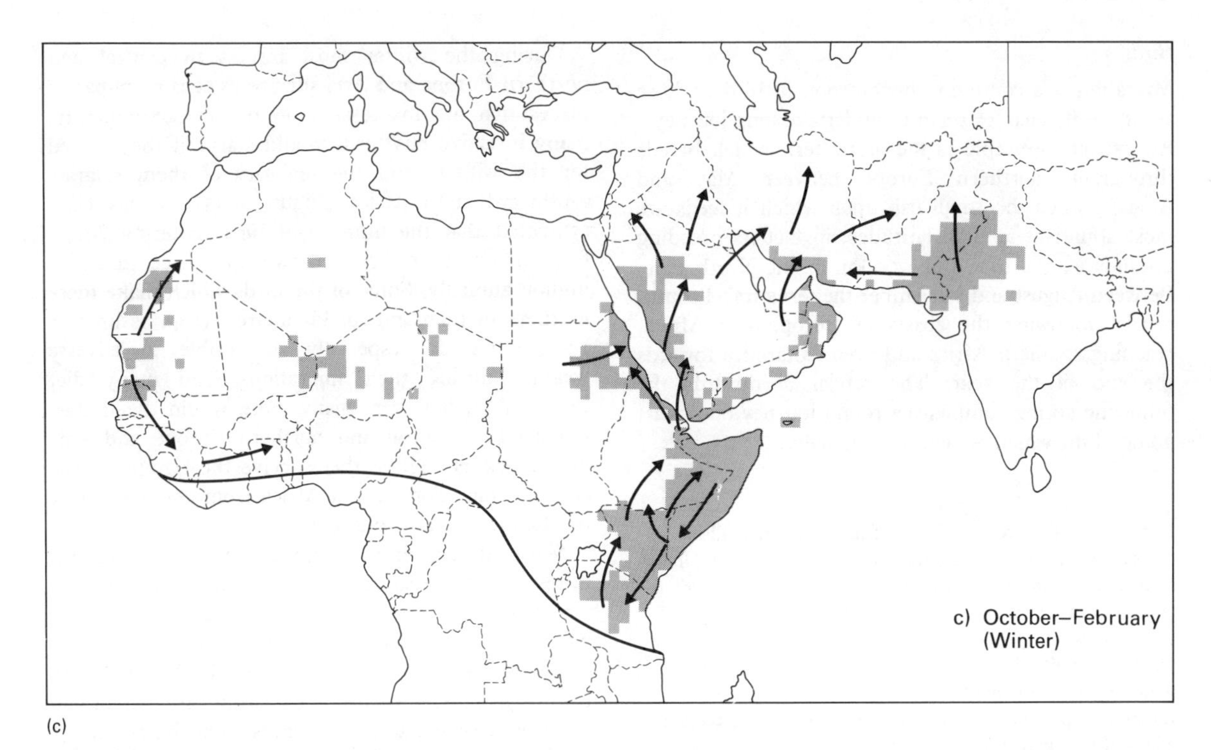

(c)

stripes, but at high densities they are dark grey or black with thin pale stripes. High densities of caterpillars are followed by emigration of the resulting moths shortly after emergency (cf. desert locusts).

A characteristic of the biology of armyworm moths is the sudden appearance of millions of caterpillars on grasslands or cereals followed a few weeks later by their disappearance as they pupate in the soil. After a further 2–3 weeks, adults emerge and may migrate considerable distances with the prevailing winds, resulting in similar larval infestations elsewhere. Regular outbreaks occur progressively later from south to north (Table 6.1). But in September to early November the north-east monsoon blowing across Somalia, Ethiopia, Sudan and Kenya brings to the south armyworm moths originating from late infestations in the north.

Table 6.1 Outbreak periods of adult armyworm in eastern Africa, showing the northward trend in the starting time

	From	To
Ethiopa	March–April	August–September
Uganda	January–February	May
Kenya	December–January	June–July
Tanzania	November–December	May–June

The painted lady butterfly

This attractive insect is found in all continents except South America. In North Africa, large numbers breed in December to March, then during March and April adults migrate northwards across Egypt into Turkey and the Ukraine, whilst others from Morocco and Algeria migrate into the Mediterranean area and thence into northern Europe. Some adults reach Finland, Scandinavia, Iceland, and even north of the arctic circle, but fail to breed there. In some years millions migrate northwards into Europe whereas in other years many fewer migrants reach the temperate regions. In Europe, eggs are laid in July and some adults hibernate but most migrate southwards in September, across Europe into North Africa, some continuing as far as West Africa.

Many other butterflies migrate regularly within Africa. *Catopsilia florella* is a common yellow butterfly that is often seen in considerable numbers flying close to the ground in one direction; though sometimes they fly several hundred metres above the ground. Populations south of the Sahara migrate during the severe dry season, December to February, southwards into the Congo basin and to the region of Lake Tanganyika, where they breed. In March to May populations in southern Africa migrate to the wetter eastern coastal areas and northwards into Central Africa.

Birds

Migration is a common phenomenon in birds, whose ability to fly enables them to undertake long journeys. An extreme example is the arctic tern which breeds throughout northern Europe between May and August, when the small fish upon which it feeds are most abundant at high latitudes; in fact its breeding grounds extend to within 700 km of the North Pole. Between August and November the terns travel southwards, following the coasts of Europe and Africa, reaching southern Africa and even Antarctica towards the end of the year. The return northwards the following spring completes a round journey of at least 20 000 km, which is repeated annually.

Figure 6.35 The breeding range of the Eurasian swallow (a) is extensive, and when it migrates south it occupies almost all of subsaharan Africa. It is sometimes extremely abundant (Figure 6.32). The nightingale (b) is famous for its song, especially where it breeds, but it also sings in Africa, usually from the middle of a thick bush. Both its breeding and wintering areas are much smaller than those of the swallow: the most important wintering areas are in the moist savanna zones. (See Figure 2.6.)
(Moreau, R. E., 1972. *The Palaearctic African Bird Migration System*, Academic Press: London.)

a) Eurasian swallow

breeding range

winter range

b) Nightingale

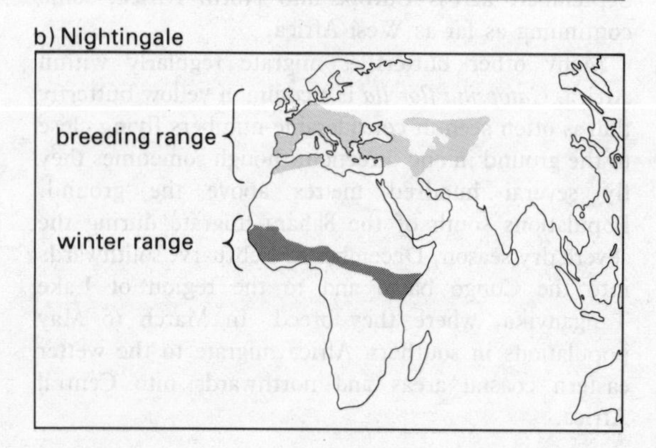

During the winter most insects in central and northern Europe and Asia survive as eggs or pupae in places that are inaccessible to birds. Consequently, many insectivorous birds would starve if they stayed for the winter, and the majority of them migrate southwards into Africa (Figure 6.35). It has been estimated that the number of birds entering Africa from the north is between two and three thousand million annually. Some of the birds which make these journeys of thousands of kilometres weigh as little as 10 g. They are especially susceptible to adverse weather during their migrations, and many die. Presumably, however, many more would die if they remained in central and northern Europe and Asia during winter. Those that survive the rigours of the autumn migration stay in Africa from about October to March. (See also Fig. 6.29).

Many African birds also undergo regular migrations in response to seasonal variations in their food supply, which in turn are usually related to rainfall. One which has been studied extensively is the red-billed quelea (sometimes called black-faced dioch: it has both features). This is a small seed-eating bird which can cause extensive damage to crops. The birds' movements depend upon the patterns of rainfall, which are very variable in the drier regions that they normally inhabit (Figure 6.36). This unpredictability complicates the control of queleas, because it is hard to know where and when to expect them.

Mammals

For a given distance travelled, flying requires the least effort, and walking or running the most. Swimming is intermediate. Not surprisingly then, the long-distance migrants amongst mammals are ones that swim, notably whales (Figure 6.37). Some bats undergo seasonal migrations, particularly the larger fruit bats, but the distances covered are usually quite short, sometimes only a few kilometres, and rarely more than a hundred. Better known are the annual migrations of some of the large herbivores, particularly those of the wildebeest, zebra and several other species in the Serengeti region of northern Tanzania. The enormous herds of these animals are a world famous spectacle (Figure 6.38). The food specializations of the various species were outlined on pages 120 and 123 (see also Figure 5.25). Optimal feeding conditions are only found in one part of the region at any particular time of the year. However, the rainfall patterns are different in the northern, southern and western parts of the Serengeti region (Figure 6.39). Consequently, by following the rains, each species is able to obtain near optimal conditions throughout the year.

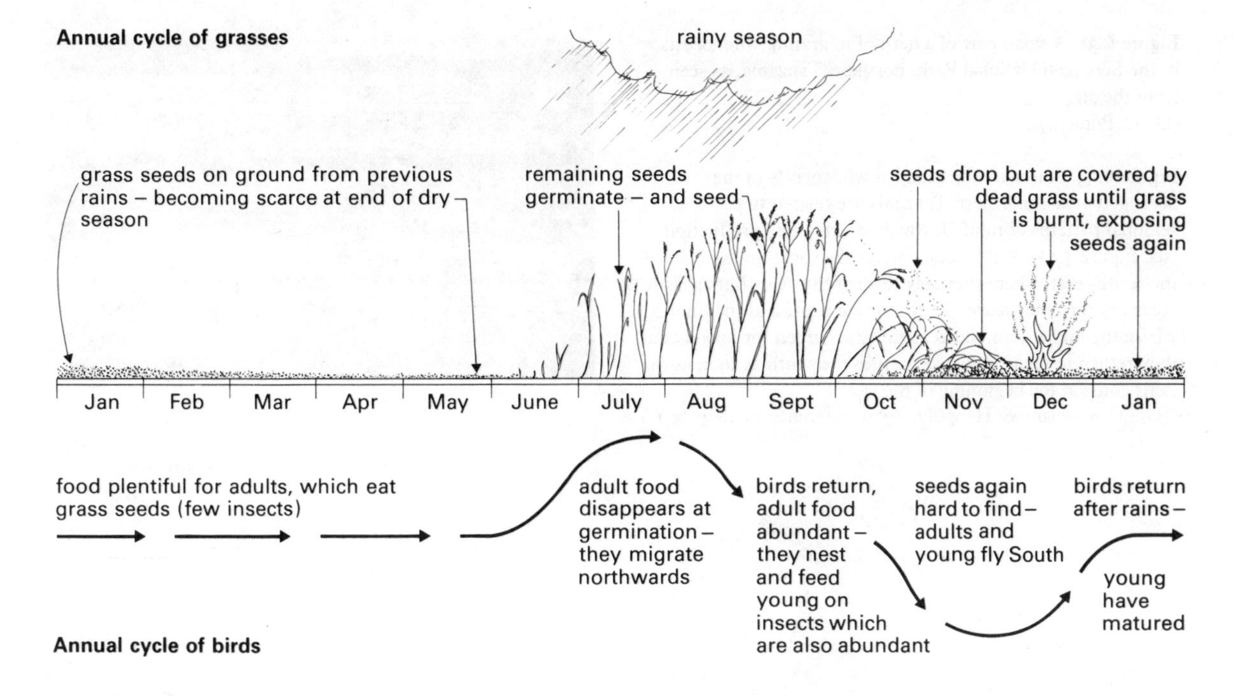

Figure 6.36 The effects of seasonality on the migrations of the red-billed quelea in Chad. The adult birds feed almost exclusively on grass seeds, which are abundant in the savannas after the rains. But as the rains move southwards (see pages 6 and 8), usually about June, any remaining seeds germinate, and from then until about September, when the new crop of seeds begins to fall, the birds would find no food in the area. They respond by moving northwards to areas where the rains passed earlier and the grasses are already seeding. Later, the birds follow the rains back south again. On this journey they may stop to breed, since by then conditions are ideal, with abundant seeds as well as insects on which to feed their young.

As the grasses dry out, they drop their seeds, but these become covered by a thick mat of dead grass. The seeds are then hard to find, so the birds again migrate southwards, to more favourable areas. They return after fire has removed the dead grass, exposing the seeds.
(Based on Ewer & Hall, 1978, reference in chapter 7.)

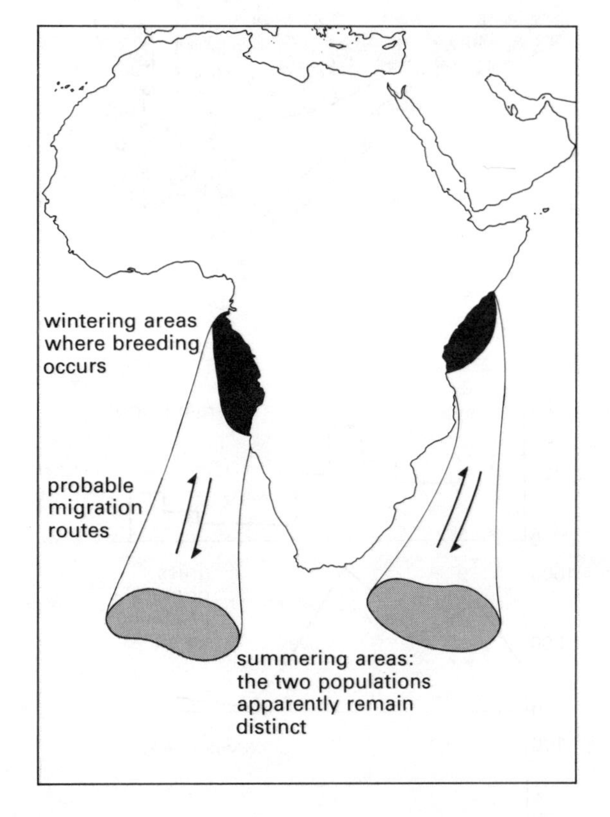

Figure 6.37 Humpback whales migrate from the Southern Ocean to their breeding areas off the African coast. Their food in the south consists of abundant crustaceans known as krill, but from about November to April the seas there are too cold, whereas tropical waters are favourable to the new-born whales (which are about 4 m long and weigh 1.3 t! – adult humpbacks average 11 m in length and 29 t in weight). The whales feed very little whilst in tropical waters, living mainly on their stored food reserves (blubber). The two populations are thought not to mix, and although in the south they are always far from land, in the breeding areas they are coastal, occasionally entering harbours.
(Partly from Grzimek 1972–75, *Grzimek's Animal Life Encyclopedia*, Van Nostrand Reinhold, New York.)

Figure 6.38 A small part of a herd of migrating wildebeests in the Serengeti National Park, northern Tanzania, as seen from the air.
(D. E. Pomeroy).

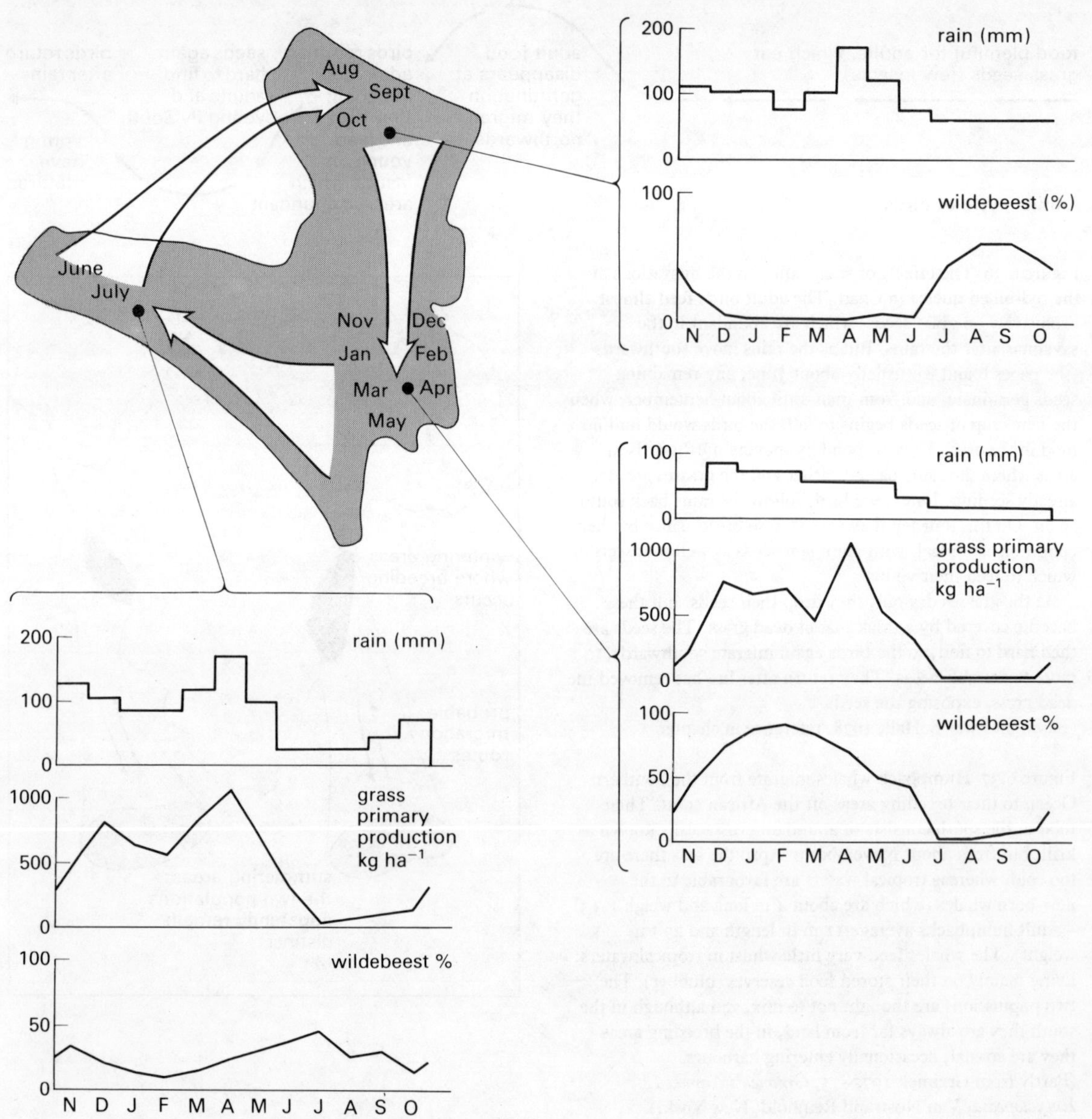

Figure 6.39 Seasonal migrations of wildebeests in the Serengeti area of northern Tanzania are related to the seasonal patterns of rainfall, which affects grass production (see Figure 5.25). Wildebeests prefer the short-grass plains of the south–east, where they stay for half the year, but as those areas dry out, they leave. They migrate west, and then north, where the higher rainfall keeps the grass green for longer, but they return to their preferred areas in the south as soon as the rains come at the beginning of the year.
(Based on Delany & Happold, 1979, reference in chapter 1.)

Suggested reading

In 1962, Klopfer published a book entitled 'Behavioural aspects of ecology' and this remains a useful introduction, but the subject has been one of increasing interest since then, and amongst recent texts, Krebs & Davies, 1978, 1981, give much more interesting accounts.

Several of the texts on animal behaviour and ethology are useful to ecologists, for example Immelman, 1980, and Manning, 1979, who both discuss social behaviour and territoriality. Social behaviour is dealt with in greater detail by Wilson, 1971, 1975: the former specificially concerning insects. Wilson developed the concept of 'sociobiology', a topic of much interest in recent years, and some of the most influential papers on this subject are reprinted in Clutton-Brock & Harvey, 1978, who provide useful commentaries.

Excellent examples of animal coloration are seen in Skaife, 1979, a beautifully illustrated book on African insects; more general reviews are Cott, 1975, Fogden & Fogden, 1974, and Wickler, 1968, all of which provide nicely illustrated examples of mimicry and protective coloration. Sheppard, 1975, in his book on natural selection and heredity, also gives clear accounts of protective coloration and mimicry, especially amongst African butterflies.

The sense of smell is highly developed in mammals, and Stoddart, 1976, describes well how they make use of it. A useful account of bird song is to be found in Campbell and Lack 1985. Foraging strategies are discussed by Krebs & Davies, 1981, Crawley, 1983, and especially by Townsend & Calow, 1981.

Migration, dispersal, homing and navigation are described exhaustively by Baker, 1978, but accounts in a shorter version (Baker, 1982) and in Gauthreaux, 1980, are more straightforward. Another readable book, extensively illustrated, is edited by Baker, 1980. These works cover migration by all the main groups of animals, and the last includes seed dispersal too.

Essays and problems

1 Review territorial behaviour in animals.

2 Describe and contrast strategies of courtship and mating in invertebrates, birds and mammals.

3 Describe the patterns of seasonality of some named animals in your own country, and discuss the timing of reproduction in these species.

4 Why are some species of animals social, yet most are not?

5 Write an essay on insect migration and its importance in pest or vector management. (Suggested references are Baker, 1982, D. L. Gunn & R. C. Rainey, 1979. *Strategy and Tactics of Control of Migrant Pests*, The Royal Society, London and Service, M. W. 1980. *International Journal of Biometerology*, **24**, 347–53.

6 Referring particularly to either birds or mammals, give an account of how animals communicate with others of their own species.

7 Choosing mainly local examples, discuss how coloration and pattern are used by animals to minimize predation.

8 What are the relative advantages and disadvantages of sound and odour in animal communication?

9 Describe territorial behaviour in the impala and Uganda kob. (Suggested references are Delany & Happold, 1979, reference in chapter 1, Kingdon, 1971–82, reference in chapter 5 and Leuthold, W. 1966. *Behaviour*, **27**, 214–57). How would you test the hypothesis concerning kob leks that was outlined on pages 139–140?

References to suggested reading

Baker, R. 1978. *The Evolutionary Ecology of Animal Migration*, Hodder & Stoughton: London.

Baker, R. (ed.) 1980. *The Mystery of Migration: The Story of Nature's Travellers through the Cycle of the Seasons*, Macdonald and Jane's: London.

Baker, R. 1982. *Migration-paths through Time and Space*, Hodder & Stoughton: London.

Campbell, B. and Lack E. 1985. *A Dictionary of Birds*, Poyser.

Clutton-Brock, T. H. & Harvey, P. H. 1978. *Readings in Sociobiology*, Freeman: San Francisco.

Cott, H. B. 1975. *Looking at Animals*, Collins: London.

Crawley, M. J. 1983. *Herbivory*, Blackwell: Oxford.

Fogden, M. and Fogden, P. 1974. *Animals and their Colours: Camouflage, Warning Coloration, Courtship, Territorial Display, Mimicry*, Peter Lowe, London.

Gauthreaux, S. A. (ed.) 1980. *Animal Migration, Orientation and Navigation*, Academic Press: New York.

Immelman, K. 1980. *Introduction to Ethology*, Plenum: New York.

Klopfer, P. 1962. *Behavioural Aspects of Ecology*, Prentice-Hall: New Jersey.

Krebs, J. R. & Davies, N. B. (eds). 1978. *Behavioural Ecology: an Evolutionary Approach*, Blackwell: Oxford.

Krebs, J. R. & Davies, N. B. 1981. *An Introduction to Behavioural Ecology*, Blackwell: Oxford.

Manning, A. 1979. *An Introduction to Animal Behaviour*, (3rd edn). Edward Arnold: London.

Sheppard, P. M. 1975. *Natural Selection and Heredity*, Hutchinson: London

Skaife, S. H. (revised by J. Ledger) 1979. *African Insect Life*, (2nd edn). Country Life: London.

Stoddart, D. M. 1976. *Mammalian Odours and Pheromones*, Studies in Biology, No. **73**, Edward Arnold: London.

Townsend, C. R. & Calow, D. 1981. *Physiological Ecology, an Evolutionary Approach to Resource Use*, Blackwell: Oxford.

Wickler, W. 1968. *Mimicry in Plants and Animals*, Weidenfeld & Nicolson: London.

Wilson, E. O. 1971. *The Insect Societies*, Belknap Press: Harvard.

Wilson, E. O. 1975. *Sociobiology: the New Synthesis*, Belknap Press: Harvard.

7 Animal abundance

Chapter 5 and 6 covered the various components of an animal's environment, including the interactions of individuals with their own and with different animal species. Now we take a more quantitative approach, focussing on populations rather than individual animals. The basic parameters of population change are births and deaths, immigration and emigration. As animal populations grow in size we shall see how their innate capacity for increase, $\mathbf{r_m}$ *can be derived by simple calculations. We will then examine the actual rates of increase,* **r**, *of some real populations to see how well the theories describe the growth of natural populations.*

It is a matter of everyday experience that the size (and hence density) of populations vary from place to place, and at different times. For example, the numbers of some animals fluctuate cyclically during the 24-hour day, other animals show marked variations in seasonal abundance whilst many species experience 'good' and 'bad' years.

We shall distinguish between population control and population regulation, the latter referring to changes in population size that are largely determined by population density. We will show how climate, acting in what is called a density-independent way can cause high mortalities, and is sometimes the most important factor in controlling (i.e. changing) population size. Other factors such as predation and competition working in a density-dependent manner are frequently important in regulating populations whose densities are close to the carrying capacity. But predation and competition can also regulate populations below the level that would be obtained if food or space were regulating their size. Many habitats are neither stable nor permanent but stability is important in determining what kinds of animal inhabit it. Some species, conveniently described as **r**-*selected, because of their high innate capacities of increase (*$\mathbf{r_m}$*), are best adapted to temporary or unstable habitats. Others, the* ***K***-*selected species, have populations which are near the carrying capacity (****K****) and these are characteristic of the more stable habitats.*

There is still controversy as to what causes populations to change in size and how populations are regulated. For instance, are density-independent, or density-dependent factors the main cause of changes in population size? We will summarize theories and show how models have grown to be important in the study of animal populations.

Introduction: Populations and their parameters

Ecologists are much concerned with the numbers of the organisms they study. When they say that some species are more abundant than others, they mean the populations are bigger. All the individuals of a particular species living in a certain area, and at a particular time, are referred to as a **population**. The area could be as large as the whole biosphere, as when we say that the human population of the world in 1980 was about 4240 million people. But ecologists usually consider populations of much smaller areas, for instance the population of frogs in a pond, or of weevils in a grain store.

The human population of the world can be divided geographically into numerous smaller populations some of which are still very large, as in the great land-mass of Eurasia. At the other extreme, there are many oceanic islands with a few tens or hundreds of people. Such a division of the whole population into local populations is characteristic of most species of plants and animals, as well as humans. Obvious examples are species inhabiting freshwater pools, or patches of forests, but the phenomenon is more widespread than this. Sometimes some contact may be maintained between local populations by dispersal and migration, but at other times local populations of a species become isolated from each other, with no inter-breeding between them (page 94). When this happens, the flow of genes between the various populations ceases and eventually natural selection will be likely to produce differences between them. This is because the environment in the various localities will almost certainly differ somewhat, so that slightly different genotypes will be favoured in each.

Extinction is not infrequent in small isolated populations, usually because of habitat changes, but the whole species becomes extinct only in the much rarer event of all populations dying out. Such species extinctions do occur, however; indeed, it is believed that of all the animal species which have ever inhabited the earth, about 99 per cent are now extinct. Since we still have several million species alive today, the number of species that have become extinct must be formidable.

There are many practical difficulties in studying and measuring populations. For example, many species have extensive distributions, such as the aardvark in Figure 5.1, and an ecologist studying such a species can consider only a small part of its population. Other animals, e.g. corals and termites, form colonies in which individual organisms cannot exist independently. Similarly many perennial grasses and other vegetatively-propagating plants grow in clumps, and one cannot adequately define separate individuals. In cases such as these, measuring and counting colonies or clumps is usually more logical than describing the populations in terms of individuals composing them. We can, for instance, easily observe the density of termite mounds in a field.

In describing populations, we commonly refer to an animal's **abundance** which is taken to mean the size of the population, but this is misleading unless the amount of space it occupies is also considered. For example, if fifty fish occur in a small pond and another fifty in a large pond, it is clear that both populations are the same size, but ecologically the situations are different because the fish are more crowded in the smaller pond than in the larger one. Thus, in addition to population size, a most important population parameter is the **population density**, defined as the number of individuals per unit area. The unit can be a square metre or a square kilometre, depending upon such factors as the size of the animal and of its habitat. It is occasionally more appropriate to measure the population density as the number of individuals per unit of space, for example aphids in the air, fish in ponds, or arthropods in the soil, although sometimes fish densities, especially for bottom dwellers, are better given as numbers m^{-3}. Soil organisms such as earthworms are usually estimated as numbers m^{-2} of the soil surface.

Properties of populations

Each of the individuals that comprises a population has various attributes, such as age and sex. There may also be different morphological forms within the population, such as winged and wingless insects or different castes as in ant and termite colonies. Whole populations also possess properties, such as age-structures, sex-ratios, density, and the rates of birth, death, immigration and emigration (Table 7.1). These last four are sometimes described as the basic parameters of populations: a change in any one of them causes a change in population size. If, on the other hand, the population size remains constant for some time, then

Table 7.1 Properties of an individual and a population measured a) at one point in time and b) through the life of the individual or cohort

	Individual	*Population*
(a)	age	age – structure
	size	size – structure
	sex	sex – ratio
	state of maturity	proportion in each stage of development
	state of reproduction	proportion breeding
	place in space	range
	no equivalent	density
	no equivalent	dispersion pattern
	no equivalent	birth rate (= natality)
	no equivalent	death rate (= mortality)
	no equivalent	immigration rate
	no equivalent	emigration rate
	no equivalent	population growth rate
(b)	life span (longevity)	average life span (mean longevity)
	number of eggs produced	mean fertility
	number of young produced	mean fecundity
	no equivalent	survivorship curve.

it follows that:

$$\text{Natality} + \text{Immigration} = \text{Mortality} + \text{Emigration}$$

i.e Potential causes of population increase = Potential causes of population decrease. (Natality and mortality being the rates of births and deaths respectively.) In nature, however, population size never remains constant for very long.

We shall consider how populations behave or are expected on theoretical grounds to behave, in the simplified conditions created in a laboratory, then discuss what happens in more complex natural environments. Later (pages 200–203) we present some theories of population regulation which have been vigorously debated amongst ecologists for many years and which still generate considerable controversy.

Basic population processes

Populations usually begin with a very few organisms, but increase their numbers with time. Now some habitats are inevitably temporary, for example, freshwater pools, which when filled with water are rapidly colonized. Some of the new colonizers may withstand desiccation, which is a constant threat to temporary water collections. For example, eggs of *Aedes* mosquitoes can remain dry but viable for many months or even years and yet hatch when soaked, and larvae of the chironomid fly, *Polypedilum vanderplankei*, can withstand severe drying for months, but revive when pools become reflooded. When a field is planted with crops, it has the potential to support new populations of the crops' pests, which after harvesting are eliminated or greatly reduced. New buildings are potential habitats for cockroaches and bed-bugs, which can be distressingly successful in establishing large populations within a short time. These habitats are admittedly more permanent: they may stand for many years. In contrast, some habitats change very little, and populations of species inhabiting rainforests, for example, may persist for very many generations.

Population growth

If a population were to grow continuously, with nothing to limit its size, the increase in numbers would follow an exponential curve like that in Figure 7.1(a). The shape of the curve – in particular its steepness – varies with the species (Figure 7.1(b)). Under ideal conditions, even elephant populations would rise exponentially: Charles Darwin calculated that a pair of them could multiply to 19 million in 750 years. But bacteria dividing every 20 minutes would reach a similar population in about seven hours! The bacterium is said to have a much higher **innate capacity for increase** than does the elephant. The symbol r_m is used for the innate (or intrinsic) capacity for increase, which was defined in 1954 by the Australian ecologists Andrewartha & Birch as 'the maximal rate of increase attained at any particular combination of temperature, moisture, quality of food, and so on, when the quantity of food, space, and other animals of the same kind are kept at an optimum and other organisms of different kinds are excluded from the experiment'. In effect, this innate capacity for increase depends upon the amount by which the birth rate exceeds the death rate. Box 7.1 shows how r_m can be calculated. We shall see shortly how values of r_m can be used.

Knowledge of the innate capacity for increase is valuable because we can use it to estimate how fast a population would grow under a given set of environmental conditions such as climate and certain food resources. But, of course, populations do not increase indefinitely since all organisms would ultimately be limited by the availability of their resources, which themselves cannot be infinite. (The converse, however, is not necessarily true: as we saw in chapter 5, the availability of resources is not the only component of environment which can limit the size of a population.)

If we were to place some maize weevils in a bag of maize grain in a reasonably warm, dry place, from

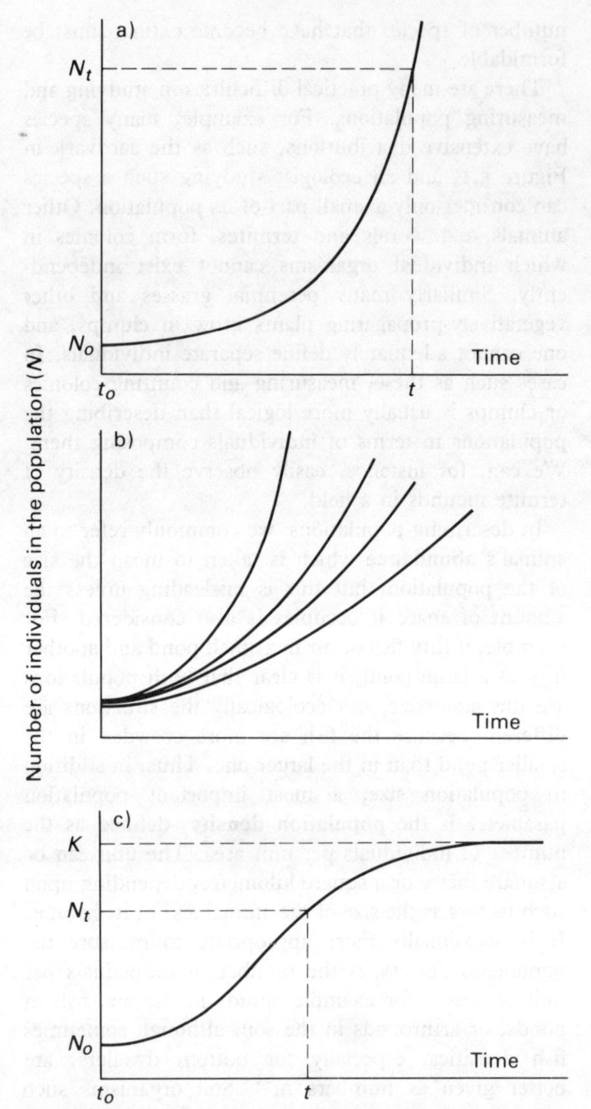

Figure 7.1 (a) Under constant favourable conditions, the size of a population of living organisms will increase exponentially. The exact rate depends upon both the species and the conditions, but as long as conditions do not change, the size of the population N_t at any particular time, t, can be related to the original number of organisms N_o at time t_o by the equation $N_t = N_o e^{r_m t}$.

(b) The slope of the curve shown in (a) varies with the species and the particular set of conditions. When these conditions are constant and favourable, a value of r_m, the innate capacity for increase, can be determined for that set of conditions. This figure shows curves differing only in their values of r_m, steeper curves have greater values of r_m.

(c) When the supply of resources is kept at a certain level, the population increase follows the form of a logistic curve, whose parameters are shown here, and featured in equation (5). K is defined as the carrying capacity of the environment for the species under consideration.

Box 7.1 Calculation of r_m

The innate capacity for increase is related to time by the exponential equation:

$$N_t = N_o e^{r_m t} \quad (1)$$

The meaning of the terms N_t, N_o and t can be seen from Figure 7.1(a); and e is the exponent of natural (Naperian) logarithms. It is a constant whose value is 2.718. We can simplify equation (1) by using natural logarithms (i.e. logs to the base e). The equation then becomes:

$$\log_e(N_t) = \log e(N_O) + r_m t \quad (2)$$

or, in terms of common logarithms (logs to the base 10):

$$\log_{10}(N_t) = \log_{10}(N_o) + \frac{r_m t}{2.303} \quad (3)$$

(2.303 is $\log_e$ of 10).

In practice, we can evaluate r_m by keeping a population of animals under a particular set of conditions. By re-arranging equation (3) we have:

$$r_m = \frac{2.303[\log_{10}(N_t) - \log_{10}(N_o)]}{t} \quad (4)$$

Suppose, for example, that we are studying a species of fly. We introduce a fertile pair of flies into a cage whose environment we are keeping favourable and constant. One hundred days later we observe that the population has risen to 200. What is the value of r_m under these circumstances? We have $N_o = 2$, $N_t = 200$ and $t = 100$ days. Hence, using equation (4):

$$r_m = \frac{2.303[\log_{10}(200) - \log_{10}(2)]}{100}$$

$$= \frac{2.303 \times 2.00}{100}$$

$$= 0.046$$

Note that the particular value of r_m depends upon the units of time, in this case, days.

which predators and diseases were excluded, their population would begin to increase exponentially, as in Figure 7.1(a). But after a short time the weevils would begin to respond to various effects of crowding. To maintain the population we should have to start supplying more food. Supposing that we now did this, whilst keeping all other conditions constant, we should find that the growth of the population began to level off, as in Figure 7.1(c). A curve of this shape, resembling a tilted letter S, is called sigmoid, and the particular form of a sigmoid curve that describes the growth of a population in an environment with limited but constantly renewable and renewed resources, is the logistic curve. It can be defined mathematically as:

$$N_t = \frac{K}{1 + e^{a(-r_m t)}} \quad (5)$$

Here, N_t, t, r_m and e have the same meanings as in Figure 7.1(a) and in equation (1) in Box 7.1. The other two terms require some explanation. a is a constant for a particular curve, changing its values moves the curve as a whole to left or right. K is described mathematically as the asymptote – or more simply as the size of the population at the point where it levels out. We have described r_m (page 176) as the innate capacity of an animal to increase its numbers, but the actual rate of population increase, termed r, is often less than r_m. When a population is decreasing in size, r becomes negative, that is we have a declining population. Under optimal environmental conditions the actual growth rate, r, approaches the theoretically calculated r_m. Figure 7.2 shows a logistic curve for a particular set of data. Notice that the actual rate of increase, r, declines as the population increases in size, until it levels off (here birth rates equal death rates) after which the population remains the same size. That is, there is a maximum number of animals which can be supported by a particular supply of resources and this number is called the **carrying capacity** (K).

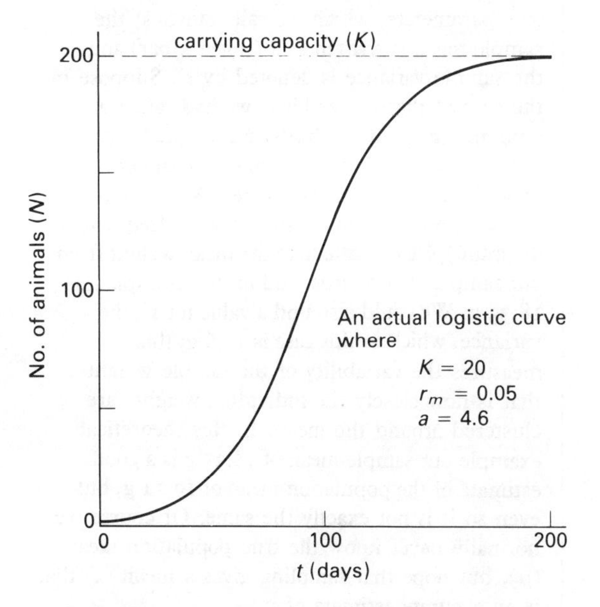

Figure 7.2 Here is an example of a logistic curve, in which $r_m = 0.05$, K (carrying capacity) = 200, and the number of organisms N_t at time $t = 70$ is 50. The constant a was found to be 4.60 (see equation 6). The relationship between r and the potential growth rate, is as follows $r = r_m (K-N)/K$, where N = population size.

Box 7.2 Statistics, parameters and mathematical models

Statistical analysis has become such a valuable tool to ecologists, and its use is now so widespread, that it is advisable to be acquainted with at least the simpler statistical procedures; especially as they are less difficult to understand than is generally believed. A few general points are introduced here.

It is well known that if we measure almost anything on any species of organism we are likely to find that the results vary. A good example is animal's weights (more correctly, masses).

The continuous line in this figure shows the result of plotting the individual weights of an imaginary population of millions of a species of weaver bird. In theory, if we weighed every bird, we could calculate the true mean of the whole population and its variance. True values are denoted by Greek letters, μ (mew) for the mean and σ^2 (sigma) for the variance; all true population properties such as these are called parameters. In this imaginary population we have calculated the mean weight (μ) as 19.24 g and its variance (σ^2) as 1.35 g. But we can very rarely weigh all individuals of a population and thus have to take a sample, and make estimates of the true parameters, which we call statistics; the sample mean is written as $\bar{x}$ (called x bar) and the sample variance is denoted by s^2. Suppose in the case of the weaver birds we had taken a random sample of 100 birds and weighed them individually we could then plot a frequency-diagram as shown in the figure. (A class-interval of 1.00 g was used in constructing the frequency-diagram.) We can calculate the mean weight from our sample of 100 birds, and in this example it is 18.33 g. We could also find a value for s^2, the variance, which in this case is 1.46 g; this measures the variability of the sample weights, that is how closely the individual weights are clustered around the mean. In this theoretical example our sample mean of 18.33 g is a good estimate of the population mean of 19.24 g, but even so it is not exactly the same. Of course we normally never know the true population mean (μ), but hope that sampling gives a mean ($\bar{x}$) that is an accurate estimate of it.

Unfortunately the use of particular letters for particular statistics, is not completely standard. Occasionally, for example, you will see m instead of $\bar{x}$ for the mean, which, incidentally, is mathematically the same as the average.

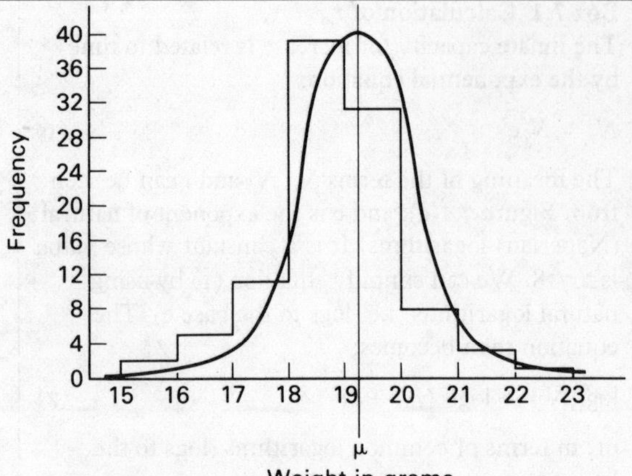

Mathematical models, such as the logistic curve, are useful in helping us to describe more accurately a predictable process. A model is a simplified version of the real situation, and thus assists us to understand the real situation. If it is too simple, the model becomes unrealistic; whilst if it exactly describes the real situation it is itself unnecessary.

The logistic curve (equation (5) on page 177) certainly does not describe real populations exactly (see, for example page 179). But it does describe very well the general shape of population growth under its given conditions, i.e. limited resources and an environment which is otherwise constant and favourable.

The most serious limitation of the logistic curve, and others like it (such as the exponential growth model, equation (1) in Box 7.1), is that they predict only a single possible outcome. Thus given values for each of the other statistics, a single estimate of K can be made. Such a 'one-answer' model is called deterministic. There are other models which more accurately reflect ecological realities. These are the stochastic models which take account of variability, and they yield answers in terms of probabilities. A stochastic model for population growth, for example, could take the form of a prediction that the value of K in a given set of circumstances would fall between (say) 180 and 220, with 95 per cent probability. Obviously stochastic models are necessarily rather more complex than deterministic ones, but in principle they are not hard to understand: see, for example, Krebs, 1978, pp. 199–204. This book, together with some others recommended as being good introductions to statistics for ecologists, are listed under Suggested reading at the end of this chapter.

Carrying capacity is a variable: it depends upon the species, and the supply of resources, and all the other components of the species' environment. When the latter are optimal, the carrying capacity is determined solely by the supply of resources, especially food (see also page 182).

Fitting a logistic curve

Fitting a logistic curve is not difficult. Suppose we had found that the population of maize weevils of the previous section settled down at 200, we could use this as our estimate of K, the carrying capacity. We should evaluate r_m for the prevailing conditions by experiment. By observing the initial stages of the population growth we should obtain values for N_o and N_t for several values of t. Then, using equation (4) in Box 7.1 we could estimate r_m for each value of t, and thus obtain a mean value for r_m.

Equation (5) can be re-arranged too:

$$a = \log_e \left(\frac{K - N_t}{N_t} \right) + r_m t \quad (6)$$

This is how Figure 7.2 was constructed; given that $K = 200$ and $r_m = 0.05$, then the constant a was found to be 4.6.

The values of r_m and K change with environmental conditions, but given a particular value of r_m we can predict K. At least one set of values for t and N_t would have to be found by experiment, for each set of conditions. Then K could be calculated as follows:

$$K = N_t(1 + e^{a - r_m t}) \quad (7)$$

There is an opportunity for you to try this yourself in the 'Essays and problems' section at the end of this chapter.

The construction of models is becoming increasingly common in ecology because they are useful in both describing ecological processes and predicting the probable outcome of events; the logistic curve for example is a model of population growth. The use of models generally, and in population ecology in particular, is a topic of Box 7.2. The actual values of a population at any time depend on the relative values of the birth and death rates, that is, so long as emigration equals immigration. This leads to the notion of population **stability**, that is, population processes interact to maintain the population at *about* the same size. The opposite is population instability, which leads to a population becoming either extinct or extremely large. Box 7.3 shows how the various

Box 7.3 Population growth

Diagrammatic representation of how changes in birth and death rates can by themselves lead to unstable population processes. In (a) and (b) the birth and death rates remain the same at all population densities, in (c), (d), (e) and (f) one of them varies.

(a) birth rate exceeds death rate, and population explodes; (b) death rate exceeds birth rate so population decreases to extinction. (c) here there is a potential equilibrium at point x_1 where birth rate equals death rate, but the population is unstable because any slight increase or decrease in birth rate results in population exploding or becoming extinct. (d) as above a point is reached where birth rate equals death rate, but again unstable as any departure of death rates leads to population explosion or extinction. (e) a stable population is reached because death rate increases both below and above the birth rate, and so equilibrium is reached; (f) again a stable condition is reached because of decreasing birth rates leads to equilibrium (Adapted from D. Rogers in A. Youdeowei and M. W. Service 1983. *Pest and Vector Management in the Tropics* Longman: London.)

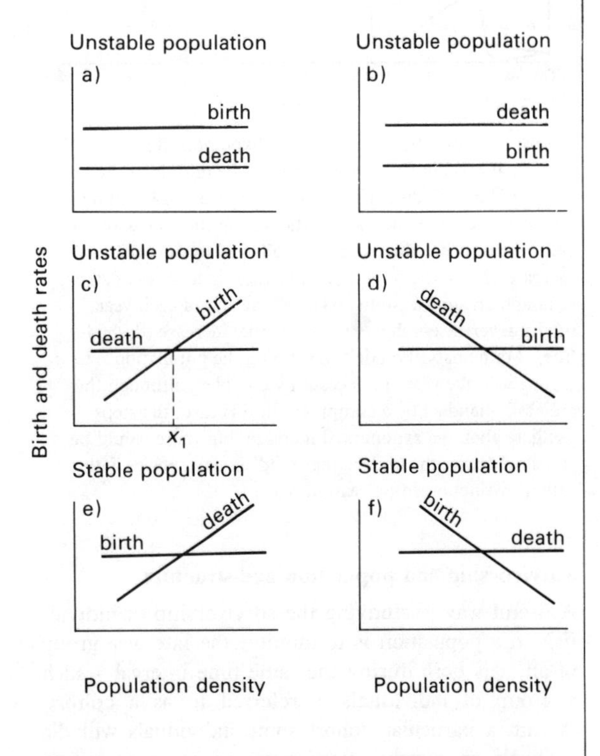

interactions of birth and death rates can lead to stability or instability.

The curves of population increase that we have looked at in Figure 7.1(a)–(c) make the assumption that births are occurring continuously – it is the excess of births over deaths that generates the population increases. But many animals have distinct breeding seasons, producing young only once or perhaps twice a year (pages 160–164). For these species, increases can occur only periodically, as in Figure 7.3.

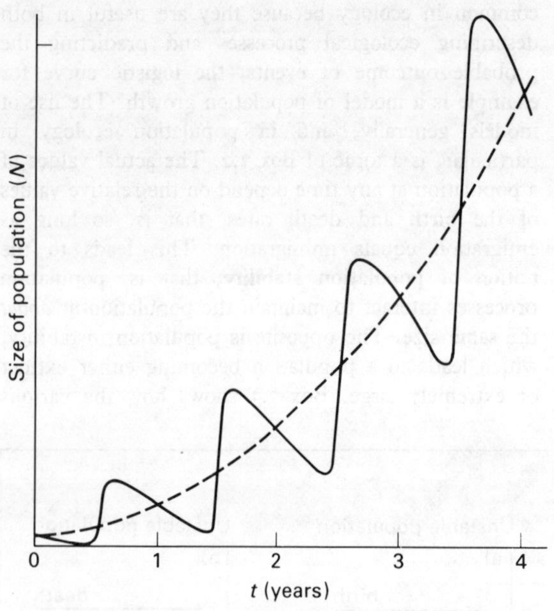

Figure 7.3 Population increase of a hypothetical insect with an annual breeding season which is in the middle of the year. The numbers of adults rise steeply as a result of breeding, but then decline as individuals die during the course of the year; the next breeding season is followed by another increase. Thus, the population increases in a series of steps, although an annual count, taken at the end of each year, might nevertheless show an exponential increase (dashed line). Mathematical models describing the population size throughout the year are necessarily complex, although they are easily handled by a computer. In this case, the steps oscillate about an exponential increase, but there would be steps, whatever the underlying trend, in any species with a rather distinct breeding season.

Survivorship and population age-structure

A useful way of studying the survivorship of individuals in a population is to monitor the fate of a group of animals born during the same time interval – such a group of individuals is referred to as a **cohort**. Within a particular cohort some individuals will die at birth or shortly afterwards and some as young adults, whilst others may survive to old age. If the decrease in numbers of animals during the life of a cohort is plotted, usually on a logarithmic scale, against time, you obtain a survivorship curve. This is called an **age-specific curve**, or graphical representation of a life-table, because it is based on the fate of individuals in a real cohort measured throughout its life, and hence numbers at one age can never be greater than at a previous age. The shape of this curve indicates how mortality changes with age of the individual. Many species, especially invertebrates and marine fish, produce large numbers of offspring most of which die in early life. When illustrated graphically this produces a concave survivorship curve – termed a type III curve (Figure 7.4). In contrast a few animals, modern man being one, have a high proportion of those born surviving to senility, which leads to a type I curve. In other species, there is a more or less constant rate of mortality in all age-groups leading to a linear decline in survivorship when survivors are plotted on a logarithmic scale (type II survivorship curve). It is often very difficult to give meaningful answers to such questions as 'how long do animals live?' The average length of life is a misleading statistic for a species in which 90 per cent of the individuals die within a few weeks of birth, whilst of the other 10 per cent, some may live for several years.

Another method of representing survival data is shown in Figure 7.5, where the percentages in each age-class of a population are shown in the form of a pyramid. The numbers of survivors are usually fewer in older age-classes but not always (see 3-year-olds Figure 7.5(c)). This is a **time-specific** life-table, because it is based on a sample of a population taken at one point of time, where generations are overlapping. Careful study of Figure 7.5 shows how such information could be used to predict future trends in populations, especially of long-lived species. Where a species has a continuous breeding season, as do many insect pests, it becomes very difficult to analyse the death rates in the different age-classes because they overlap, but the same general principles apply. Various methods are available for detailed analyses of demographic data; some references on this topic appear in the Suggested Reading section at the end of this chapter.

Carrying capacity

Carrying capacity refers to the notion that under a given set of environmental conditions a habitat is able to support only a finite number of animals (pages 177–179). In some situations the carrying capacity is never reached, but in others populations increase in size until they reach the carrying capacity, and may

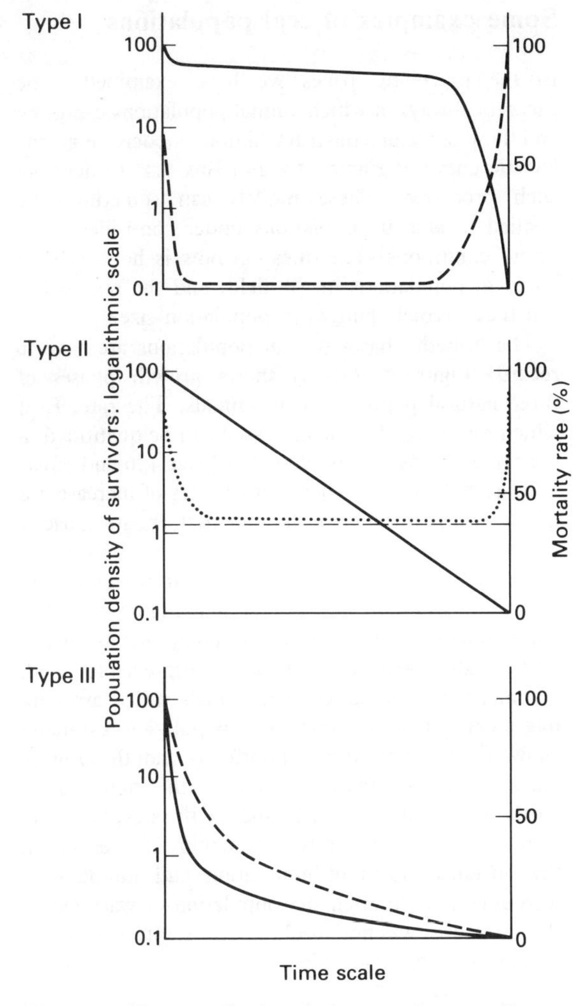

Figure 7.4 These are the three basic types of age-specific survivorship curves, based on a cohort followed through its life. The number of survivors at a particular time are shown in solid lines, whilst dashed lines indicate mortality rates, as percentages of the population dying in unit time, usually years for types I and II but months or days for type III.

Type I In a few species, such as larger endotherms and marine organisms, a fairly high proportion of offspring survives to adulthood. After an initial period of infant deaths, the mortality rate falls to near zero, only increasing again in old age, as in modern man in developed countries.

Type II In many species of birds and in most flowering plants, mortality rates do not change with age, at least in adults. On a semi-logarithmic plot – as here – this produces straight lines for both survival and mortality curves. In birds, however, mortality is invariably higher when young and may increase again with senility (dotted line).

Type III In the vast majority of species, (especially invertebrates) mortality is very high at the start of life, and very few individuals survive to breed. But the mortality rate drops with age and may be quite low in adults.

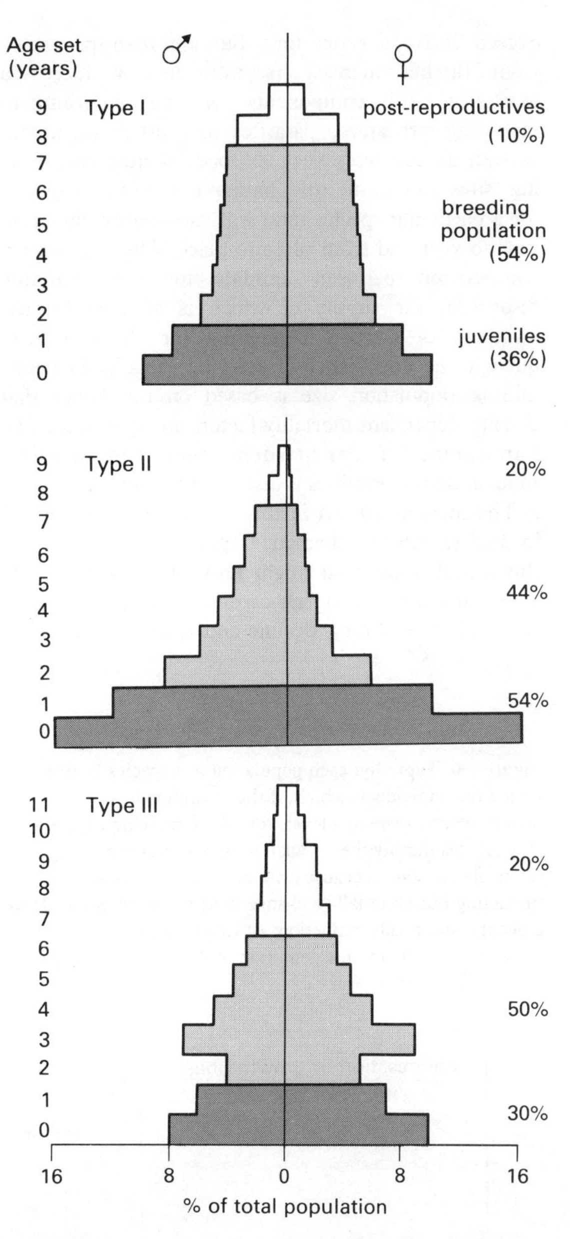

Figure 7.5 These pyramidal diagrams show time – specific mortalities of 3 hypothetical populations of the same species living under different conditions and the percentage of the total population (divided according to sex) in each cohort.

I Stable population, with a type I mortality (Figure 7.4). Sex-ratio approximately 1 : 1.

II Population with type II mortality. Sex-ratio equal at birth, but with higher mortality in males than females. In the 6-year olds the ratio females: males is 2.4 : 1.

III Population whose mortality rate changed markedly. There were either few births three years ago (age-set 2 or 3-year-olds), or an unusually heavy mortality later within that cohort. A greater proportion reach senility, as in type I; there are more females than males at all ages.

exceed it for a short time but are then prevented from further increase in numbers by increased mortality. All components of the environment including predators, parasites and other organisms as well as resources such as food, shelter and nesting sites determine the habitat's carrying capacity for a particular species, and will vary seasonally, from year to year and from place to place. There is usually competition between animals for the available resources, the supply of which is affected by the numbers of animals competing for them. Consequently the importance of carrying capacity in determining population size is based on the belief that density-dependent mortality factors are operating. The importance of density-dependent and density-independent mortalities is discussed on pages 192–198.

The ideas discussed in this section are summarized in Figure 7.6, a schematic representation of how a theoretical population might grow, become relatively stable around its carrying capacity, and then, due to some adverse factors, decline and become extinct.

Figure 7.6 Typically, each population of a species begins with a few individuals which, if the environment is favourable, increase to a level around the carrying capacity (K). At this density the population is shown as remaining relatively constant, because favourable conditions are prevailing but eventually a change in its environment leads to a decline and finally extinction. At times r is positive (black). See page 177 for definition of r (Delany & Happold, 1979, reference in chapter 1.)

Some examples of real populations

In the preceding pages we have examined some theoretical ways in which animal populations can grow and how we can construct simple models (e.g. the logistic curve of Figure 7.2 and Box 7.2) to describe such processes. These models can sometimes be applied to animal populations under controlled laboratory conditions. The question now is how do they apply to populations in the field, and the real test is can they predict changes in population size?

The growth phases of real populations are easy to record. Figure 7.7(a)–(c) shows growth phases of three natural populations of animals. The rate, r, at which these populations increased can be quantified in the same way as r_m was estimated (page 176 and equation (4) of Box 7.1). These actual rates of increase are nearly always lower than the potential, innate capacity for increase, r_m.

The relatively stable phase is what we expect to observe in an established population (Figure 7.6). The word 'relatively' emphasizes the point that no population is absolutely constant for any appreciable length of time. Even in species of mammals which are non-migratory, such as warthogs, population estimates show fluctuations from month to month (Figure 7.8(a)). The fluctuations seen in the figure arose partly from changes in birth and death rates, but there were also unavoidable errors in making the census. In this instance, rates of emigration and immigration were low and although the population of warthogs in the study area was not absolutely constant, it remained within fairly narrow limits.

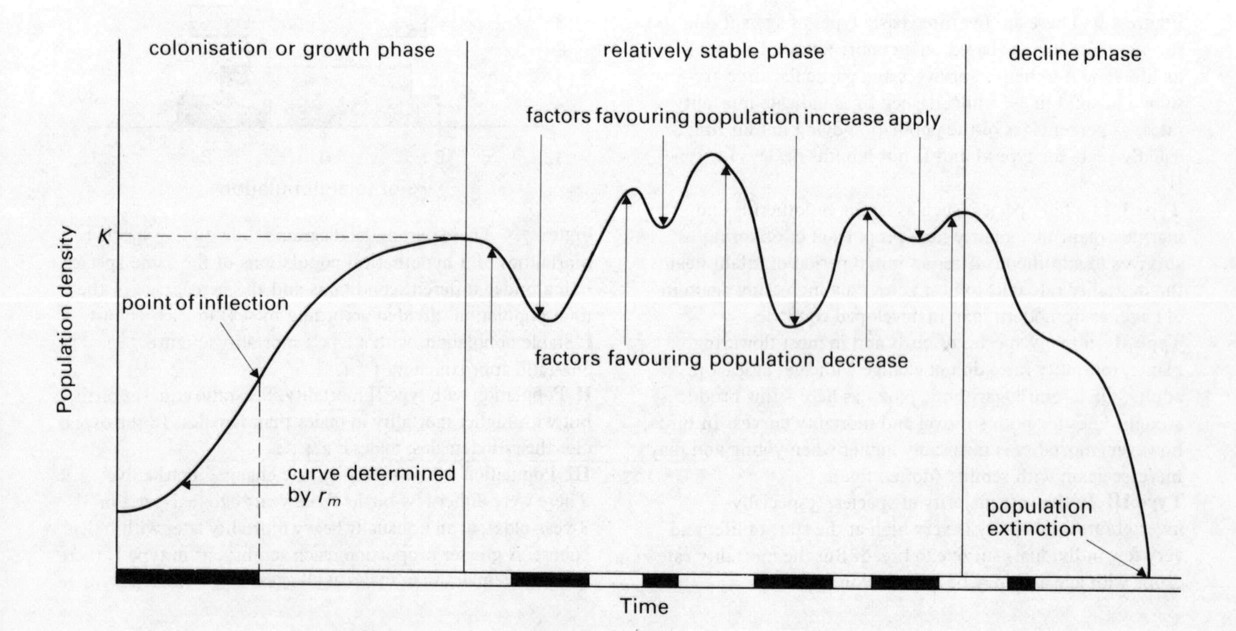

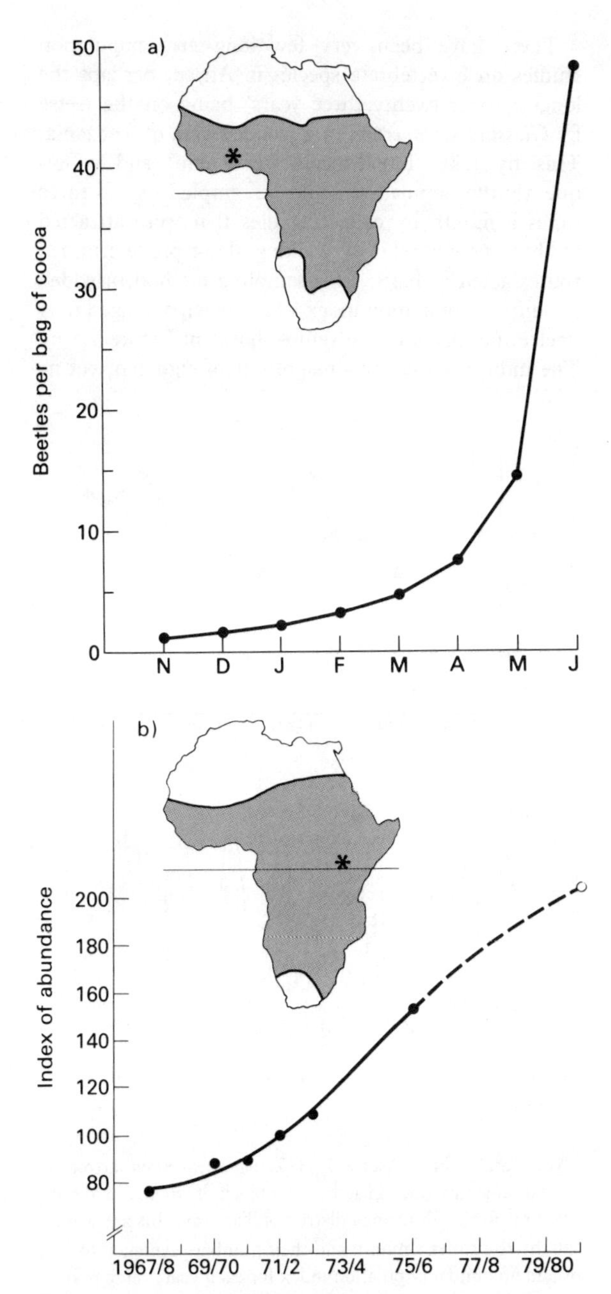

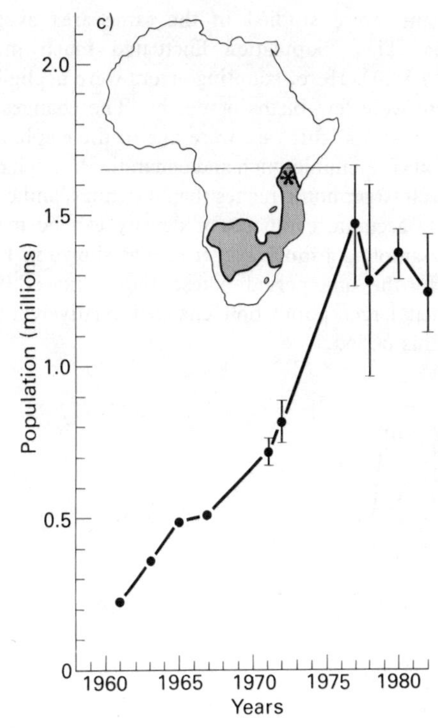

Figure 7.7 Three examples of population increase. The inset maps show the distributions of the species concerned, and the stars show the localities where the data were collected.

(a) The cigarette beetle can be a serious pest of stored products, partly because it has a high innate capacity for increase (r_m). The graph shows the increase of a population which had infested stored cocoa at Kumasi, Ghana. Within seven months they had multiplied more than 100-fold. The generation time of this species is 4–6 weeks; r_m is highest when the relative humidity is between 20 and 30 per cent and the temperature between 30 and 35 °C.
(Graph modified from Ewer & Hall, 1978; based on Cranham, J. E., 1960, *Bulletin of Entomological Research*, **51**, 203–22; additional data and map from Hill, 1975, reference in chapter 5)

(b) Marabou storks occur throughout tropical Africa, especially in drier areas where they are commensal with man. Increasing urbanization has led to more waste from abattoirs and increased numbers of rubbish tips. The marabou population has responded by increasing in many areas. The index of abundance in the graph is based upon the numbers of nests in a series of breeding colonies in Uganda and western Kenya (except for the final point where data for only one colony were available). Abundance is relative to the 1971–72 season, taken as 100. Each breeding season starts towards the end of the year and extends into the first few months of the next.
(Data from Pomeroy, D. E., 1977. *Journal of the East African Natural History Society & National Museum*, **31**, 1–11, and unpublished. Map from Mackworth-Praed & Grant, 1980, reference in chapter 3.)

(c) Wildebeest are large antelopes which are famous for forming herds, sometimes of enormous size. At the end of the last century, a viral disease known as rinderpest was accidentally introduced to tropical Africa, probably from the north. Cattle populations were greatly reduced over vast areas of eastern Africa, and many wild ungulates suffered too, including buffaloes and wildebeest. Gradually, the disease was controlled in cattle by immunization. It appears that rinderpest cannot maintain itself in wild ruminants, and by 1963, soon after its disappearance from cattle, it had also disappeared from wildebeests. The graph shows the recovery of the wildebeest population, slow at first but with increasing rapidity. Vertical lines for years 1971 onwards show one standard error. (Sinclair *et al.*, 1985. *Oecologia*, 65, 266–68. Map from Dorst & Dandelot, 1970, reference in chapter 5.)

Elephants were studied in the same area as the warthogs. Their population fluctuated much more (Figure 7.8(b)). Here, counting errors were negligible and there were few births or deaths. The changes in population size in this case were due to the elephants' movements (i.e. immigration and emigration). Elephants have much larger home ranges than warthogs, and consequently accurate estimates of density can be made only by sampling a much bigger area as shown in Figure 7.9 for the same period. These studies showed that this much larger population changed relatively little during this period.

There have been very few long-term population studies on invertebrate species in Africa, perhaps the longest (over twenty three years) being on the tsetse fly *Glossina swynnertoni* in a wooded area of Tanzania. This fly takes blood-meals from man, and consequently the population could be sampled two to three times a month by collecting flies that were attracted to three people who by walking along predetermined routes acted as baits. This sampling method provided a monthly population index which when averaged over twelve months gave the values shown in Figure 7.8(c). The indices varied by a magnitude of eighteen, yet no

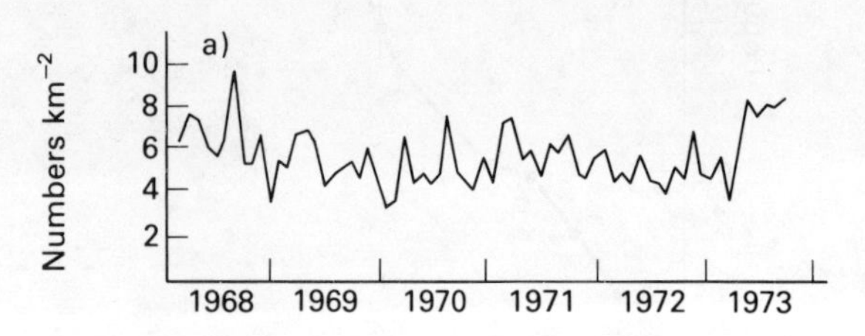

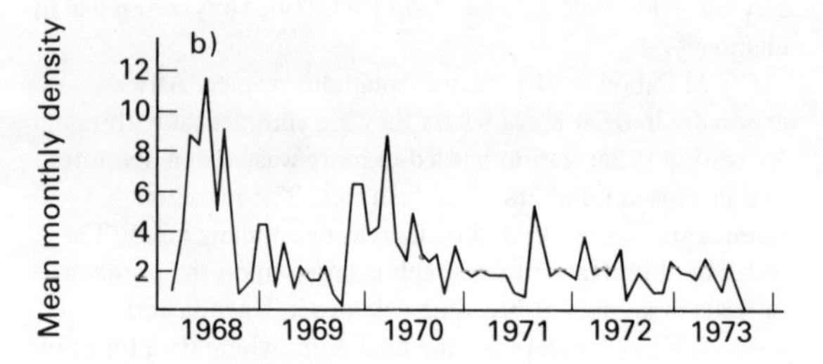

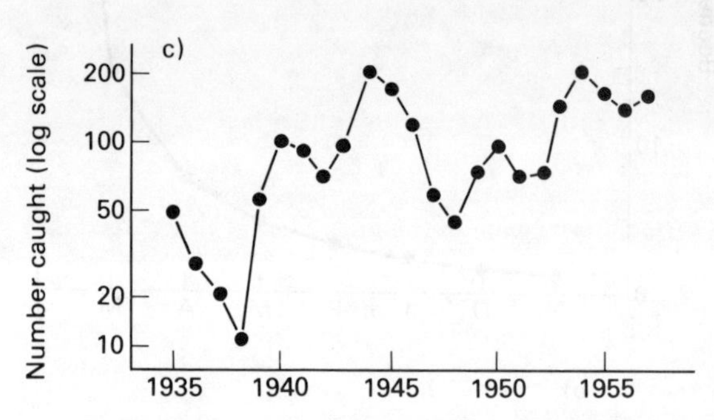

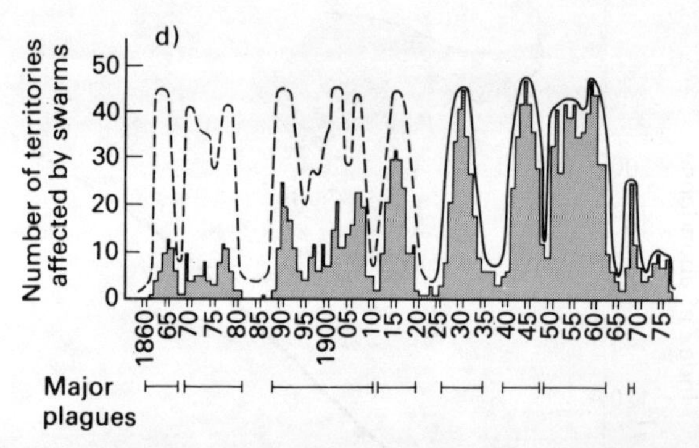

Figure 7.8 Examples of two relatively stable populations, and two that are not.

(a) Estimates of the warthog population of the Mweya peninsula, Queen Elizabeth (Rwenzori) National Park, Uganda. The area of the peninsula is 4.4 km^2. Numbers over a six-year period fluctuated comparatively little: about 85 per cent of all estimates fell between four and eight animals per km^2.

(b) The elephant population in the Rwenzori National Park as a whole varied relatively little between 1968 and 1973 (see Figure 7.9) but within a small area there were much greater fluctuations. This is seen here in the elephant data for Mweya peninsula, which is only about 0.3 per cent by area of the whole Park.

(Data for (a) and (b) from Eltringham, S. K., 1980. *African Journal of Ecology*, **18**, 53–71 and Eltringham, S. K. & Malpas, R., 1980. *African Journal of Ecology*, **18**, 73–86.)

(c) Catches of the tsetse fly, *Glossina swynnertoni*, attracted to man in standardized catches in a block of woodland about 115 km^2 in the Shinyanga district of Tanzania. Insects were caught 2–4 times a month and their numbers averaged to obtain an annual population index for each year; catches are presented here on a logarithmic scale. Peak populations occurred irregularly in the years, 1940, 1944, 1950 and 1954 and lowest catches were in 1938, 1942, 1948, 1951–52 and 1956.

(Modified from Glasgow, J. P. & Welch, J. R., 1963. *Bulletin of Entomological Research*, **53**, 129–37.)

(d) Major plagues of the desert locust, 1860–1977. The undulating line above the histogram demarcates plague and recession periods from 1925–77, while the dashed line suggests the infestation periods from 1860–1924 when records were incomplete.

(From Ashall, C. & Chaney, I., 1979. *Span*, **22**, 98–100.)

factors, artificial or natural, could be identified as causing these fluctuations. Inability to find the cause, does not, however, mean there were not ecological reasons for these changes in population size, rather that they were not discovered. This amplitude in population size (eighteen times), is not, however, large, for there are several instances of mammal populations varying 100-fold between different years and much greater annual fluctuations occur in insects such as locusts. In these species, the wide population fluctuations indicate that no stable phase exists. This situation is well illustrated by the sporadic migrations which are characteristic of locust populations. There are three important species of locust in Africa: the desert locust, migratory locust and red locust. Each species has a range which includes an outbreak area, where it can always be found, and an invasion area (Figure 6.33). Locusts are usually absent from the invasion area, but at times when their enviroment is particularly favourable, populations erupt in the outbreak areas, resulting in dense populations which stimulate emigration into the invasion areas. Invasion areas thus experience years with no locusts, interrupted by occasional years with very high numbers. Figure 7.8(d) shows years of major plagues of desert locusts; a large number of countries had swarms during 1941–61, reflecting large population explosions during these years.

The decline phase of a population (Figure 7.6) occurs at the end of each eruption with species such as the locust. Another example is seen in Figure 7.9, showing a dramatic decline in an elephant population which was attributed to a sharp increase in the market price of ivory. That in turn led to heavy poaching of elephants in the Queen Elizabeth (Rwenzori) National Park, and elsewhere in Africa.

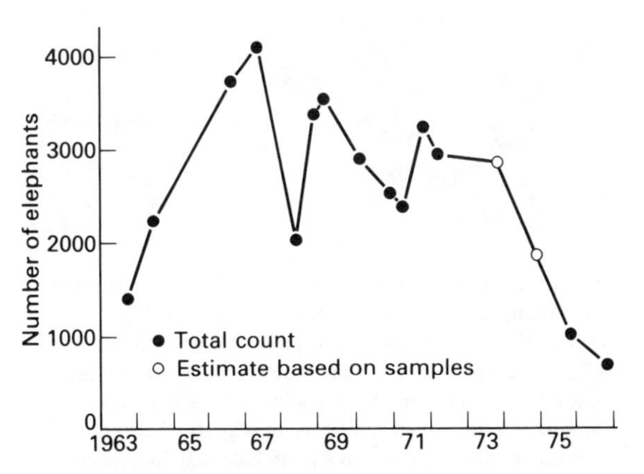

Figure 7.9 The elephant population in Queen Elizabeth (Rwenzori) National Park, Uganda, from 1963–76. From 1963–72 the fluctuations in numbers were mainly due to movements of elephant herds into and out of the National Park. But there was then a steep decline from an estimated population of 2864 in late 1973 to only 704 in late 1976. (Redrawn from Eltringham, S. K., & Malpas, R., 1980. *African Journal of Ecology*, **18**, 73–86.)

Changes in population density

Changes in terms of place

A study of any species of animal will invariably show that the density of its population differs from place to place (Figure 7.10). We also recognize this when we talk about a species showing habitat preference, because this implies that some habitats are more favourable than others. Naturally, we expect the species to be more abundant in the more favourable habitats. It follows that a study of such differences will help us to identify important components of the species' environment. Thus, the abundance of the termites in Figure 7.10 depends, amongst other things, upon temperature (Figure 7.11).

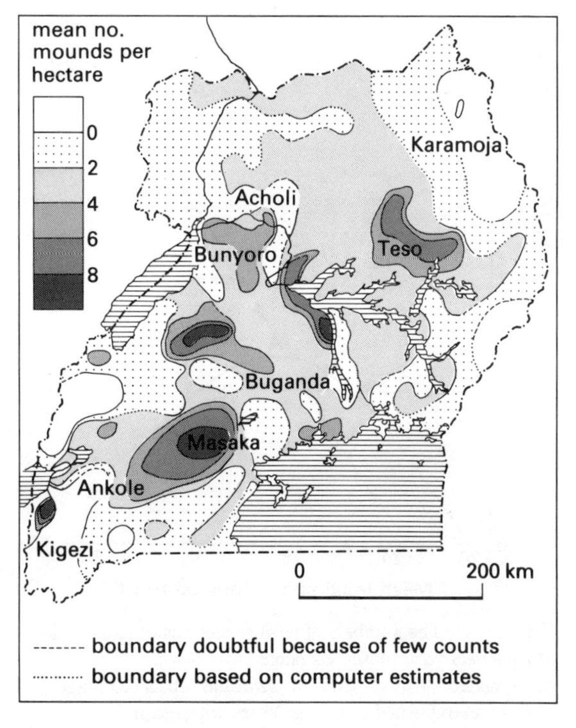

Figure 7.10 Geographical variations in population density, as illustrated by the distribution of large termite mounds in Uganda. The mounds are built by two species, *Macrotermes herus* and *M. bellicosus*; the map shows the average density of mounds.
(From Pomeroy, D. E., 1977. *Journal of Applied Ecology* **14**, 465–76.)

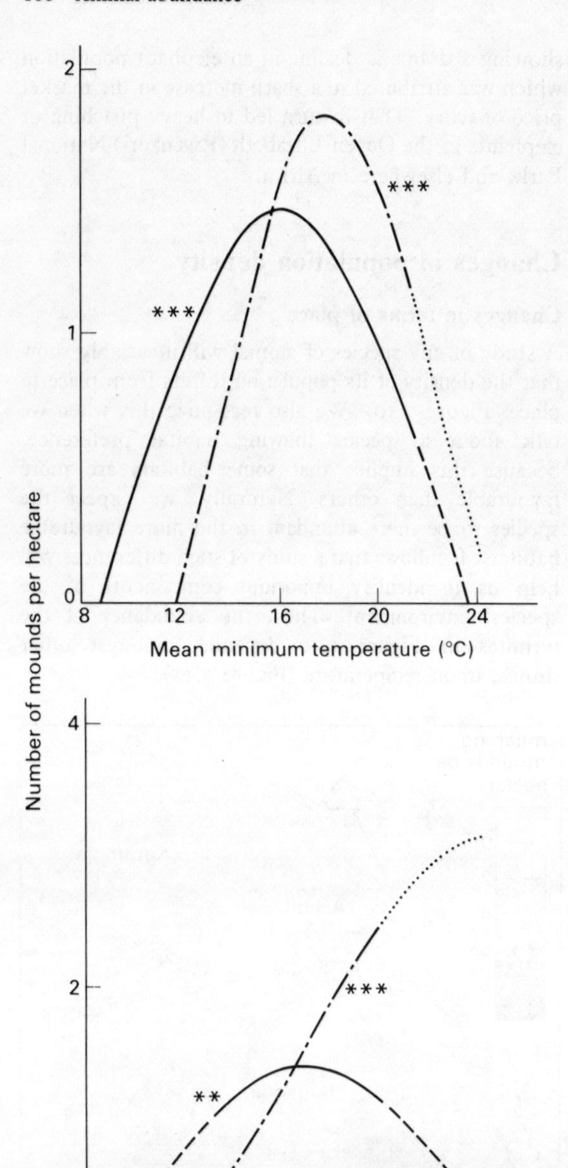

Figure 7.11 The numbers of termite mounds of two species, *Macrotermes herus* (observed range ———, extrapolated -------) and *M. bellicosus* (observed range -.-.-., extrapolated . . . .) vary with mean temperature, which itself is closely related to altitude. *M. bellicosus* is most numerous in hotter areas, and *M. herus* where it is relatively cooler. The agreement between the fitted curves and field data, on which they are based, were highly significant (**$P < 0.01$ *** $P < 0.001$). (Modified from Pomeroy, D. E., 1978. *Journal of Applied Ecology*, 15, 51–63.)

Other examples of habitat preference are shown by the three mosquitoes involved in yellow fever transmission. *Aedes africanus* is the species principally responsible for transmitting yellow fever virus amongst monkey populations, and it is most often encountered in forests because its larvae occur in water-filled tree holes, Whereas *Ae. simpsoni*, which is an important vector of the disease from monkeys to man, occurs in cultivated land around villages because its larvae are found in water-filled leaf axils of banana plants. In contrast, *Ae. aegypti*, the mosquito responsible for transmission of yellow fever from person to person, is found predominantly in villages and towns because it breeds in man-made habitats such as domestic utensils and water-storage containers (Figure 7.12).

Habitat preferences, and thus local variations in species abundance, are also well known in tsetse flies, which are responsible for spreading sleeping sickness and animal trypanosomiasis (nagana). Tsetses are divided into two major groups according to their habitat preferences. Firstly, we have the riverine

Figure 7.12 Water-storage pots such as these support large populations of the mosquito, *Aedes aegypti*, which is responsible in Africa and the tropical Americas for spreading yellow fever from man to man. The breeding sites are thus typically man-made and the mosquito is referred to as a domestic species. In some regions of Nigeria where water is very scarce in the dry season a household may have about forty water-filled pots. In the absence of more normal habitats, such as ground pools, the important malaria vector *Anopheles gambiae* also breeds in water-storage pots. This atypical behaviour caused by seasonal lack of preferred breeding places is important because in the dry season malaria transmission in the area is maintained by such peridomestic breeding.
(M. W. Service.)

species (the *palpalis* group) living and breeding in the relatively humid and sheltered environment of vegetation growing along rivers, streams and lake-shores. These species feed commonly on animals associated with this type of habitat, e.g. warthogs and reptiles including probably monitor lizards. The other group of tsetses, the savanna flies (the *morsitans* group) can tolerate rather drier conditions and usually feed on antelopes. They are encountered more frequently away from riverine vegetation, and as the group name suggests are more common in savanna areas.

A very important aspect of spatial distribution is dispersion, meaning how the individual members of the population are spaced in relation to each other (pages 132–135). Thus, whilst some species are social, occurring in swarms, herds or flocks, others are solitary. This topic was considered in more detail in chapter 6, mainly in relation to sociality and other factors that cause individuals to aggregate.

Population changes with time

The variations in estimated population densities seen in Figure 7.8(a), (b) are rather irregular. They were caused by a variety of factors, including sampling errors. But some species exhibit more or less regular changes in population density in response to some component of their environment, such as climate.

Seasonal changes in numbers

Many tropical animals have well-defined seasonal peaks of abundance. Some have several peaks in a year as shown in Figure 7.13(a) for a cocoa pest, but probably most have a single peak in numbers as shown by a mosquito in Figure 7.13(b). Frequently, the monthly patterns in abundance are similar in most years. Fluctuations in population density may result from (a) the seasonality of mating and production of offspring; (b) seasonal reduction in numbers as the food supply drops (as in Figure 7.13(a)); (c) increased mortality from natural enemies; (d) reduction in places to shelter, or in which to live, e.g. small temporary aquatic habitats (as in Figure 7.13) and (e) migration or dispersal (as in Figure 6.29) or a combination of factors. The frequency of population peaks is usually annual, but not necessarily so (as in the examples in Figure 7.13(a) where there were several in a year).

The butterfly *Acraea encedon* is widely distributed in tropical Africa but tends to occur in isolated colonies. In Sierra Leone, its population is highest at the end of the rainy season and the beginning of the dry season, and lowest at the end of the dry season and beginning of the rains. The most likely reason is that its larval food-plant, *Commelina*, dies back during the dry season but flourishes during the rains. However, in Uganda the distinction between wet and dry seasons is less sharp and the butterfly population does not seem to be limited by the availability of *Commelina*. There are, however, variations in the numbers of male butterflies, and since these are always fewer than the females there are times when there are so few that some females remain unmated and thus fecundity is reduced. This periodic scarcity of males (the cause of which is unclear) seems to be the cause of fluctuations in the butterfly's population density in Uganda. Here is an example of quite different factors affecting population densities of the same species in different places.

The armyworm moth is a major pest of grasses including cereal crops in much of tropical Africa. There are often sudden outbreaks of the moth during the rainy season (pages 167 and 169). In southern and central Tanzania outbreaks occur mainly in December–January, whereas in Kenya they are usually between March and June. These differences apparently arise from different times of emergence and dispersal of successive moth populations. Newly emerged adult moths have been observed flying downwind within two hours of emergence, and this together with the sudden occurrence of increased catches in light traps some 40–50 km downwind, supports the idea that armyworm outbreaks in an area result from invasions of migrants.

Two moths of the genus *Achaea* (*A. catocaloides* and *A. lienardi*) show very marked but irregular patterns of migration and swarming in parts of West Africa. In the Legon area of Accra, Ghana, these two moths are at times so abundant after the first rains in March and April that they become a nuisance. They rest in vast numbers amongst vegetation, on buildings and sometimes enter houses. 'Moth plagues' lasting about a month were recorded in 1969, 1972 and 1973, and in some years it was estimated that there were about a million moths in the Legon area. They do not breed in the infested areas; the probable breeding sites being forested areas north of Legon, where following a build-up in their population there are large-scale migrations. Similar mass migrations by *A. lienardi* have been reported occasionally from East Africa, for example, they were observed swarming in Mombasa, Kenya in 1970.

Migrations also influence local population densities in fish. In the wet season, adult fish belonging to the families Cyprinidae and Clariidae disappear from Lake Victoria and swim up rivers to reach the swamps where they breed. Amongst birds, and to a lesser extent mammals, there are also many species which undergo seasonal migrations (pages 165 and 172).

The response to climate may vary geographically, as with the butterfly *Acraea encedon* mentioned earlier.

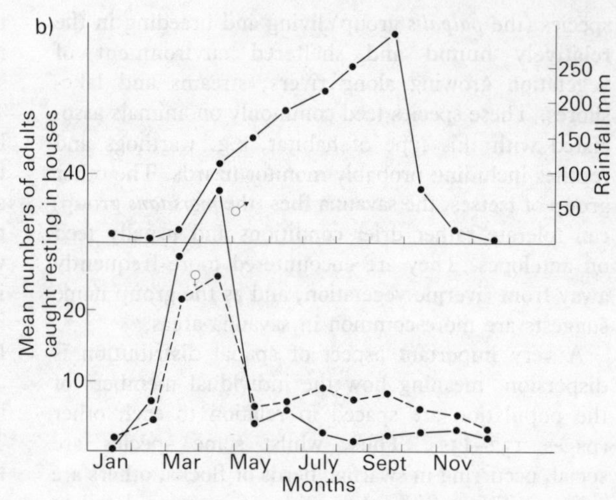

Figure 7.13 (a) On a Ghanaian cocoa farm the population of a moth pest, *Characoma strictigrapta*, was monitored throughout the year. Trees were selected at random and all caterpillars found in one hour were collected. Samples were taken weekly when populations were high, but twice weekly at other times. Results are shown on the lower graph; the upper one shows the percentage of trees on which new buds and leaves were appearing (called flushing)
From Akotoye, N. A. K. & Kumar, R., 1976, *Journal of Applied Ecology*, **13**, 753–73.)
(b) The seasonal incidence of both sexes of the man-biting mosquito *Culex quinquefasciatus* in the Kaduna area of Nigeria, as shown by the mean number of adults collected from houses. Populations rise dramatically at the beginning of the rainy season before rapidly crashing: the explanation is this. The immature stages occur in small, polluted collections of water, and at the beginning of the rains such larval habitats are common. As the season continues the small habitats become flushed out by torrential rainfall leading to a sudden decrease in breeding places and decline in population.
(Modified from Service, M. W., 1963 *Bulletin of Entomological Research*, **54**, 601–32.)

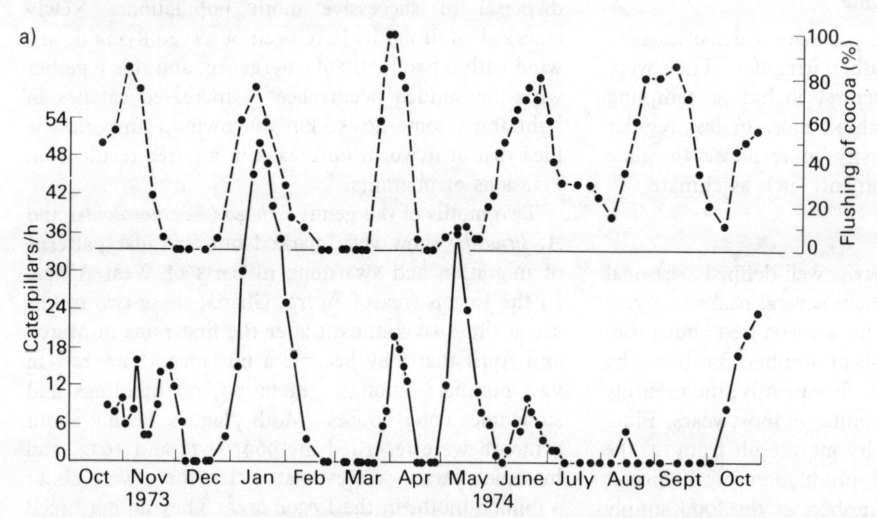

Another example is the common toad *Bufo regularis*, which breeds over most of its range in the wet season, but in Sierra Leone it breeds in the dry season, and may do the same in Liberia, Cameroun and other areas of high rainfall. There are several possible explanations. *B. regularis* breeds in shallow pools, which in Sierra Leone are commonest in the dry season, because they become flushed out during high rainfall. It has also been observed that extreme wetness depresses amphibian activity and there is some evidence that during the rains ovarian development is inhibited. Another possible reason for dry season breeding is that in Sierra Leone food supply may be better in the dry season than during the torrential rains. Whatever the reasons it is clear that climatic conditions affect the breeding activity of *B. regularis* and that in Sierra Leone the population is greatest in the dry season.

Shorter cyclic variations in activity and abundance

A few animals are active both during the day and night, for example some endoparasites of man and certain soil arthropods. A more conspicuous example is the elephant, which feeds during much of the day and night. Impala and several other large mammalian herbivores are also active both at night and during the day. Populations of many animals, however, show a daily or diel periodicity. The term **diel** means a 24-hour day and **diel periodicity** refers to events repeated daily. In moths, cockroaches, frogs, bats and various other mammals as well as birds such as owls, activity is greatest at night, although in many animals activity is greatest during the hours of daylight especially if vision is important in their lives. There are species of marine plankton, such as crustaceans and jellyfish which exhibit marked diel vertical migrations, moving towards the surface at night and descending to lower depths by day. They are followed by fish which feed on them. Obviously, changes in activity can effect the abundance of animals in a particular habitat.

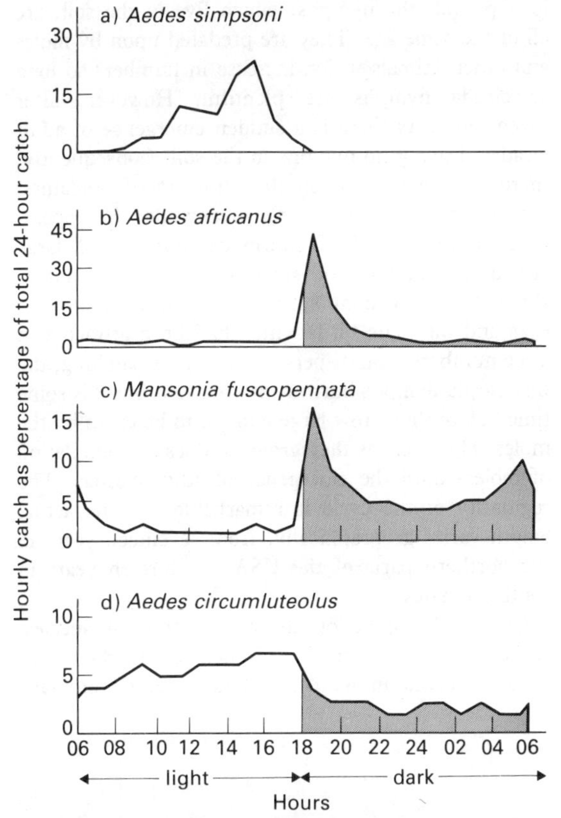

Figure 7.14 Examples of different activity cycles of four African mosquito species. These graphs show the percentage of females arriving each hour of a 24-hour day at a suitable host animal to bite and take a blood-meal. Clearly *Aedes simpsoni* is principally a diurnal feeder, while in marked contrast *Mansonia fuscopennata* and *Aedes africanus* bite mainly at night; *Ae circumluteolus* although biting mainly during the day nevertheless continues to feed throughout the night.
(Modified from Gillett, J. D., 1971. *Mosquitos*, Weidenfeld & Nicholson: modified (a) from Gillett, J. D., 1969. *Annals Tropical Medicine and Parasitology*, **63**, 147–56; (b) from Haddow, A. J. *et al.* 1947 *Bulletin of Entomological Research*, **37**, 301–30; (c) from Haddow, A. J. & Gillett, J. D., 1958. *Annals of Tropical Medicine and Parasitology*, **52**, 320–25; (d) from Haddow, A. J., 1960. *Bulletin of Entomological Research*, **50**, 759–79.)

Another kind of diel periodicity is seen in some birds. For example, in the hot, dry areas of the Sudan and Sahelian savannas, queleas visit their feeding grounds in the early morning and late afternoon. During the heat of the day they shelter at the roosting sites which are also used at night. Hence, although the population of queleas in the general area stays the same, the numbers of birds at any particular site vary greatly during the day.

As well as species whose activity is greatest by night or by day – nocturnal and diurnal respectively – are the crepuscular species which are active at dawn and dusk. Examples include birds such as nightjars, and adults of various mosquitoes.

The biting patterns of mosquitoes can be surprisingly complex (Figure 7.14). An interesting example is the behaviour of *Mansonia fuscopennata* as observed in the Mpanga forest, Uganda. During the day, females of this mosquito bite man and other animals at ground level, whereas at heights of 9 m and above there is virtually no biting activity. However, from about 1800 to 2000 h biting is frequent at heights between 27 and 36 m. Apparently this mosquito makes daily vertical migrations, ascending in the evening and descending towards dawn. However, why it does so is a mystery.

Monthly or lunar periodicities are much less common than daily ones, but striking examples are shown by some aquatic insects in Lake Victoria. Most famous is the lakefly, *Tanytarsus balteatus*, which is a species of chironomid fly. These emerge in swarms shortly after each new moon sometimes forming dense clouds up to one kilometre across (Figure 7.15).

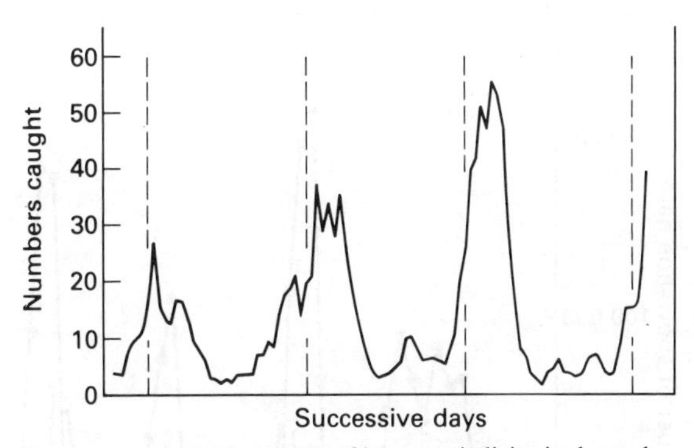

Figure 7.15 A good example of lunar periodicity is shown by the chironomid fly, *Tanytarsus balteatus*, whose larvae are detritivores in the mud at the bottom of lakes. The numbers of adults caught in a light trap sited at Jinja, Uganda on the shores of Lake Victoria over 100 consecutive nights were highest on nights following a new moon (shown by vertical dotted lines), demonstrating a lunar rhythm of adult emergence from the lake.
(From Corbet, P. S., 1958 *Nature*, London, **182**, 330–31.)

Cycles of more than a year

Everyone is familiar with the fact that there are 'good' and 'bad' years for pests – years when they are scarce or common. In some species the year to year fluctuations appear to be fairly regular, and they are referred to as 'cycles'.

The best-documented cycles are those of the lemming, a small rat of northern European moorlands and tundras. Regularly, every three or four years, there is a 'population explosion' of lemmings, and huge numbers emigrate from their crowded subalpine breeding sites to lower ground. Occasionally they attempt to cross inlets of the sea and many are drowned. Records of lemming cycles have been kept continuously since 1826, although the phenomenon was first reported in 1579. Similar cycles, but of various lengths, are known from several other Arctic mammals. Population cycles of a species may be in phase over large areas, or the phases may differ geographically. Sometimes the cycles of predators are of similar duration to those of their prey, as with the snowshoe hare and its predator the Canadian lynx (Figure 7.16).

Another famous cycle is shown by a North American cicada, *Magicicada septendecim*. The individuals comprising the population show a remarkable degree of synchronization. They can take as long as seventeen years to reach maturity, but despite this extraordinarily long period, the nymphs, which live in the soil, are all of the same age. They are predated upon by moles and other animals which increase in numbers so long as cicada nymphs are plentiful. However, after seventeen years there is a sudden emergence of adult cicadas, leaving no nymphs in the soil. Consequently, there is a sharp drop in the numbers of predators, such as moles (Figure 7.17). Meanwhile, the appearance above ground of enormous numbers of large cicadas provides a food surplus for predatory birds. But within two months, the adult cicadas mate, lay eggs and die. The birds which had been attracted in large numbers soon disperse again. After hatching, the tiny cicada nymphs burrow into the soil, but it is some time before they grow large enough to be eaten by the moles. However, as they grow, so does the population of moles, until the emergence of adult cicadas. The regularity of the cycle is remarkable, but its actual length varies geographically, from seventeen years in the northern parts of the USA, to thirteen years in southern states.

Cycles of population density in tropical species, especially forest animals, are much less marked than those observed in temperate and subarctic animals.

Figure 7.16 Population cycles of two North American mammals. Yearly variations in the population of the snowshoe hare, *Lepus americanus*, (.–.–.–.) and its predator the Canadian lynx, *Lynx canadensis* (o–o–o–o). Notice that the hare's population has peaks at regular intervals of about ten years, while the lynx population usually peaks about two years later. Until 1903 the data on hare numbers were based on furs collected by trappers and sold to the Hudson Bay Company, but from 1903–35 census data were compiled from replies to questionnaires. The lynx numbers were derived from fur returns until 1911, then from a variety of sources. Despite the population estimates being based on different methods of assessment it seems likely that they are comparable from year to year.
(Based on MacLulich, D. A., 1937. *University of Toronto Studies in Biology Series*, No. 43, 5–136.)

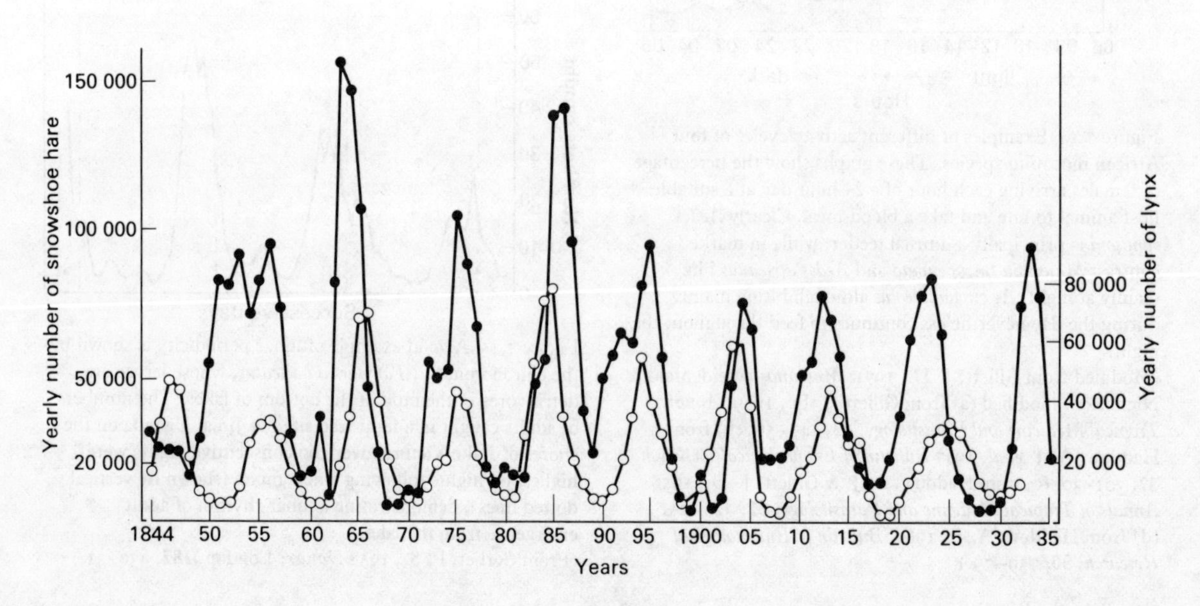

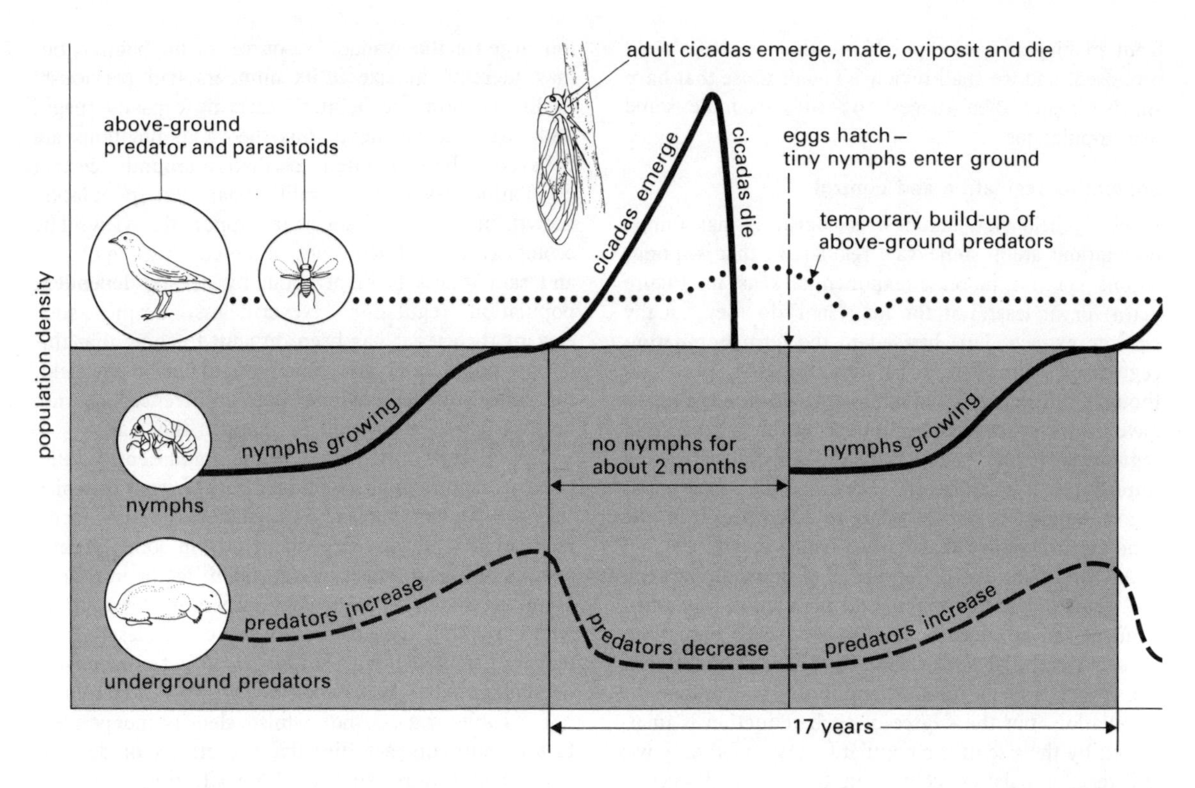

Figure 7.17 Periodical population cycles of the North American cicada, *Magicicada septendecim*. This diagram illustrates the ecological relationships between the cicada and its principal predators below and above ground. For a full account of this 17-year cycle see page 190
(Modified from Lloyd, M. & Dybas, H. S., 1966. *Evolution*, **20**, 133–49.)

Analyses of census data have suggested that elephants may go through five- and fifty-year population cycles, but such claims are poorly substantiated. Short-term studies (eight years) on Nigerian rodents in the Gambari Forest showed that population densities varied by a magnitude of only about four and a half times during this period. More marked annual cycles are, however, recorded in the desert locust (Figure 7.8(d)), and in the tsetse fly, *Glossina swynnertoni*. At Shinyanga, Tanzania, investigations over twenty three years showed some periodicity in the numbers caught, a few years of high catches being followed by several years of small catches, the largest being eighteen times greater than the smallest (page 184). This, in fact, is a comparatively small amplitude of variation, and most other long-term studies on insect populations have shown much greater fluctuations. It was noted that low catches in 1934 and 1944 coincided with years of sunspot minima and high catches in 1937 and 1947 were in years of maximal sunspot activity. On this basis it was predicted that catches would increase after 1948, and they did so. It was also predicted that catches would decrease after 1955, but this did not occur. Thus relationships to solar activity seem unlikely (and would have been hard to explain in any case), but neither could fluctuations of tsetse flies be related to any other component of the flies' environment.

What determines population densities?

This question, of what it is that determines the densities of populations is of major significance to ecological theory. It has practical importance too; an understanding of the underlying principles of population ecology would help enormously in the task of managing animal populations. These considerations have led to many studies being made, usually observational in the field but experimental in the laboratory. The lack of suitable experiments in the field (mainly because of problems in carrying them out) means that none of the various theories has been adequately tested in nature, at least not in Africa. Moreover, as there are believed to be several million species of animals in the world, one might suspect that a single set of principles would not apply to them all. This, though, has not prevented ecological theorists from making the attempt. Many theories have been proposed, and we shall review a few of those that have survived, and then (pages 193–198) examine some real populations.

Population regulation and control

A widely held view amongst ecologists is that animal populations are in some way 'regulated', that is, populations do not increase exponentially as in Figure 7.1(a) or at least not for long, nor do they usually become extinct. This has led to the term **population regulation** being used to describe the processes thought to determine population size. Some ecologists have, however, objected to its use on the grounds that regulation implies some outside control which by implication is purposeful. However, the term population regulation has become so widespread in the ecological literature that its use seems inevitable.

An important distinction needs to be made between the terms population *control* and population *regulation*. Control can be more or less equated with reduction. Thus insecticidal spraying or a severe drought may cause very large mortalities, and in this way reduce the population, but the degree of such reduction is unaffected by the size of the population. We shall only use the term population control in this sense. Processes such as those that change population size without regulating it are caused by density-independent factors, that is, mortality is acting independently of population size. In contrast, the term population regulation is best restricted to describe the decrease, or increase, in population size resulting from density-dependent processes. There is a so-called 'negative feed-back' into the population, that is when population size is increasing, various natural checks on population growth tend also to increase in severity, and as a consequence cause larger mortalities. When, however, population density is low, these natural restraints to population growth are less severe and allow the population to increase in size.

The concept of **density-dependent** and **density-independent** mortalities is well illustrated in Figure 7.18. Here factor A is a cause of mortality which kills a certain proportion of the population (70 per cent in this case), regardless of whether the population density is low or high, thus there is no 'feed-back'. In contrast, mortality caused by factor B increases progressively with the density. Here there is a negative feed-back; as the population grows factor B acts with increasing severity – an essential feature of any regulatory system.

So, we have the theory of population regulation, by which a population is kept close to its carrying capacity. The population is prevented from becoming too large for the available resources of the habitat, but may increase in size if its numbers had previously declined below the habitat's carrying capacity (pages 180–181). According to this theory, populations are believed to be in a state of oscillation around a certain population level; the 'stable phase' of population growth in Figure 7.6 seems to support this view. The ecologist, A. J. Nicholson, accepted this hypothesis and said it was brought about by density-dependent population regulation. Nevertheless, despite such elegant theories it has been difficult to show that this is how populations are actually regulated in the field, but failure to demonstrate their occurrence does not necessarily mean they do not happen this way.

The concept of density-dependent and density-independent factors appears simple, but the ways in which they act can be complex. For instance, the same factor may act in a density-dependent way in some circumstances and as a density-independent factor in others. Here, as so often in ecology, we must be cautious when making generalizations. Nevertheless, it is usually true that living (biotic) factors act in a density-dependent way whereas non-living (abiotic) factors do not (Table 7.2). And whilst density-independent factors normally act directly, the effects of density-dependent factors, such as those affecting fecundity, may not become apparent until the next generation.

Figure 7.18 Suppose two factors, which we will call A and B, cause death in an animal population. Now when factor A operates it causes a 70 per cent mortality rate regardless of whether the population density is high or low, and thus mortality is *density-independent*. In contrast, few, if any, animals are killed by factor B when the population-density is low, but at times or places when the density is high, up to 100 per cent are killed; here *density-dependent* mortality is operating. Note that density-dependent mortalities are not necessarily linear, as in this case.

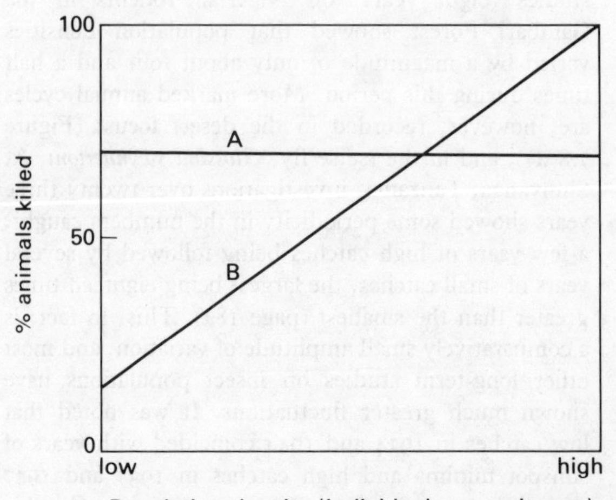

Table 7.2 Ways of classifying animals' environments

Looked at from the animal's point of view (see pages 94 and 95)	Examples	Considered as part of an ecosystem
Climate	Temperature Humidity Wind Light	Density-independent factors
Resources	Food* Nesting sites Drinking places Space	Factors which can act in a density-dependent way
Other individuals of the same species Animals of other species	Mates* Competitors* Parasites* Predators* Competitors*	

* These are sometimes called biotic factors (page 192), the others being non-living or abiotic.

The term delayed density-dependent factors is used to describe such situations.

We shall proceed, on pages 193–199, by considering various aspects of population regulation. Then on pages 200–206 we will outline some of the attempts which have been made to generalize principles of population regulation.

Factors whose effects can be density-dependent

Predators and their prey

On pages 125–126 we considered predation as a component of a prey animal's environment. Now we discuss the ways in which predators may control the numbers of their prey and mention other close interactions that develop in predator–prey systems. Firstly, we will devote some time to general ecological considerations of predator and prey strategies then discuss more specific predator–prey relationships.

The idea that lions must regulate the numbers of antelopes, or that spiders reduce the population of flies, seems so obvious that it was not until comparatively recently that it was seriously questioned. This is because animal populations are surprisingly difficult to study in the field, and further, the relationship between predator and prey really needs long-term studies. There are rather few African studies that meet these requirements, and consequently we shall range more widely in selecting cases that illuminate predator–prey interactions.

(a) *Lions and their prey*. Lion populations have been monitored in the Serengeti ecosystems of northern Tanzania for many years, the most detailed records being from 1966–69 and 1974–77. Most lions were territorial and lived in resident groups known as prides (page 138), but some nomadic lions had no fixed territories; they followed the migratory herds of ungulates, especially wildebeest (pages 170, 172). During the period between 1966–69 and 1974–77 the population of resident lions in the Serengeti study-area increased by about 50 per cent. These lions took a variety of prey species, whose numbers were also increasing. An indication of the improvement in lions' food supply was that more of their cubs survived (Figure 7.22).

During the period when the numbers of resident lions were increasing, the less numerous nomadic lions declined in numbers, although there was a continuous increase in the numbers of wildebeest, their principal prey (Figure 7.7(c)). The explanation appears to be that during this period of increased prey numbers the nomadic lions had joined the resident prides. Thus, the populations of both groups of lions were affected by prey density, but there was no evidence that lions affected the numbers of prey. On the contrary, the lions were concentrating on the old, the sick and the young, most of which would have died anyway – the

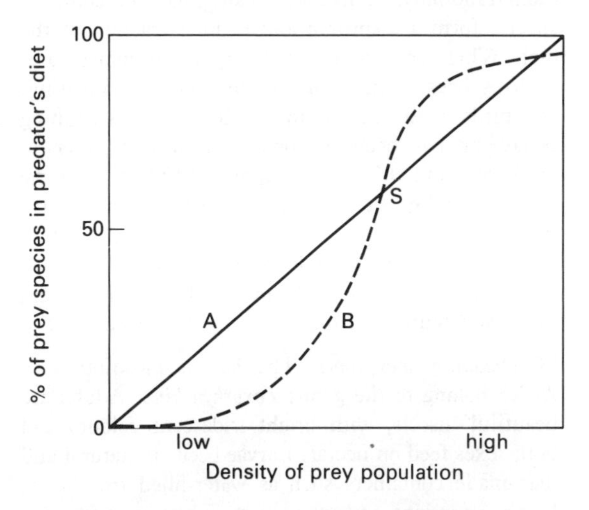

Figure 7.19 The graph shows the percentage of a particular prey species taken by a predator in relation to the prey's density. The simplest situation is shown by line A: as the density of the prey population increases, it forms an increasing proportion of the predator's diet. But it sometimes happens that the predator *switches* its preference as indicated in curve B. Here, few if any of the prey are taken by the predator at times when prey density is low. As the prey population increases, the predator switches to it and a stage S may be reached when this species of prey forms a disproportionately high percentage of the predator's diet. See also Fig. 6.26. (Based on Hassell, 1976.)

latter from intraspecific competition for resources such as food, with the result that relatively few of them would have reached maturity. We will return later (page 196) to consider what does limit the ungulate population since apparently it is not predators.

(b) *Snowshoe hares*. The population cycles of snowshoe hares in Canada were mentioned earlier (page 190). Careful study of Figure 7.16 shows that the number of the predatory lynxes lagged behind those of their prey, typically by a year or two. An increase in prey apparently led to an increase in predators – as with the lions. Did the increase in lynxes cause the subsequent crash in the hare population? Presumably not, because hare cycles, including crashes, were found to occur in places without lynxes as well as in places with them.

(c) *Great tits*. Great tits are small insectivorous birds, widely distributed in the woodlands of the Palearctic region. They feed their young principally upon caterpillars and the particular species of caterpillar eaten is related to caterpillar abundance at the time. But the relationship is not a simple one. Caterpillars which are rare form an even lower proportion of the tits' diet than would be expected if the birds were catching them randomly. Conversely, caterpillars of common species form a disproportionally high fraction of the diet. When one species of caterpillar becomes rare, perhaps because most of the survivors are pupating, the birds then change to another. This **switching** behaviour has been demonstrated in a number of predators, and although the causes may vary with the species, the effect is always similar (Figure 7.19, 7.26(c), (f)). The tendency to concentrate on the commonest species only so long as it remains common means that the effect on the prey is density-dependent (see also Figure 7.24 and search images (page 160).

(d) *Predatory mosquitoes*. The largest mosquitoes in Africa belong to the genus *Toxorhynchites*. Adults are beautiful insects, with bright iridescent colours and both sexes feed on nectar. Larvae occur in natural and man-made containers such as water-filled tree holes, bamboo sections and water pots. They are voracious predators, killing larvae of other mosquito species. The commonest African species, *T. brevipalpis*, was introduced into Hawaii, Samoa and Tahiti in the hope that it would reduce populations of mosquitoes which were disease vectors, but the introduction was not successful. The reason is that, although *Toxorhynchites* larvae are rapacious predators, they are usually unable to substantially reduce and maintain low prey populations. This is partly because *Toxorhynchites* eggs cannot withstand desiccation, and their small container-type larval habitats are very liable to dry out. When reflooded with water *Toxorhynchites* populations take a considerable time to re-establish their former level, whereas most of the mosquitoes upon which they prey have both drought-resistant eggs and shorter life-cycles. Thus, when reflooded, these mosquitoes can recolonize the habitats quickly and complete the aquatic phase of their development before *Toxorhynchites* have re-established themselves. On the other hand, there is no doubt that these predatory mosquitoes can destroy large numbers of mosquito larvae, each killing some 150–200 mosquito larvae during its development.

An interesting phenomenon is that, about two to three days prior to pupation, *Toxorhynchites* larvae kill all prey larvae in their habitat, although they may not be eaten. *Toxorhynchites* larvae are also often cannabalistic prior to pupation. It has been suggested that both types of behaviour, by reducing the chances of survival of other members of the species, increase their own survival rates.

These four examples have concerned predators taking animal prey in the field. The next cases illustrate the value of laboratory experiments, and the convenience of invertebrates as subjects.

Laboratory experiments

Predatory mites are one of the more conspicuous groups of predators amongst the smaller invertebrates of many terrestrial habitats. Many other mites are herbivorous. The experiments whose results are shown in Figure 7.21 concern one species of predatory mite and its prey, a herbivorous mite feeding on oranges. When oranges were readily accessible to the herbivorous mites, their numbers increased rapidly – but the same happened to their predators (Figure 7.20(a)). Soon, the predators had eaten all their prey and consequently died of starvation. When, however, the oranges were made less accessible and most of their surfaces rendered unsuitable for feeding, a different sequence of events occurred (Figure 7.20(b)). The prey species now had difficulty in finding food whilst the predatory mites had trouble finding the scattered populations of herbivorous mites, and spent much potential feeding time in searching for them; they experienced a relative shortage of food (page 108). Some prey populations were not discovered, others were found only after they had reproduced and some young already emigrated, perhaps to found new colonies.

This example illustrates what seems to be a very important point. That is, so long as there are some places where prey remain inaccessible to predators, then populations of both species may persist indefinitely. Such situations can arise in at least two ways.

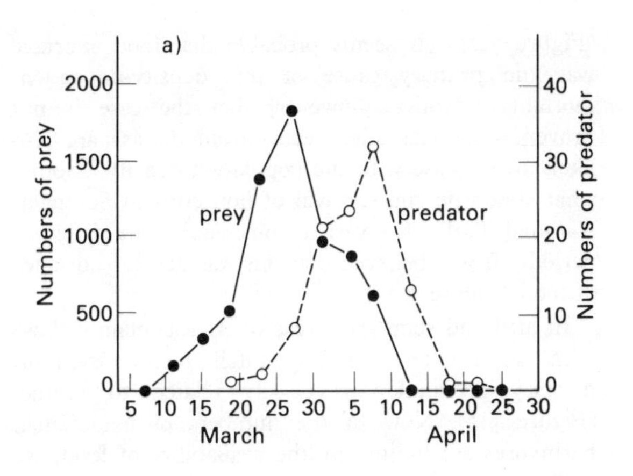

Figure 7.20 Experiments with two species of mites, one a predator and the other the prey (a herbivore). In (a) the prey could easily find the oranges that they ate, and the predators easily found their prey. The experiment resulted in the extinction of both populations. In a second experiment (b) only small and scattered patches of orange skin were exposed, and the prey had difficulty in locating them. The predators also had trouble finding the prey, so that some survived to breed and their populations persisted for much longer than in the simple environment of experiment (a). (From Ewer & Hall, 1978, reference in chapter 1.)

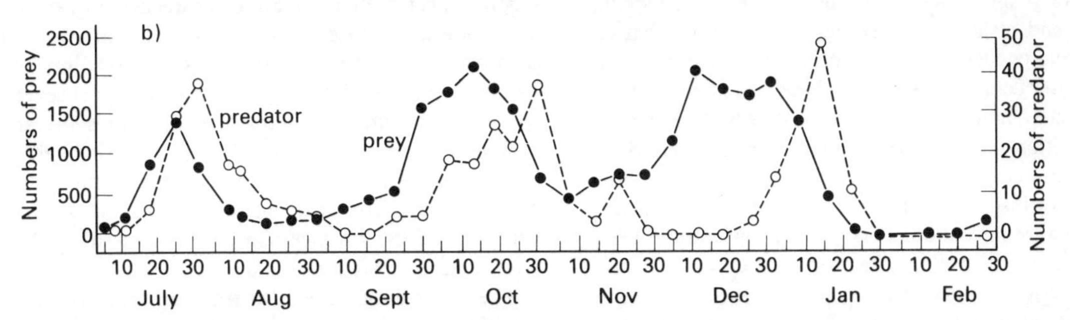

The first was illustrated by the herbivorous mites, where under certain conditions populations were scattered, so that, by chance, some escaped detection, at least for long enough for them to reproduce. Alternatively, there may be physical hiding places (refuges) where the prey are safe from the predator. For example, domestic cats can easily catch lizards in the open, but some lizards manage to escape by sheltering under rocks and in crevices. So that although some are caught by cats, or other predators, the refuges allow the population to avoid extermination. There are many other methods that have evolved to protect prey from predators and some of these are described on pages 125 and 126.

So far we have considered predators and prey in the conventional sense. Similar relationships exist between parasitoids and their hosts. (Parasitoids, despite their name, are really a type of predator; females lay their eggs on or near other insects. The larval parasitoids then develop within the host, eventually killing it just before it pupates. Most parasitoids are host specific.) Both predators and parasitoids kill their prey, whereas parasites do not usually kill their hosts, and herbivores do not generally kill their food-plants. Both parasites and herbivores resemble predators and parasitoids in that they can act as density-dependent agents in regulating population size. The general case is that of the effect of the consumer upon the consumed, of one trophic level on that below it. For instance, intensive grazing by livestock causes plants ('prey') to use a proportion of their energy reserves to replace the parts eaten. Thus, they have less energy available for reproduction, fewer offspring are produced, and their populations consequently decline, a good example of a delayed density-dependent effect. In due course, there is a change in the composition of the pasture, unpalatable plants gradually increasing as the palatable ones decline. In other instances, plant predators such as fruit-eating herbivores destroy a number of seeds, but at the same time they assist in seed dispersal which may result in colonization by the plant of areas where there is less competition for light or soil nutrients.

Resource limitation and competition

All organisms require certain things, which we call resources, from their environments (pages 106–112). Resources are not unlimited, so if populations which are near the environment's carrying capacity increase or the resources decrease a stage may be reached when resources become inadequate so there is no longer sufficient for all. We then say that competition is occurring. Resources can act as a density-dependent factor because their shortage becomes more severe as the demand increases.

Interspecific competition is difficult to demonstrate in nature and so we do not really know much about its frequency in natural ecosystems (pages 120–123), although it can undoubtedly happen. Although in the laboratory it is not too difficult to demonstrate that interspecific competition (pages 119–120) leads to the elimination of one species by another – an example of the competitive exclusion principle – it is much harder to show the existence of interspecific competition in natural populations. There seem to be several reasons for this, for example, as has already been stressed, it is unlikely that two or more species will have identical requirements, and thus compete for the same resources (pages 61–64). Another point is that in the field we may be looking at the end result of interspecific competition, that is, observing populations of the species that have survived, the poorer competitors having already been eliminated. We may, however, expect to detect interspecific competition where a species has been introduced. A much-quoted example comes from Hawaii where three species of insect parasitoids, all belonging to the genus *Opius*, were introduced to control an agricultural pest, the oriental fruit fly. The first species to be introduced was not very efficient at parasitizing the fruit fly and when a more efficient parasitoid was introduced, the first was virtually eliminated. Finally, a third parasitoid species was established and this caused a much higher percentage of parasitism and both earlier species were eliminated. (Interspecific competition is discussed in more detail on pages 118–124).

In contrast to interspecific competition, intraspecific competition is more commonly observed and often intense: the large numbers of offspring produced by many species, relative to the numbers required to maintain populations near their optimum size, makes this inevitable at times. (Not always, however, because climatic disasters and predators may kill many offspring.) The defence of a territory can be considered as a means to ensure that at least some members of the population will receive an adequate food supply, and there is some evidence that territory size is related to food requirements (page 108). Three further examples will serve to illustrate some ways in which resources can affect animals' numbers.

The increase in wildebeest in the Serengeti area has been described already (Figure 7.7(c)). At first the increase represented a recovery from earlier years of drastic reductions by the exotic cattle virus, rinderpest. Subsequently a number of years with above-average rainfall helped to sustain the increase. Studies in these later years showed that the mortality rate was much higher at times when food supply was low (Figure 7.21). It seems probable that food shortage was the primary cause of this density-dependent mortality. (Notice, however, that the case is not proven – for example, deaths from disease are also likely to increase with the population density.) Somewhat similarly, the survival of lion cubs in Serengeti National Park also varied considerably in different periods. It was believed that this was due to shortages of food (Figure 7.22).

In arid and semi-arid areas of Africa, mainly those with an average annual rainfall below 700 mm, primary production is closely related to rainfall (Figure 3.5). Now, if the numbers of mammalian herbivores are limited by the availability of food, we might expect their numbers to be related to rainfall, as is indeed the case (Figure 7.23). This is a further example of a classic density-dependent relationship. Notice here, that although at most of the sites several species of mammalian herbivores were present, and competition probably occurred, much of it is likely to have been intraspecific, as in the case of the wildebeest. A probable explanation is that most wild herbivores differ (to a certain extent, at least) in their food preferences (pages 120 and 122).

The third example is also from a harsh arid area, the Namib desert in southern Africa, where average annual rainfall is less than 100 mm. A severe drought

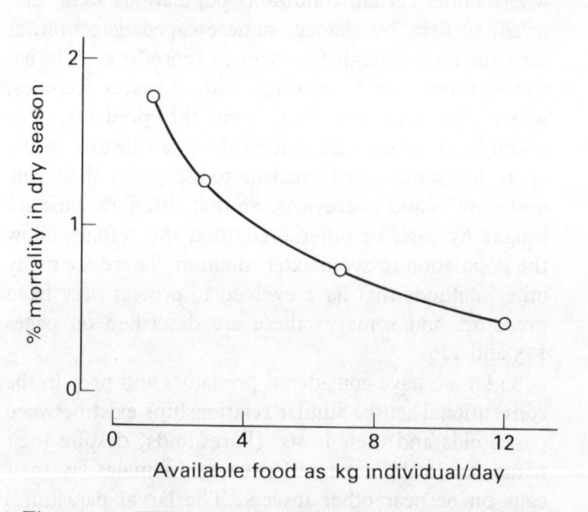

Figure 7.21 Wildebeest mortality in Serengeti National Park, Tanzania, in relation to food supply. Most deaths occur during the dry season, when the net primary production is zero and the herbivores have to survive on what grew during the rains. Here, monthly mortality rates in the dry season are plotted against the amount of food available to the wildebeest, measured in kg individual^{-1} day^{-1}. Each point represents the mean for one year. (Drawn from data of Sinclair & Norton-Griffiths, 1979.)

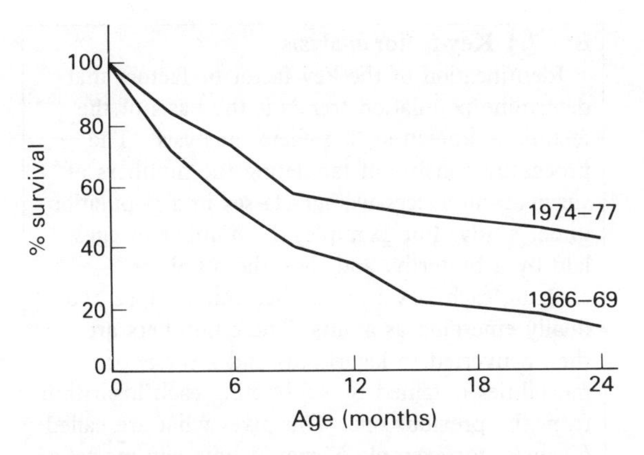

Figure 7.22 Numbers of lion cubs surviving up to the age of two years, in two prides in Serengeti National Park, Tanzania. During 1966–69, only 14 per cent of the cubs reached an age of two years. But in the wetter period 1974–77, about three times as many (44 per cent) were still alive after two years. This difference was probably caused by increase in the prey populations on which the lions fed. When prey is short, there is no food left for the cubs after the adults have eaten.
(Sinclair & Norton-Griffiths, 1979.)

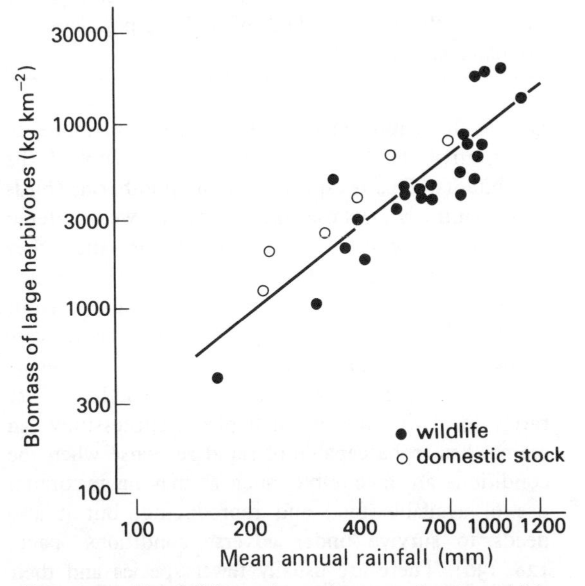

Figure 7.23 Rainfall determines primary production, which in turn limits the population of herbivores. In this graph the herbivores were large mammals, and their populations were estimated in terms of biomass per square kilometre. When converted to logarithms (note the scale) the biomass was closely related to rainfall. Notice that all six values for domestic stock are above the fitted regression line, suggesting that these pastoral areas were over-stocked. Most of the values plotted were based upon more than one species of mammal, especially in the wildlife areas. (Modified from Coe, M. J., *et al.*, 1976. *Oecologia*, **22**, 341–53.)

which occurred the year previous to a study on the recovery of several populations of three rodent species was responsible for causing reproductive failure and reducing their population densities to very low levels. During the study period rainfall was higher and plants grew vigorously allowing populations to recover rapidly. Exceptional rates of increase were found in one population of the Namaqua gerbil, which approximately doubled its population each month. Rates of increase varied considerably between different colonies of the same species, much depending on their reproductive success and survivorship. In most of the three rodent species, rates of population increase were positively correlated with birth rates and inversely with survival rates. In other words, survival rates were lowest during times of rapid population growth, and higher when populations were declining.

Thus as in other studies on desert rodents, survival rates were generally highest during adverse environmental conditions. This paradoxical situation probably arises because during hot, dry times the rodents spend less time on the surface searching for mates and food, but longer seeking refuge in their burrows and thus are less exposed to predators. This interesting study clearly shows that rainfall in causing a lack of resource – food – was responsible for the rodent population crashing. But when high rainfall caused an abundance of food there was no resource limitation on population growth and populations increased explosively to high levels. Although density-independent factors resulted in large changes in population size, density-dependent factors also operated in the population, for example birth rates were inversely correlated with survival rates.

Density-independent factors

There is no dispute amongst ecologists about the potentially important effects of climate on populations, or that these effects are density-independent by definition. In fact, climate and other density-independent factors are often very important in causing changes in population size. For example, many butterflies will fly and mate only in sunny weather, and if, as in Britain, the mating period is relatively short, continuous dull weather substantially reduces the numbers of matings and consequently the numbers of eggs laid. In this case, densities of the butterfly populations are not responsible for their reductions in size. Similarly, eggs of many *Aedes* mosquitoes are laid on damp leaf-litter and debris at the edges of the water line of aquatic habitats. Now, if some time later the water level drops, then many eggs remain unflooded and are unable to hatch. Thus, here is an example of weather, acting in a density-independent way, being responsible for variation in population size of *Aedes* mosquitoes.

In these examples the most important factors influencing population size were sunshine and rainfall, and can be regarded as the 'key factors' responsible for population changes (see Box 7.4). But we must remember that although weather in these cases was the main factor causing changes in population size, being density-independent, it cannot regulate population size (see page 192).

Before proceeding further it is worth stressing again that the various factors which affect population densities interact in many ways, some obvious, but others more subtle and complex. There can also be more than one way of looking at a situation. The dependence of mammalian herbivores upon primary production, and the limiting effect that this can have on their populations, is clear. But equally it could be argued that the animals' populations are limited by rainfall albeit indirectly, as implied by Figure 7.23. For instance, since a shortage of grass during drought periods causes mortality in the herbivores, its effects could be described as density-independent, because a shortage of grass affects populations irrespective of their density. But such shortages can also result in populations aggregating in a few favoured places where some grass remains. In these places there is likely to be severe competition for the grass, and thus mortality in this sense could be regarded as density-dependent. Hence the distinction between density-independent and density-dependent processes is not always clear, at least initially.

This leads to two useful conclusions. Firstly, that it is more important to understand the processes involved in a particular situation than to simply categorize and label them. Secondly, it can be misleading to draw conclusions from correlations, particularly when only two variables have been considered.

Box 7.4 Key-factor analysis

Identification of the key factor or factors that determine population trends is the basis of the technique known as 'key-factor analysis'. The procedure consists of tabulating the numbers of survivors in successive age-classes in a population under study. For example, the number of eggs laid by a butterfly, and then the numbers entering each larval instar, becoming pupae and finally emerging as adults. These numbers are then converted to logarithms and a series of mortalities obtained by subtracting each logarithm from the previous one. This gives what are called k-values, for example k_0 may denote egg mortality and k_4 mortalities of 4th-instar larvae, adding these gives total generation mortality (K). The k-values for the different age-classes and for total mortalities (K) obtained for several generations are plotted so that the k-value contributing most to K, can be identified visually. The key mortality factor is the mortality that contributes most to the animal's total mortality. The ecologist tries to determine what causes the mortality. The analysis of population mortalities by the k-factor technique is one of the better methods of studying population dynamics.

Some evolutionary aspects

The stability of an ecosystem is a significant factor in determining which species inhabit it, and how common they are. Some species are able to survive in unstable ecosystems which experience large and sometimes unpredictable changes, such as ponds and deserts. Others exploit the nearly constant environments of stable ecosystems like rainforests and large lakes. An insight into these matters is helpful in comparing population dynamics in different ecosystems.

Ecosystem stability

Stable ecosystems are those whose physical environment remains more or less the same for lengthy periods. Animals (and plants) living in such places are likely to experience intensive intraspecific competition, and also possibly some interspecific competition. The lack of fluctuation in the environment contributes to the relative stability of populations of many of the inhabitants. This is especially so of endotherms (birds and mammals), whose physiology is well-buffered from their physical environment. For instance, they are relatively independent of variations in temperature (pages 102–105). Stable ecosystems also favour species diversity (pages 65–66), the classic case being lowland tropical rainforests.

Conversely, in unstable ecosystems, conditions are rarely ideal. To live in such places successfully, an animal has to be capable of rapid response when the conditions are favourable, such as making maximum use of available food and reproducing, but it also needs to survive under adverse conditions (pages 128–130). There are usually fewer species and their abundance is subject to wide fluctuations. Here again endotherms have an advantage, since they can regulate their internal environments over a much wider range of ambient temperatures than most exotherms, and because of this their population size fluctuates less. In contrast, many exotherms survive low temperatures in temperate zones by entering into a state of dormancy, few endotherms use this strategy.

Ecosystem stability is discussed in greater detail in chapter 4.

r- and K-selection

Species inhabiting unstable ecosystems are subject to rather different selective pressures from those of more stable ecosystems. Ecologists studying animals in rainforests have come to expect that these animals will live longer, grow more slowly and produce fewer offspring, than animals living in savannas. In fact a whole series of adaptations are associated with the degree of ecosystem stability. Species in unstable ecosystems are subject to what has come to be called *r*-selection, the term reflecting their high innate capacity for increase, r_m (pages 176–177). In stable ecosystems, where many species have populations at or near the carrying capacity (*K*), their adaptations are typically a product of *K*-selection. This has led to the idea that there are two extreme types of life-history strategies, one mainly characterized by *r*-selection and the other by *K*-selection pressures. Typical *r*-selected animals, are the opportunistic colonizers which often live in temporary and other less favourable habitats (Box 7.5). Many insects are *r*-type strategists especially pests, such as mosquitoes, locusts, armyworm moths and aphids. Their populations often increase explosively, then crash to low numbers, but they are very resilient and can increase their numbers quickly. This ability to recover rapidly from low population numbers, is one of the reasons why such pests are often difficult to control.

At the other extreme are the *K*-type animals which live in more permanent unchanging habitats. Many birds and mammals are *K*-type species as are a few insects, such as tsetse flies (*Glossina* spp.), cicadas and the large blue butterflies of the genus *Morpho* which are common in South America.

Of course, no ecosystem is absolutely stable: there are degrees of stability. Hence few species exhibit all the characteristics of *K*-selection, whilst few of the species from unstable ecosystems have every *r*-selected characteristic. Rather, the features typifying *r*- and *K*-selection can be thought of as extremes of a continuum, with most species being somewhere in between, so you can say a certain species is more '*r*' or more '*K*' than another. Thus a forest tree has the typical *K*-selected slow growth rate and long life, but most trees produce enormous numbers of seeds, more typical of *r*-selected species. Most weaver birds breed once a year and produce only two or three young; but their close relative *Quelea quelea* (page 161) has evolved a short generation time. Its reproductive strategy has evolved from the broadly *K*-selected pattern of most weavers towards an *r*-selected one, and it resembles insect pests in its ability to increase in numbers rapidly at times of food abundance.

Ecosystems undergoing successional changes are by definition unstable, and their early colonizers typically

Box 7.5 A summary of features of typically *r*- and *K*-selected species

Typical of *r*-selection	Feature	Typical of *K*-selection
unpredictable or ephemeral	**physical environment**	constant or predictably seasonal
variable and usually a fraction of the carrying-capacity	**population size**	fairly constant, always at or near the carrying-capacity
often high, mainly caused by density-independent factors	**mortality-rate**	usually low, often caused by density-dependent factors
type III – short life-span	**survivorship** (Figure 7.4)	type I or II – relatively long life-span
efficient – long distance dispersal and/or migratory	**mobility**	offspring disperse only locally; non-migratory (but exceptions, e.g. some birds)
little and infrequent	**competition: intra- and inter-specific**	intense and frequent
rapid development, high r_m	**consequences of natural selection**	slow development, low r_m
reproduction early in life, once only,		delayed reproduction, repeated every season,
producing many small eggs – hence high productivity		with few large eggs or young – hence high efficiency

show many features of r-selection. After a time, if a more stable environment is established, K-type selected species are more favoured than r-type selected ones. Consequently the populations of species having K-selected characters increase at the expense of the less competitive r-selected species.

Scramble- and contest-competition

We have seen that interspecific and intraspecific competition are more frequent in K- than r-selected species (Box 7.5); here we are concerned with intraspecific competition of which two major types can be recognized. When individuals compete for food or certain other resources, the competition may be passive, or it may be contested, such as fought for; the former is referred to as **'scramble-competition'** and the latter **'contest-competition'**. An example of scramble-competition is caterpillars competing for food plants. All caterpillars eat as much as they can until the food is depleted, but when this happens, there may be insufficient for their life-cycle to be completed, and consequently the entire population may die. Scramble-type competition often contributes to large fluctuations in population size.

In contrast, when there is active competition between animals for food, such as between predators for their prey or scavengers for carcases, the dominant individuals usually get the food while the remainder have very little or none, and may die. Here the fittest survive, but at the expense of the weaker animals dying of starvation. There may also be competition for breeding sites or other resources (cf. Figure 7.22).

Theories of population regulation

For at least fifty years ecologists have vigorously debated the different ways in which they think the sizes of animal populations are determined, and even today there is no general agreement on how this happens. Amongst early proponents were A. J. Nicholson, and H. G. Andrewartha and L. C. Birch, but many others have joined in the controversies. There was a profusion of population theories between 1950 and 1960, and two rather different points of view were presented in major textbooks which both appeared in 1954. One was 'The Distribution and Abundance of Animals' by the Australian scientists Andrewartha and Birch; the other was 'The Natural Regulation of Animal Numbers' by D. Lack of England.

Andrewartha and Birch based much of their writing on their experiences with insects in South Australia, which has a highly seasonal climate with hot, dry summers and cool, moist winters. They concluded that population size is limited mainly by a shortage of time during which the population can increase (i.e. when r is positive, there being more births than deaths). The lack of time was caused mainly by the shortness of the favourable season – at most times the weather was less than ideal as required for r to approach r_m. They also concluded that population size could be determined, though less often, by relative or absolute shortages of resources such as food, shelter or resting-places. But they stressed that intraspecific competition for a resource could operate only when the population size was sufficiently large to affect the supply adversely, whereas, in reality, many populations remain at low densities for most of the time, so that competition is infrequent.

They also concluded that the distinction between density-dependent and density-independent factors is often not simple, because most mortality could be affected by density at some stage. For example, it might appear at first that adverse weather can cause only density-independent mortality, because its affects all populations equally, irrespective of their size. But during a drought, when there are fewer places where animals can drink, fierce competition may happen amongst the few that do remain, resulting in a form of density-dependent mortality. Andrewartha and Birch argued that the distinction between the two types of mortality was often unclear, and hence the terms would be better abandoned in favour of a 'components of environment' approach (page 94). This is based upon actual descriptions of the factors affecting populations, rather than labelling them with abstract terms. They also made a strong case against the use of terms such as 'balance' or 'ultimate limits' in discussions of animal populations, pointing out that such concepts are not susceptible to scientific experiments – and hence are not science.

Andrewartha and Birch's book was noteworthy as being the first general ecological work to emphasize the importance of rigorous scientific methodology, and to demonstrate the value of thorough statistical analysis in interpreting ecological data.

David Lack's book was based largely upon his extensive knowledge of birds. Many species of birds are known to have relatively constant populations from year to year, and this led Lack to argue strongly for density-dependent factors as being the key to understanding population dynamics. Although he considered that density-dependent processes were essential to population regulation, it has to be pointed out that he failed to prove it.

A more quantitative case for the overriding importance of density-dependent factors was made by a third

Australian ecologist, Nicholson, whose main work – by a remarkable coincidence – was also published in 1954. However, Nicholson had begun to develop his ideas much earlier, and an important paper published in 1935 with his co-worker Bailey foreshadowed a line of thinking which has gradually developed into the most general approach of the 1980s. This is to base ecological theory on mathematical models.

Nicholson used models of competition and predator–prey relationships as a basis for his arguments. He wrote of populations having an equilibrium level, or balance, but did not think of this level as being constant. Rather, he believed that there was a dynamic equilibrium level which altered according to changes in the environment, i.e. changes in carrying capacity (although he did not use that term). Thus, in the rainy season when food is abundant, populations have a higher equilibrium-level than in the dry season, when the equilibrium level is much lower. Nicholson believed that density-dependent mortalities regulated population size, although other factors which are not density-dependent, notably weather, could greatly affect the level at which the population was regulated. Some ecologists argue that there is no really satisfactory way of determining whether or not a population is stable, for example does stability imply a small variance about the mean population size, if so who decides what is small?

The earliest attempts to model the relationships of predators and prey had been those of A. J. Lotka and V. Volterra. (Here was another remarkable coincidence of timing. Lotka, an American and Volterra, an Italian, did not know each other, yet produced their ideas at almost the same time (1925–26)). They started from the assumptions that (a) the birth rates of predators would increase as the population of their prey increased, and (b) the mortality rates of prey would increase as populations of the predators increased. Their arguments and equations have become classics in the literature of ecology (Box 7.6).

One of the assumptions implicit in the Lotka–Volterra equations is that individuals from both predator and prey populations breed continuously: in other words that they have no distinct breeding seasons. Nicholson

Box 7.6 The Lotka–Volterra equations

Imagine a simple ecosystem with two species, a predator and its prey. We will call their population sizes at a particular time, P for the predator's population, and Y for that of its prey. Now suppose that their populations are not constant, but change with time, t. The conventional way of representing a rate of change is as a differential equation; so we write $\frac{dY}{dt}$ meaning 'rate of change of Y with time'.

To see how the population of the predator might be affected by the prey we shall make several assumptions (these are necessary to keep the model relatively simple):

1 The intrinsic rate of increase, r_m, of the prey is the principal driving force. The rate for the predator is assumed to be unimportant – it never obtains enough food to achieve its intrinsic capability. But,

2 The death rate of the predator, D, is important because when there are too few prey, predators die.

3 The proportion of the prey caught by the predator, C_1, is assumed to be constant; it is low if the prey are good at escaping. The efficiency with which the predators catch their prey (C_2), is also taken to be constant.

Now we can write an expected rate of change of the prey population:

$$\frac{dY}{dt} = r_m Y - C_1 YP \quad (1)$$

This is equal to the prey species' initial population, times its intrinsic rate of increase, less the numbers that are predated. The latter obviously depends on the numbers of prey, numbers of predators, and the predation-rate, hence we have $C_1 YP$.

Similarly, changes in predator population are:

$$\frac{dP}{dt} = C_2 YP - DP \quad (2)$$

That is, the predator's population is determined by the amount of food obtained from its prey (which will allow its own population to increase), less its losses from deaths, (D) times P. (The term $C_2 YP$ simply combines the numbers of prey and predators with the utilization of prey.)

Equations (1) and (2) are commonly called the Lotka–Volterra questions. Theoretically equations can be used to predict actual changes in numbers of predators and prey with time, in a two-species ecosystem. In practice, the numbers of both species would oscillate, with the predator population reducing the numbers of prey until it begins to suffer a shortage of food itself. As the predators starve and die, the pressure on the prey population is lessened, enabling them to increase again, and so on (Figure 7.24(a)).

worked in a highly seasonal climate and he rejected the Lotka–Volterra view in favour of a model in which both predator and prey have fixed breeding seasons, leading to discrete generations. His model was developed with co-worker V. A. Bailey from observations on parasitoids searching for host eggs. All models need to make some simplifying assumptions, or they would become too complex. In the Nicholson–Bailey models, the main assumptions are that (a) the parasitoid searches randomly for hosts, which are randomly dispersed; (b) the parasitoid has a limited range – its 'area of discovery'; and (c) the host, in this case an egg, does not respond, e.g. by flight. The model can be generalized by considering parasitoids as equivalent to predators, and hosts as equivalent to prey. Whereas the Lotka–Volterra model predicts stable oscillations in the numbers of both predators and prey (Figure 7.24(a)), the Nicholson–Bailey model predicts an unstable relationship. Any external factor, such as adverse weather killing some of the prey, creates larger oscillations, leading to either a population explosion of the prey, or the extinction of both predator and prey (Figure 7.24(b)).

The Lotka–Volterra and Nicholson–Bailey models fit very few actual sets of data, for instance most observations have shown that population size remains considerably more stable than predicted by these models. This is mainly due to the models' oversimplified assumptions, and the omission of important variables such as hunting and handling times of predators, immigration and emigration, and intraspecific competition. These models can also be criticized because they are derived from formulae that do not take into consideration any components of the animals' environment except the population densities of the two species. The result of these rather simplistic models is that coupled oscillations of abundance arise only from direct predator and prey interactions, leading to the prediction of much greater population instability than actually occurs.

Nevertheless, the idea that models can provide useful insights into population fluctuations is still valid. Subsequently ecologists have tried to build models which are more elaborate, but at the same time more realistic. From 1960 onwards, an ever-increasing number of models has been proposed, and some of these are beginning to describe the behaviour of actual populations quite accurately. In doing so, the models require data on all components of the animals' environments, and consequently the models themselves increase in complexity. Figure 7.25, for example, presents the predictions of a rather more sophisticated model than we have considered, although it is much less complex than some.

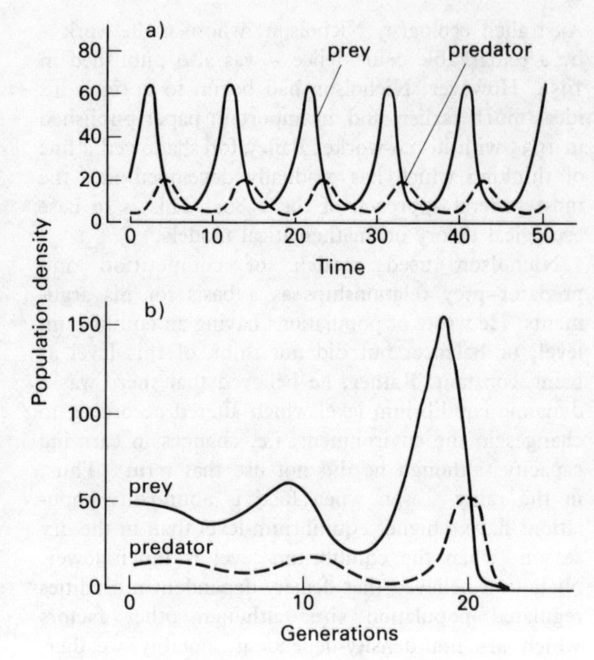

Figure 7.24 (a) numbers of predators and prey as predicted by the Lotka–Volterra equation and (b) by the Nicholson–Bailey model. For the Lotka–Volterra model, initial densities of twenty prey and four predators were taken, C_1 was set at 0.1, C_2 at 0.5 and r_m at 1.0. The oscillations continue indefinitely with peaks in numbers of predators following those of the prey (cf. Figure 7.16). The more complex Nicholson–Bailey model is based upon discrete generations. It has more than one possible outcome, depending upon the values of such variables as rates of increase and search area of the predator, or parasitoid. The case illustrated is an unstable one, with oscillations of increasing amplitude leading to extinction of both predator and prey after 22 generations.
(Figure 7.24(a) from Krebs, 1978; Figure 7.24(b) from Colinvaux, 1973, references to both in chapter 1.)

No discussion of predator–prey interactions would be complete without at least a brief reference to the effect of prey densities on the prey death rate inflicted by predators. That is the change in numbers of prey attacked by a single predator in unit time when the initial prey density is varied. The variations in prey consumption arising from such different situations have been referred to as 'functional responses'. Theoretical responses as well as actually observed ones are illustrated in Figure 7.26.

The advent of computers (and of ecologists who are not shy of mathematics) has made the use of complex models quite practicable. At present, the greatest constraint to the development of really efficient models is a shortage of reliable field data. Any model is only

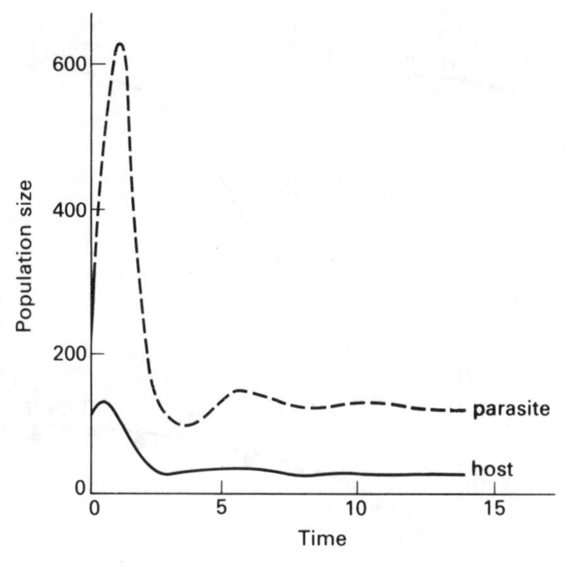

Figure 7.25 This figure is based on a more complicated model than the Lotka–Volterra one depicted in Figure 7.24 and concerns host–parasite numbers (notice that parasite populations can be greater than those of their hosts, an unlikely situation with predators and their prey). Like the Lotka–Volterra model, this one is based on mathematical formulae, but in this case they allow variable birth and death rates of host and parasite, including the possibility that hosts may die either from parasitic infections, or from other causes. It assumes that the parasites are aggregated, and incorporates a term for this (*k* of the negative binomial series), and it also incorporates a measure of transmission efficiency. This particular model, unlike that shown in Figure 7.24(b), exhibits oscillations which are dampened and give way to stable equilibria in populations of both host and parasite. (From Anderson, R. M. & May, R. M., 1978. *Journal of Animal Ecology*, **47**, 219–47.)

as good as the information that is put into it. Nevertheless, it will probably not be too long before ecologists are making reasonably accurate predictions of future population trends in some of the more important species, and in this way contributing more to conservation and pest management.

Conclusions

Animal populations behave in ways which are usually complex; the variations in population size often seem to be erratic and difficult to explain. In this chapter we have discussed some of the principal causes of population changes, and taking these into consideration, but also incorporating relevant information published elsewhere on population dynamics, we now summarise our conclusions on the growth and regulation of animal populations.

1 When introduced to a new and favourable habitat, all species increase in numbers geometrically at first, but the rate of increase soon slows and, in an ideal environment (with replenishment of resources) reaches an equilibrium level (Figure 7.1(c)).

2 In practice, environments do not remain ideal, and population densities vary in both space and time, sometimes seasonally or cyclically, but often in less regular ways. The degree of variability differs considerably, but populations in 'stable' habitats are less variable in their size (Figures 7.6, 7.8).

3 Fluctuations in numbers are a consequence of interactions between the animals and the components of their environment.

4 Population densities of exotherms, the so-called cold-blooded animals, tend to show greater variability than those of endotherms; exotherms tend to have more *r*-selected characteristics, while in endotherms *K*-selection is more pronounced. This is presumably because endotherms maintain a fairly constant internal environment: consequently they are less influenced than exotherms by outside changes, especially of weather.

5 The differences between exotherms and endotherms partly explain the controversy which grew up between the 'density-dependent school' of ecologists such as Lack, who studied birds, and the 'climatic school', following Andrewartha and Birch, who studied insects. Present-day ecologists have come to accept that there are many sorts of interactions between animals and their environments, so it is unreasonable to expect that all populations will behave similarly. This is not to say that general principles do not exist: rather, they operate to different degrees with different species, and sometimes their significance varies, within a species, and with time and place as well. Figure 7.27 is an imaginary illustration of some of these facts.

6 The predator–prey relationship is a good example of how a general principle can be developed. In an ecosystem with only two species of animals, one predator and one prey, population sizes sometimes agree quite well with mathematical predictions (Figures 7.25, 7.26). Under these circumstances, there is clear density-dependent regulation of prey numbers and predator numbers. Complications arise because nature is more complex. For instance:

(a) Even a small change of climate will be likely to alter the pattern of abundance, because the responses of the two species will differ.

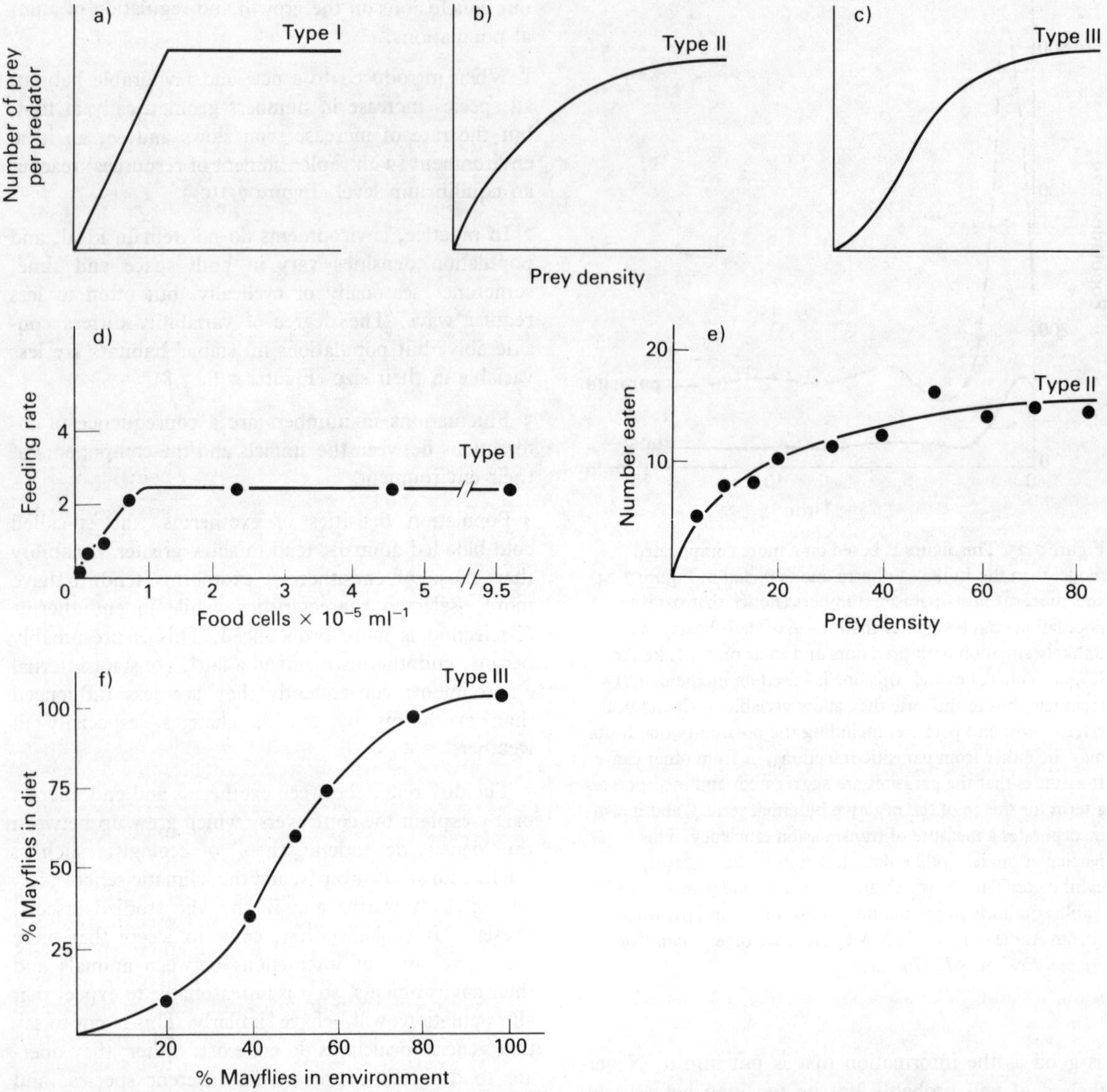

Figure 7.26 Predator–prey functional responses. (a) – (c) Firstly the three theoretical types of functional response showing the numbers of prey attacked by a single predator when prey densities change (After Holling, C. S., 1959. *Canadian Entomologist*, **91**; 293–320.). (d)–(e) Actual responses of predators to different prey densities; (d) different numbers of a yeast cell consumed by a water flea, *Daphnia*, by filter feeding. Below 10^{-5} consumption is proportional to density of yeast, but above this density the water flea is unable to increase its feeding rate because it cannot swallow all the yeast it filters, so an abrupt plateau is reached (Modified from Rigler, F. I., 1961. *Canadian Journal of Zoology*, **39**, 857–68.). (e) Feeding rate of a 10th-instar nymph of a damselfly, *Ischnura elegans*, on *Daphnia*, showing that as prey increases the predation rates increase less and less and finally approaches a plateau (After Thompson, D. J., 1975. *Journal of Animal Ecology*, **44**, 907–16.). (f) The functional response of an aquatic hemipteran, *Notonecta glauca* when given the choice of two prey species – the isopod, *Asellus aquaticus* and mayfly nymphs, *Cloeon dipterum*. At first, the predator took a disproportionately high number of mayflies, but as these became scarcer due to predation it took less because it had switched to feeding more on the alternative prey – the isopod. (See also Fig. 7.19). (After Lawton, J. H. *et al.* 1974 in *Ecological Stability* (eds) M. B. Usher & M. H. Williamson, Chapman & Hall: London.)

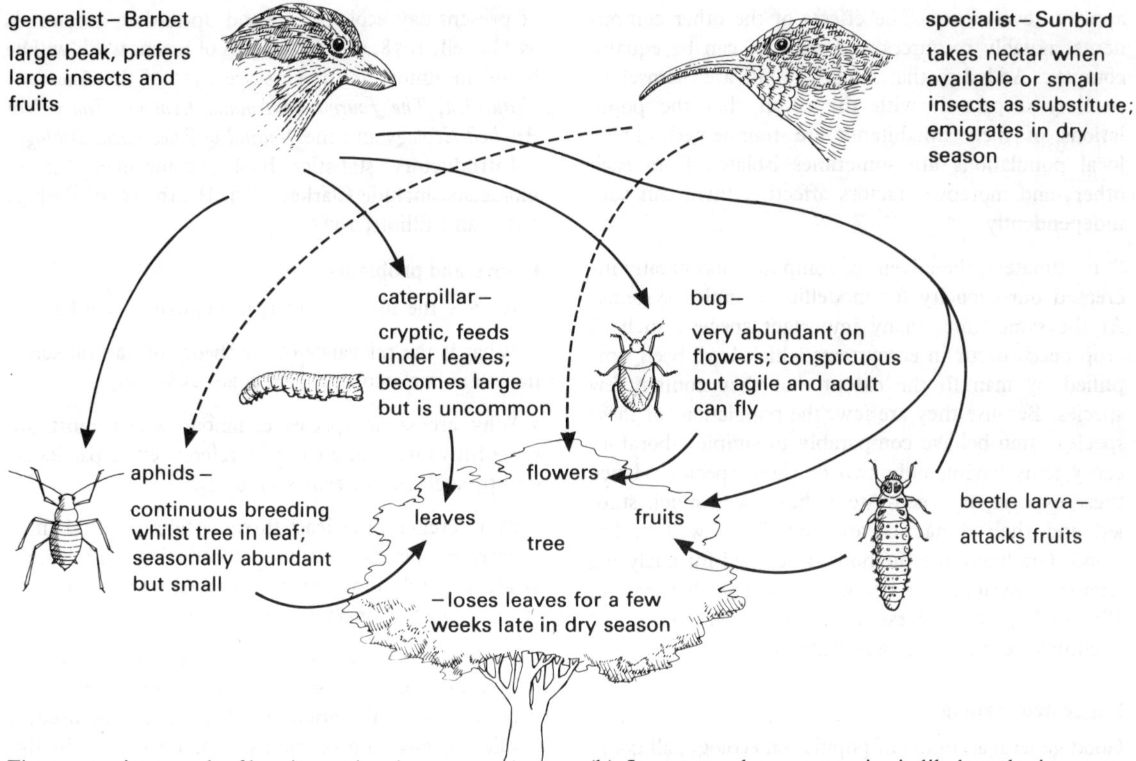

Figure 7.27 An example of how interactions between species can affect their populations. Even an apparently simple ecosystem is not really simple at all. Increasing the number of species leads to an exponential increase in the number of possible interactions between them.

In this imaginary case, there are four herbivorous insects, preyed upon by two insectivorous (i.e. predatory) birds. However, none of the insects is available throughout the year, so the birds are obliged to respond accordingly. At times there is no food for the sunbird, which emigrates to another ecosystem for a few months every year, but the barbet, survives during this period by feeding instead on fruits. Thus, at times, aphids for example, have no predators, whilst at other times they are heavily predated. The caterpillars are uncommon and the 'optimal foraging strategy' (pages 157 and 159) of the barbet is that it searches for them only when they reach a large size, i.e. just before they pupate, because at that time one caterpillar provides 100 times as much food as one aphid. The barbet prefers fruits when they are available, but if it finds a beetle larva in the fruit it will eat that too. However, since it cannot tell where the larvae are, its consumption of them is largely a matter of chance.

Other variables affecting numbers of the insects are: (a) other trees of the same species are not quite synchronized – so there are always some in leaf somewhere; (b) other species of tree provide alternative food for the bug and beetle, but not for the caterpillar or aphid; (c) climate – different species react more or less to climatic change. Thus, a model to predict the likely effects of given numbers of predators on the insect prey will be complex.

(b) In nature, the prey species is likely to be inaccessible to the predator at certain times, i.e. it might be sheltering in a refuge.

(c) Predator and prey vary in their ability to disperse and found new populations.

(d) Many predator–prey models are based on the assumption that both prey and predator are randomly dispersed, and have random encounters, whereas in most ecosystems both prey and predators are patchily dispersed. Many ecologists believe that aggregation and non-random searching by predators are important clues to understanding population stability.

(e) Some predators, and especially parasitoids, are host-specific, but most prey species have more than one predator. Most predatory species can switch from one prey species to another and do so for a variety of reasons (Figures 7.19, 7.26).

(f) The prey species taken by a particular predator often depends upon their respective ages, sex and previous experience.

7 Our discussion of some aspects of predator–prey relationships indicates the sort of complexities involved in studies of population dynamics, although not all

apply to every case. The effects of the other components, notably resources and climate, can be equally complex. Added to that, most habitats are themselves patchily dispersed, with the result that the populations of their inhabitants are fragmented. These local populations are sometimes isolated from each other, and moreover factors affecting them can vary independently.

8 Fortunately, the advent of computers has greatly increased our capacity for modelling complex systems. At the same time, many important species, such as crop pests, occur in ecosystems which have been simplified by man to the extent that they contain few species. Because they are few, the populations of these species often behave comparably to simple laboratory ecosystems having only two or three species. Hence these experimental ecosystems have been much studied and their dynamics are now fairly well understood. Gradually the methods developed for analysing simple ecosystems are being extended so that eventually we hope to understand more complex ones, and eventually even forests and coral reefs.

Suggested reading

Good general accounts of population ecology, all going into greater detail than this book, can be found in: Colinvaux, 1973, part 3; Emmel, 1976 chapters 4–6, 8, 11; Krebs, 1978 part 3; and Whittaker, 1975 chapter 2. A unified approach to the ecology of plants and animals is found in a profusely illustrated book by Begon & Mortimer, 1981. Other useful accounts are the shorter books of Hassell, 1976, and Solomon, 1976.

More advanced treatments of the subject, dealing particularly with ecological modelling, include May, 1981 and Maynard-Smith, 1974; whilst Christiansen & Fenchel, 1977, and the book edited by Halbach & Jacobs, 1979, contain discussions on a variety of related topics.

References to the pioneering works of Andrewartha & Birch, 1954, Lack, 1954, and Nicholson, 1954, are given in the list below; the various approaches were reviewed and discussed quite briefly in an appendix by Lack, 1966. Andrewartha & Birch, 1984, have greatly revised their original ideas, although the analysis of components of the environment remains their central theme.

Further actual examples of population studies in Africa can be found in Ewer & Hall, 1978 book 2; and in the detailed studies from Serengeti in Sinclair & Norton-Griffiths, 1979, as well as in various journals, especially the *African Journal of Ecology*.

Population modelling, including predator–prey and host–parasite models, is one of the most active fields of present-day ecology. A good up-to-date account is by Hassell, 1978, other examples of recent work will be found in almost any issue since 1970 of the *American Naturalist, The Journal of Animal Ecology, Journal of Applied Ecology* and the *Journal of Theoretical Biology*.

Introductory statistics books, comprehensible to biologists, include Clarke, 1969, Heath, 1970, Parker, 1973, and Elliott, 1977.

Essays and problems

1 Review the use of models in population studies.

2 What is the relevance of the theory of natural selection to animal ecology? (see pages 198–199).

3 Why are some species common, whilst most are rare? Illustrate your answer by reference to a particular group, such as butterflies or birds.

4 By reference to at least three vertebrates, and three flowering plants, outline contrasting life-history strategies and comment on the extent to which each species can be considered as *r*- or *K*-selected.

5 Calculate the average value of r for the wildebeest population in Serengeti between 1961 and 1977 (Figure 7.7(c)). Equation (4) (Box 7.1) can be used; it applies to any rate of increase (or decrease). In this instance, r probably approaches r_m, since conditions were particularly favourable. The data are these:

Year	Population estimate
1961	250 000
1963	350 000
1965	450 000
1967	480 000
1971	750 000
1972	850 000
1977	1440 000

(Take year 1961 as t_o, so for 1963, $t = 2.0$, $N_o =$ 250 000, N_t is 350 000; etc.)
How does the estimate of 1 350 000 for 1980 compare to the predicted value, based upon your estimate of the mean value of r for 1961 to 1977? Comment upon the significance of the estimates for 1978 onwards, as seen in Fig 7.7(c).

6 Using equations given (on pages 176–178) and its box evaluate K for the marabou data in Figure 7.7(b). The actual figures are these:

Breeding season	N (population index)
1967–68	76
1969–70	86
1970–71	80
1971–72	100
1972–73	110
1975–76	166
1980–81	206

Note: (a) take 1967–68 as t_o, so for 1969–70, $t = 2.0$, etc. (and $N_o = 76$; for 1969–70, $N_t = 86$).

(b) the term ($e^{a-r_{m}t}$) is readily evaluated on a scientific calculator; the instruction book will explain how (under 'exponents').

7 Give examples of invertebrate populations whose size changes (a) over a 24-hour day, (b) seasonally and (c) yearly. Why do fluctuations occur?

8 What is the evidence that density-dependent factors regulate population size?

9 Describe and discuss the relationships that may occur between predators and their prey. What are the advantages to a predator of 'switching'?

10 Describe, with examples, the differences between intraspecific and interspecific competition. What contribution do they make to population regulation?

11 What do you understand by resource limitation? Give some examples.

References to suggested reading

Andrewartha, H. G. 1970. *Introduction to the Study of Animal Populations*, (2nd edn). Methuen: London

Andrewartha, H. G. & Birch, L. C. 1954. *The Distribution and Abundance of Animals*, University of Chicago Press: Chicago.

Andrewartha, H. G. and Birch, L. C. 1984. *The Ecological Web*, University of Chicago Press: Chicago.

Begon, M. & Mortimer, M. 1981. *Population Ecology; A Unified Study of Animals and Plants*, Blackwell: Oxford.

Christiansen, F. B. & Fenchel, T. M. 1977. *Theories of Population in Biological Communities*, Springer-Verlag & Heidelberg: New York.

Clarke, G. M. 1969. *Statistics and Experimental Design*, Edward Arnold: London.

Colinvauz, P. A. 1973. *Introduction to Ecology*, Wiley: New York.

Elliott, J. M. 1977. *Some Methods for the Statistical Analysis of Benthic Invertebrates*, (2nd edn). Freshwater Biological Association, Scientific Publication, No. **25**: Ambleside.

Emmel, T. C. 1976. *Population Biology*, Harper & Row: New York.

Ewer, D. W. & Hall, J. B. (eds). 1978. *Ecological Biology* Longman: London.

Halbach, V. & Jacobs, J. (eds). 1979. *Population Ecology*, Gustav Fisher Verlag: Stuttgart & New York.

Hassell, M. P. 1976. *The Dynamics of Competition and Predation*, Edward Arnold: London

Hassell, M. P. 1978. *The Dynamics of Arthropod Predator-Prey Systems*, Princeton University Press: New Jersey.

Heath, O. V. S. 1970. *Investigation by Experiments*, Edward Arnold, Studies in Biology, No. **23**: London.

Krebs, C. J. 1978. *Ecology*, (2nd edn). Harper & Row: New York.

Lack, D. 1954. *The Natural Regulation of Animal Numbers*, Oxford University Press: Oxford.

Lack, D. 1966. *Population Studies of Birds*, Oxford University Press: Oxford.

May, R. M. (ed.) 1981. *Theoretical Ecology – Principles and Application*, (2nd edn)., Blackwell: Oxford.

Maynard-Smith, J. 1974. *Models in Ecology*, Cambridge University Press: London.

Nicholson, A. J. 1954. An outline of the dynamics of animal populations. *Australian Journal of Zoology*, **2**, 9–65.

Parker, R. E. 1973. *Introductory Statistics for Biology*, Studies in Biology, No. **43**, Edward Arnold: London.

Sinclair, A. R. E. & Norton-Griffiths, M. (eds). 1979. *Serengeti: Dynamics of an Ecosystem*, Chicago University Press: Chicago.

Solomon, M. E. 1976. *Population Dynamics*, (2nd edn). Edward Arnold: London.

Whittaker, R. H. 1975. *Communities and Ecosystems*, (2nd edn). Macmillan: New York.

8 The ecology of man

All earlier chapters have dealt with species other than ourselves, but here we are concerned with man. We often regard ourselves as something special, and in many ways we are. No other species has extended its life expectancy and conquered many of its diseases or extensively modified its environment, to anything like the same extent, or achieved such complex social communities. But we are still animals living within the constraints imposed by our environment. Because we have been such successful animals our population has suffered few biological checks to its growth, and consequently our numbers are now approaching the world's carrying capacity.

This chapter attempts to show how, in achieving our present state, we have exploited and destroyed much of the world's valuable renewable and non-renewable resources. We argue that the human species is now at a critical phase in its history. We believe that politicians, administrators and economists, and those engaged in production such as farmers and industrialists, must be made aware of the urgent need to reshape our living strategies if ecological catastrophies are to be avoided. We realize that such an ecological education will not be easy: and that it will take time. But now is the time to start.

Origins of man

Our immediate ancestors, the early hominids called *Homo habilis*, appear to have evolved mainly in southern and eastern Africa some 4–6 million years ago. They were probably omnivorous, feeding on a variety of plant and animal foods. They gradually evolved into a taller and more erect species, culminating in the appearance about 1.5 million years ago of *Homo erectus*, who used simple tools. The transition from *H. erectus* to *H. sapiens* was gradual and cannot be precisely dated; but there were certainly people we should recognize as *H. sapiens* living 100 000 years ago.

Fire had already been discovered by *H. erectus*, at least 0.5 million years ago and he thus began, albeit on a very small scale, to modify his immediate surroundings to suit his mode of life. It seems that agriculture began some 12 000 years ago when man began to collect seeds and grow a few plants near places where he lived and sheltered. Growing plants demanded a more fixed abode for man and consequently more permanent settlements were established, and gradually villages and some form of community and social life evolved. It was not long before man started keeping animals. This development of modern man did not take place at the same pace over the entire world. Agriculture and the domestication of animals arose in the Middle East and it was probably only some 5000 years ago that it spread into North Africa, and much later into many other more southerly parts of the continent. Even today, some people still lead a nomadic existence especially in the drier and less fertile savanna areas, keeping livestock but practising no agriculture.

Man began changing his environment many thousands of years ago. At first the effects were rather localized; but within the last few hundred years of rapid industrialization man's influence has been on a much larger scale, and his activities in one place may have world-wide repercussions. For example, there has been considerable concern that the destruction of vast areas of rainforest in South America and Malaysia as well as in Central Africa, will affect world climate; and pesticides produced and used extensively in northern areas of the world have resulted in pesticide residues occurring almost world-wide. What concerns ecologists and many others is the exponential rate of the changes man is making to his environment, and this is due mainly to one phenomenon – the logarithmic rate of human population increase. Added to this are the high levels of often wasteful and needless squander of consumables by the rich nations. More people means more land devoted to growing food, greater extraction and use of the world's renewable and non-renewable resources, and this has been accompanied by the destruction of much of the environment, resulting in industrial and other forms of pollution. But not all is gloom. In many areas man's nutrition and health have improved considerably. For example a vigorous, well co-ordinated campaign has eliminated smallpox from the world, and efficient drugs now exist for the prevention and cure of many diseases. It must be admitted, however, that many tropical diseases such as malaria and cholera still take their toll, and malnutrition is actually becoming commoner in some countries.

It is the intention of this short chapter to discuss briefly various aspects of human ecology and man's place in the environment.

Components of man's environment

Although man is a rather special animal he still interacts with his environment as do other animals. In chapter 5 we pointed out that each species, whether it is plant or animal, has its own set of components of the environment and we can extend this idea to our own species. The main components of our environment are:

(a) climate;
(b) resources – renewable and non-renewable;
(c) other individuals of the same species, in this instance our fellow people;
(d) animals of other species, which may be helpful to man in providing him with food and clothing, or cause a nuisance such as pests.

We next discuss some of these in more detail.

Climate

Extremes of temperature to man, like other endothermic animals, are less important than they are to exothermic animals. But man has also developed further independence from the harshness of weather in that he protects himself from the cold by clothing and living in heated houses. Protection from high temperatures is less serious so long as there is sufficient to drink. So, whereas we have seen that climate is often an important, if not 'key' factor, in controlling population size of many organisms, comparatively few human deaths are caused by extreme climatic conditions, or even from natural disasters such as floods and earthquakes. Nevertheless, drought and other climatic extremes continue to cause grave hardships.

Resources

On pages 209–214 resources were defined as those things that an animal needs from its environment, typically food and shelter, but man has come to need more than just these two. It is convenient to divide resources into two major categories, renewable and non-renewable. Renewable resources are those that can either be used repeatedly like water which recycles indefinitely through hydrological cycles (page 52) or animal and plant materials which can be replaced by fresh growth. Non-renewable resources are usually transformed by use and hence can only be used once; oil is an obvious example.

Renewable resources

Soil

Given proper husbandry, soil can continue to support crops and livestock indefinitely, except where it is covered by buildings or roads or other structures. It is said that some of the most fertile soil in England lies under the streets of London. However, the more serious threat is soil degradation. Occasionally this is caused by increased salinity and alkalinity; for example, when agricultural land is flooded, the various dissolved minerals in the water remain when the water runs off or evaporates. Repeated flooding and drying out of irrigated areas consequently leads to accumulation of minerals making the soil more saline or alkaline. But the most serious form of soil degradation is soil erosion. In a natural forest the annual rate of soil erosion is often as little as 0.1 tonnes ha^{-1}, but if the land is cleared for farming and overstocked with animals or badly managed, then annual soil losses can be as much as 10 or even 50 tonnes ha^{-1}. Drastic changes in agricultural practices are required to stop this erosion. In particular great care is needed if slopes are cultivated as steep slopes need well constructed terraces.

Water

Water is another obvious resource man requires, but unlike animals which only need it for drinking – or not at all – man relies on water for many domestic purposes including cooking, washing and sanitation. In many developing countries, the most sought-after commodity is a reliable piped water supply. Not only is this more convenient than having to collect water from wells, rivers or ponds, but a clean water supply reduces the incidence of water-borne diseases. The importance of this is apparent in the United Nations declaration that everyone should have a proper water supply by the year 2000, which is also the year in which the World Health Organization's objective is 'health for all' – that is that no one should be suffering from malnutrition or parasitic diseases like malaria. Although highly commendable it seems to us that these objectives will not be reached by then.

Apart from domestic usage, water is needed to irrigate crops, to water livestock and in industry. A large paper mill requires as much water as a town of 100 000 people. Much of the water used domestically and by industry can be recycled after treatment, but it is costly to remove industrial wastes and pollutants. Africa has a plentiful supply of fresh water, but it is often present in the wrong place. Thus, the River Nile carries relatively little water but runs through densely populated areas, while the Congo River, which is the world's second largest in terms of discharge to the sea, runs mainly through sparsely populated areas. Water is undoubtedly a precious commodity but is only renewable if treated with respect.

Food

Another essential resource is food. Man must have a minimum intake of food to meet his metabolic requirements. A convenient indicator of the quantity of food in people's diet is their intake of energy, measured in joules (formerly calories). Apart from food supplying energy it must also supply essential vitamins, minerals and proteins in order to prevent dietary deficiencies and malnutrition. The quality of a diet is commonly measured in the daily intake of dry grams of protein. In Africa, the approximate daily intake of food in terms of energy needed by a relatively inactive person is 9 MJ (megajoules), but as much as 14 MJ for an active farmer. Similarly, the required intake of protein is about 50 g for the retired person, rising to 95 g for a mother breast-feeding a child.

At present the world produces sufficient food for everyone but this food is not evenly distributed. In developed countries large sections of the community are overfed and many people are trying to lose weight, whereas in the developing countries large numbers of people have insufficient food, although this varies greatly from country to country. Generally starvation and malnutrition are commoner in countries with low rainfall, but this is not always so. For example, Uganda has very fertile soils, but many people in southern Uganda suffer from protein deficiency because they mostly eat plantains and cassava, which are composed almost entirely of carbohydrates. Estimates of the Food and Agriculture Organization of the United Nations (FAO) show that in most African countries at least 15 per cent of the adult population receive less than 7 MJ a day, barely enough for basic metabolism, let alone work, but by the year 2000 it is hoped that their intake will be around 11 MJ a day. This will require nearly doubling the present crop production.

Probably no country feeds its people entirely on the food it produces, but some could do so. Britain, for example, could grow enough food and produce sufficient meat to support its population, although it would go without luxuries such as tea, coffee, cocoa and tropical fruits like bananas and pineapples. But at present many African countries would find it difficult to feed themselves. In fact according to the 'Regional Food Plan for Africa' published in 1980 by the Food and Agriculture Organization of the United Nations, the African region is falling behind other developing countries in food production. The 'Self-sufficiency Rate' (Production/Demand) was 98 per cent in 1962–64, 90 per cent in 1972–74 and is estimated to be 81 per cent in 1985. These figures are average values for overall food production in Africa and some countries will be doing better – others worse. One of the reasons for this poor performance is that much of the fertile soil of Africa is used to grow cash crops, such as coffee, tea, cocoa, sugar, cotton, pineapples, which are neither essential nor staple foods but are exported to obtain foreign exchange.

On the whole it is probably true to say that the economy of many developing countries relies on one or two major export commodities, but, because their value on the world market fluctuates widely, their earnings are unreliable. Yet it is with the earnings from such exports that the country pays for imports, essential or otherwise, whether it be food, oil, or cars. In 1969 only 2 per cent of the food consumed in Africa was imported, but by 1980 this had risen to 20 per cent, and this figure is still rising. Most of the imports consist of grain: they amounted to 1 million tones in 1979. Some African countries also export grain to the developed countries, where some of it is used to feed cattle and poultry rather than people. This is a wasteful practice, because many more people could be fed in the developing countries if the grain was consumed there, than by converting it to meat in the developed countries.

Energy is always lost when food is converted from one trophic level to a higher one (pages 38 and 48). Nevertheless, having said that, it is important that people have a balanced diet, and it is difficult, though not impossible, to remain healthy on a purely vegetarian diet. In addition, in dry savanna regions where it is difficult to grow crops without irrigation, people derive almost all their protein from their livestock. It is also important to remember that the vast areas covered by the oceans have considerable potential for increasing fish production, provided that governments can be persuaded to undertake efficient management of these resources. Also the development of inland fisheries would help meet the protein requirements of many poor people in the developing world.

In almost any country one can see large areas of uncultivated land, but it is a fallacy to think that a country's food production can be increased merely by converting these areas to farmland. Such land is usually not being exploited for a good reason; it may

Box 8.1
'Nothing grows for ever in nature because the environment has a limited carrying capacity for living organisms. This concept is familiar to farmers. Overstock the farm and the carrying capacity is reduced. Growth itself reduces the carrying capacity. It is self limiting.' Birch, L. C., 1980. *Habitat Australia*, **8**, 25–31.

be too waterlogged, too dry, too stony, have shallow soils, or be poor in nutrients. Increased food production is less likely to be achieved by extending agriculture to marginal lands than by increasing the husbandry and efficiency on existing land. Many tropical soils have a low nutrient status and require regular addition of fertilizers if they are to be repeatedly used and production increased, rather than decreased through soil exhaustion. This point is clearly but simply stated in terms of carrying capacity in Box 8.1.

During the 1970s, fertilizer use in Africa rose by about 8 per cent a year, and this trend is continuing. Better yields are obtained if weeds and pests are reduced, but again this is achieved mainly by the application of man-made chemicals in the form of herbicides and pesticides (Figure 8.1). The manufacture of fertilizers, herbicides and pesticides requires considerable energy. At present most of them have to be imported into developing countries and their costs are escalating. Thus, it becomes more and more expensive to grow crops. There are other difficulties, such as lack of education or high pressure salesmanship which may result in the use of excessive and unnecessary amounts of insecticides, and sometimes fertilizer, or the use of the wrong insecticides and fertilizers. There is also the problem of pests developing resistance to the more commonly used and relatively cheap insecticides like DDT. When this happens, farmers have to buy alternative insecticides such as the organophosphates, which although not so persistent in the environment as DDT and other organochlorines, are much more expensive and considerably more toxic. The result is that if they are used without strict regard to the recommended precautions, poisoning or even death may occur in farm workers, and farm products consumed by the public may be contaminated with unacceptably high pesticide residues.

The need to grow large quantities of food has led to monocultures, and vast areas are planted with bananas, pineapples, tea, coffee and other crops. Such a system, although economically sound and facilitating harvesting, often leads to complications. For example, the optimum conditions which are being provided for the crops, also provide optimum conditions for pests, whether they are insects such as aphids, or birds such as quelea, or rats. Monocultures frequently lead to pest outbreaks; conversely crop diversity and inter-cropping tend to produce stability. Intensive monoculture often necessitates the repeated application of man-made chemicals. In some countries in Central America, up to 80 kg of insecticides are applied per hectare each growing season to combat cotton pests. In addition, large quantities of artificial fertilizers may be regularly applied, and this can cause eutrophication of rivers and lakes if they get washed into the aquatic environment (page 220).

Figure 8.1 Cotton crop on a large irrigation scheme being sprayed with insecticides to reduce damage and loss by insect pests. Later in the season the crop is sprayed by aerial application. There are often fifteen sprayings a year; in fact cotton receives greater quantities of insecticides than any other crop.
(Camerapix Hutchinson.)

Irrigation is another very important way of increasing crop production. Israel, for example, has large areas of very dry and inhospitable land, but because of very efficient irrigation is able to produce regular surpluses of certain crops, which comprise valuable exports. In this case irrigation allows crops to be grown in areas where formerly agriculture did not exist, but irrigation can also improve production in areas where agriculture is already being practised. Medium- to large-scale irrigation projects occur in many African countries, enabling them to increase their production of rice, sugar, pineapples, cotton and other crops. Irrigation, however, can also bring problems. It is expensive in terms of energy and other inputs and requires considerable expertise if it is to be used on anything but a small scale. A few of the problems associated with irrigation, such as soil degradation and creation of breeding places for disease vectors, are referred to in Figure 8.2. Drip irrigation reduces the problem of flooded fields forming habitats for disease vectors by providing water direct to the soil through underground pipes, but the system is technically complex. The planning and operation of large-scale irrigation schemes needs the co-operation of agronomists, entomologists, engineers, economists, physicians and sociologists if they are to function efficiently while minimising health hazards.

Shelter

Another resource required by man is shelter, which can be regarded as renewable. The typical human shelter is a house, which provides protection against rain and extremes of temperature. But not everyone has a house, for example almost half the population of large towns in India such as Bombay sleep on the streets at night. Urbanization encourages overcrowding and slum conditions, with their associated deterioration not only in the quality of life but in health. Several diseases of man are more prevalent in overcrowded towns than they are in rural areas. In most slum areas there is inadequate sanitation, an accumulation of refuse and excreta, and numerous open cesspits, all of which leads to a proliferation of filth flies (*Fannia*, *Musca*, *Calliphora*, *Lucilia* spp.) and certain types of mosquitoes. One particular mosquito species, *Culex quinquefasciatus* (= *Cx. pipiens fatigans*), breeds in polluted waters and is an important urban vector of the filarial worms that cause the disease bancroftian filariasis. The incidence of this disease is increasing throughout much of the tropics because of increased urbanization. In much of Africa this mosquito has increased both its numbers and range during the past few decades to become the most common man-biting mosquito in the towns.

Figure 8.2 Rice being grown on an irrigation scheme in Sierra Leone. Many problems can arise from irrigation projects, such as increased soil salinity and alkalinity, and proliferation of mosquitoes leading to potential increases in vector-borne diseases such as malaria, bancroftian filariasis, and various arboviruses (i.e. *ar*thropod-*bo*rne *viruses*). In addition, such habitats are ideal for the aquatic snail intermediate hosts of schistosomiasis (bilharzia). Careful management can, however, do much to alleviate these deleterious side-effects.
(M. W. Service.)

Wood

Trees whether scattered or in forests are a natural resource of great value. Their roots and fallen leaf-litter build up a protective underlayer of humus which retains water, a valuable asset when dry conditions prevail. A layer of litter and roots also stabilizes the soil and reduces the rate of erosion. Forests have an important role in maintaining the optimum carbon dioxide levels in the atmosphere and in promoting rainfall. In addition trees can provide shade and act as windbreaks, which again assist in preventing soil erosion. Wood from trees is the most important source of domestic fuel in much of the developing world. In Kenya about 75 per cent of the country's energy requirements are provided by wood; some 20 million tonnes of raw wood and another 7 million tonnes of charcoal are consumed yearly. In The Gambia, wood has become so scarce that it takes 360 man–days to collect a year's supply of firewood for an average-sized family. In contrast, in nearby Sierra Leone there are still considerable numbers of trees and scrub vegetation around villages and this provides a ready source of firewood which is not yet in short supply. Fuelwood consumption on average in Africa is about 1.5 m^3 per person per year. In semi-arid and arid areas, cutting of trees for fuel and other purposes, is a major cause of desertification, at least as important as overstocking.

More than 1500 million people in the developing

world depend on wood for cooking. Wood is also extensively used for building, fencing, furniture and paper-making, and it is therefore not surprising that the world's forests are fast disappearing. Trees are of course a renewable resource but reafforestation has not kept pace with deforestation, which in consequence is one of the more alarming changes man is making to the environment.

Non-renewable resources

Energy

The greatly increasing human population coupled with the desire for better living conditions and material wealth has resulted in man using as much energy in the last 100 years as he had done in the previous 2000 years. But there is, of course, a very uneven distribution of energy consumption. The USA has only about 7 per cent of the world's population but uses about half the world's oil. Of the world's total energy consumption Africa's share is only 1.9 per cent, and of this small amount South Africa and Swaziland consume about half, much of it for their mining and large-scale industrial operations. Undoubtedly the need for energy in developing countries will be increasing and it is difficult to see how the world is going to cope with these ever increasing demands.

Although several of man's resources such as oil and various minerals are non-renewable, until comparatively recently no one really believed that demand would be so great as to lead to shortages. Over half of the world's energy is derived from oil, but, as we know, oil reserves are not inexhaustable. In 1980 it was estimated that known oil reserves could give us 600 billion barrels of oil (1 barrel = 160 litres), but this reserve would be depleted by about 2050–2100 AD. Reliable predictions are notoriously difficult, but there is no doubt that we are rapidly using up the world's oil supply, and likewise we are consuming vast quantities of other non-renewable resources such as coal and natural gas. Present technology extracts only about 32 per cent of the oil from an oilfield, and optimists believe that improved technologies and the discovery of new oilfields will postpone the time when oil is almost exhausted, and that, by that time, we will have solved the energy crisis by employing a variety of alternative energy sources. The pessimists do not see such an easy solution.

What are the alternative sources of energy? The one that has the greatest potential is nuclear power, but at present the method of splitting the nuclei of radioisotopes such as uranium 235 (by fission) is not very efficient and in any case produces very hazardous by-products. The real hope lies in employing thermonuclear fusion to provide power. It is the thermonuclear fusion process that exploded the devastating hydrogen bomb, but man has not yet mastered how to control and harness the enormous amounts of energy produced for peaceful purposes. When he does so this will provide enormous amounts of energy, with little or no radioactive wastes. The difficulty with nuclear power stations, in addition to safety hazards, pollution problems and the high technology required for their construction, is that building them takes about ten years and is extremely costly, but they have a productive life of only thirty years. It is unlikely that many developing countries will construct nuclear power stations in the near future. South Africa is the only country on the African continent that has a nuclear reactor – and is planning to build more – yet Africa has about 30 per cent of the world's known uranium reserves. It seems likely that African countries will continue to sell their uranium and other valuable minerals and metals to more industrialized countries.

Africa has some of the largest rivers in the world as well as numerous small- and medium-sized ones that have a tremendous head of water, offering a great potential for developing hydroelectric power. Good examples of harnessing water energy for power are the Volta, Kariba, Kainji and Aswan dams. Some believe that Africa could produce 27–40 per cent of the world's hydroelectric power, but at present only some 2 per cent of the world's total water power is in Africa. Reasons for this disappointing usage of Africa's rivers are that it is very expensive and takes a long time to build dams and install power lines to take the electricity from where it is generated to places where it is needed. Hydroelectric power has the great merit of being non-polluting.

Other alternatives to fossil fuel include wind power. In 1960 at least 185 000 windmills were operating in South Africa, pumping up water from wells or generating electricity for local use. Another alternative is geothermal power. Small geothermal units making use of hot springs at Lake Naivasha and Bogoria in Kenya are being evaluated (Figure 8.3), and one at Ol Kania is now commissioned. There seems to be considerable potential for using solar energy to supply hot water, cooking and drying facilities, and lighting in houses, schools and factories. In some countries, such as Israel, solar power panels, often called photovoltaic cells, are seen commonly on roofs of houses but their use in Africa is still rather local (Figure 8.4). Other alternative schemes for supplying power include harnessing tidal power from the oceans, and operation of cars and other internal combustion engines on alcohols obtained by fermenting crops such as sugar cane,

Figure 8.3 A natural hot-water spring at Lake Bogoria, Kenya. The great heat generated by such water can be trapped by man to provide geothermal power.
(M. W. Service.)

Figure 8.4 Two solar panels (photovoltaic cells) supplying electricity to operate a deep-freezer and recharge batteries for lighting at a zoological research camp in the Outamba-Kilimi National Park, in northern Sierra Leone. A problem encountered here was that atmospheric dust, especially in the dry season, substantially reduced the estimated light (energy) from the sun.
(M. W. Service.)

cassava, soya beans and peanuts. But a problem here is that obtaining fuel from plants could lead to an increase in the price of foods. This emphasizes the need for an overall policy on land use and priorities. Biogas generators also offer the prospects of a cheap fuel for cooking and lighting. They use anaerobic decomposition of excreta and organic refuse to produce methane, a gas which has a higher energy value than coal gas.

Minerals

Just as the earth's oil is being depleted, so are other important minerals. Looking around at all the metal used in houses, factories and cars, it is difficult to imagine that metals, such as iron and copper, can ever be in short supply, but they are non-renewable resources. However, unlike fossil fuels which are completely destroyed by burning, metals used in the manufacture of products are not destroyed and in many instances a high proportion of them can be reused by recycling. It should by now be realized that we can no longer tolerate the wastefulness of a society that throws away glass containers, metal beer cans and plastic soft drinks' bottles; we must recycle our valuable resources, whether they be metal, glass or paper. (Figure 8.5).

Human population increase

Some 12 000 years ago the world population was probably about 10 million, but by about 1650 it had increased to 500 million, by 1830 it had doubled to 1000 million, and doubled again by 1930. In 1984 the world population was estimated as about 4700 million, and conservative estimates forecast that by the year 2000 it will be 7000 million (Figure 8.6). Now an important point is that more than 90 per cent of this increase will take place in the developing countries. What about Africa? For many years there was very

Figure 8.5 A typical car dump in Accra, Ghana. Here is a good example of recycling of non-renewable resources. These derelict cars are not left to rust but are owned by various traders and provide a valuable source of spare parts ranging from motor car panels, and bumpers, to electrical components. Increasing use is made of this recycling when there is a limited supply of new parts and spares, due to importation difficulties.
(M. W. Service.)

little population increase, and even by 1900 there were only some 120 million people in Africa, but by the late 1940s the population had more than doubled to reach about 250 million. However, in 1980 there were an estimated 470 million people in Africa, this being an almost fourfold increase within eighty years (1900–1980). Now, if this exponential rate of increase were to continue there would be about 832 million people in Africa by the year 2000 (Figure 8.7).

What are the reasons for these fast rates of population increase in the world in general and Africa in particular? We saw in chapter 7 that population size is determined by four variables: the birth rate, the death rate, immigration and emigration, and it is the same with human populations. Although immigration and emigration may cause local changes in population size, (for example the exodus of 1 million people from Ethiopia into Somalia during 1980 and 1981 which increased Somalia's population by 30 per cent), this rarely contributes very much to any population increase.

The major reason for the world's population increase is not an increased birth rate, but a reduced death rate. This has been brought about by improved nutrition and health which have drastically reduced mortality, especially in infancy and early childhood. For example, a successful malaria control campaign in Sri Lanka was the principal reason that the annual death rate fell from 2.2 per cent in 1945 to 0.8 per cent in 1972. Although there was a reduced death rate, this was not counterbalanced by a decrease in the birth rate, so Sri Lanka experienced a rise in its population; but it is now one of the relatively few developing countries to have succeeded in reducing their birth rate. As another example the birth rate in Egypt was 4.2 per cent in 1945 and the death rate 2.8 per cent,

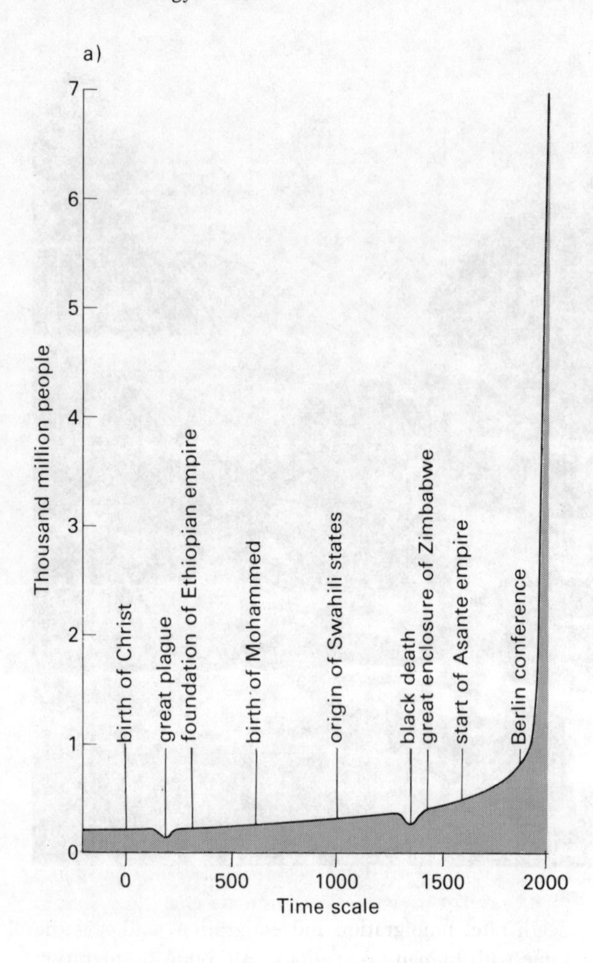

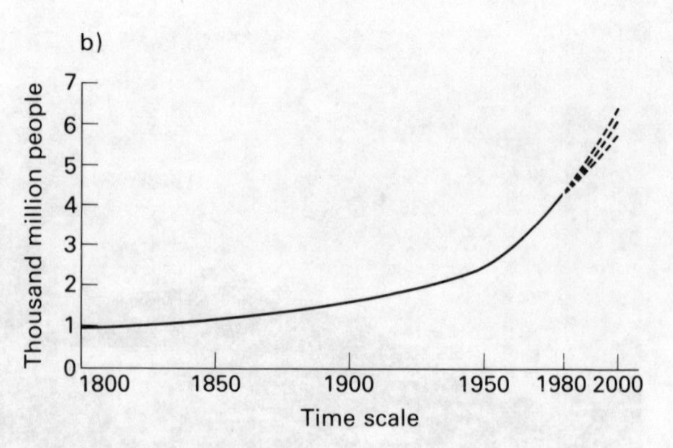

Figure 8.6 Human population growth and predictions to the year 2000. (a) World population from birth of Christ to 1980, and projection to 2000 based upon the United Nations estimates and other sources. (Redrawn and modified from Dasmann, R. F., 1975. *The Conservation Alternatives*, Wiley: New York.) (b) Actual world population increase from 1800–1980 and then United Nations predictions up to year 2000. The three different estimates (broken lines) are based on high, medium and low rates of population growth differing primarily in their assumptions of future fertility rates. (Modified from *Scientific American*, 1980.) Note how different time scales affect the visual impact, both figures show basically the same increase and projections. (You will see that in this figure the world population in 1980 is given as about 4400 million, differing from the figure of 4240 million given on page 218. These figures were taken from different sources and emphasize the difficulties of knowing accurately the present would population let alone predicting future population size.)

giving a rate of population increase of 1.4 per cent. Improved health facilities reduced the death rate to 1.6 per cent in 1972, but the birth rate remained little changed at 4.4 per cent, with the result that Egypt's growth rate was doubled to 2.8 per cent. In Kenya in 1979 the birth and death rates were 5.46 and 1.42 per cent respectively, giving a growth rate of 4.04 per cent, which is very high and means that the population will be doubled within seventeen years. Figure 8.8 shows approximate growth rates in the world.

So what is the problem? From an ecologist's point of view it is quite clear that no population can increase in size indefinitely without catastrophic results, but this is exactly what could happen to human populations. There must come a point when the world's resources *cannot* support the ever-increasing number of people, that is the **carrying capacity** has been reached. All populations can exceed the carrying capacity for short periods, but none for long. For example, if the world population continues to increase at the present rate, there would be 14 000 million people by the year 2050. This is a figure which most scientists would consider to be disastrous. If this were ever to happen there would likely be a population crash to 3000–7000 million people, caused mainly by shortage of food. But, before such a terrible event could happen, the quality of life would deteriorate. For example, most people expect more from life than just their survival to a reproductive age, they hope to live in reasonable health and comfort. However, if our numbers continue to increase, there must come a point when our finite resources can no longer supply what we have come to think of as our rights; you can imagine a time when there is insufficient supply of fossil fuels and even food. But we all know that even today many people have a very poor standard of living, and it has been pointed out that the earth's resources could not support the present 4700 million people if they were all to have the same living standards enjoyed by those of the more developed countries, or by the wealthier people in the developing countries.

Large increases in the number of people also puts

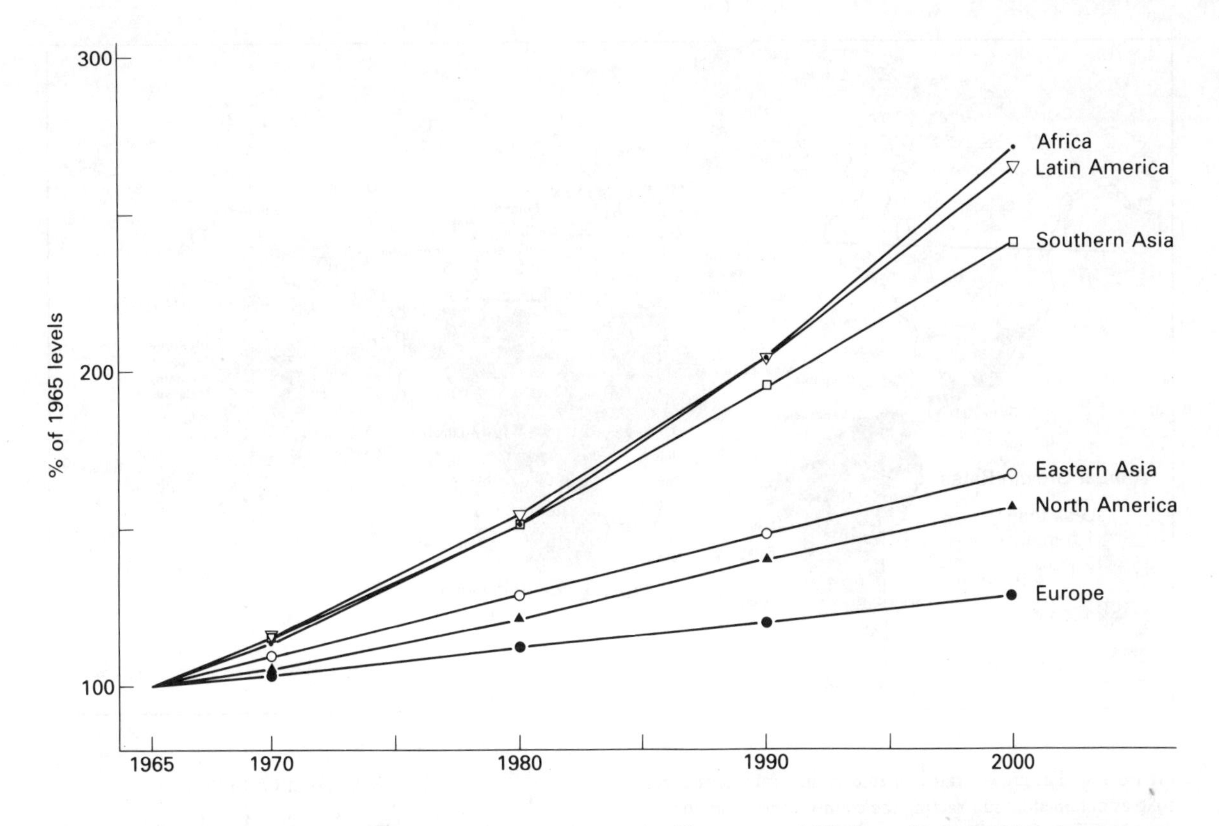

Figure 8.7 Population predictions of various regions of the world from 1965 to 2000, as a *percentage* of the 1965 levels. Based upon United Nations 'median estimates'. The population of Africa is expected to rise proportionately more than any other region, thus aggravating all population-related problems such as increased demands for food and water, health and education to a greater degree than anywhere else. (Based on figures given by Ehrlich & Ehrlich, 1977.)

a strain on a country's economy. More people mean more schools, hospitals, more food, more imports of renewable and non-renewable resources and most probably greater unemployment. So is there any solution? We can either reduce the birth rate or increase the death rate. Clearly the latter is unacceptable to virtually everyone. We need to emphasize the importance of stabilizing the growth rate to avoid great imbalance between the age-groups (Figure 8.9). But remember that reducing the birth rate leads *temporarily* to an ageing community, and both politicians and economists do not like this because it puts an economic burden on the community. Consequently they often advocate keeping the birth rate high. But this argument is unrealistic because a high birth rate *also* results in a high dependency ratio. Apart from these difficulties, there can be strong religious objections to families practising birth control and limiting their families to two children – which would maintain population size at its present level. It is therefore not going to be easy to persuade governments that they should advocate population stabilization. In China, however, where there are now an estimated 1100 million people, the government is encouraging late marriages and penalizing families having more than two children by denying them certain free social benefits. So it really becomes costly to have more than two children. This type of enforcement would not be accepted in many countries!

We appreciate that not all our readers will agree that the population growth in their country should be checked; for example, Nigeria has an estimated 91 million or more people, equivalent to about 91 km^{-2}, but who is to decide whether this is too densely populated? It is, after all, much less than that of many industrialized countries. The Netherlands, with a population density of about 426 people km^{-2}, is the most densely populated country in Europe. But there are of course differences. In general, industrialized countries have more resources (e.g. agriculture and industry) and more schools and better medical facilities to cope with their populations, whereas in many developing countries existing services can barely cope with present populations, let alone expanding ones. One of the major motives for a large family is security

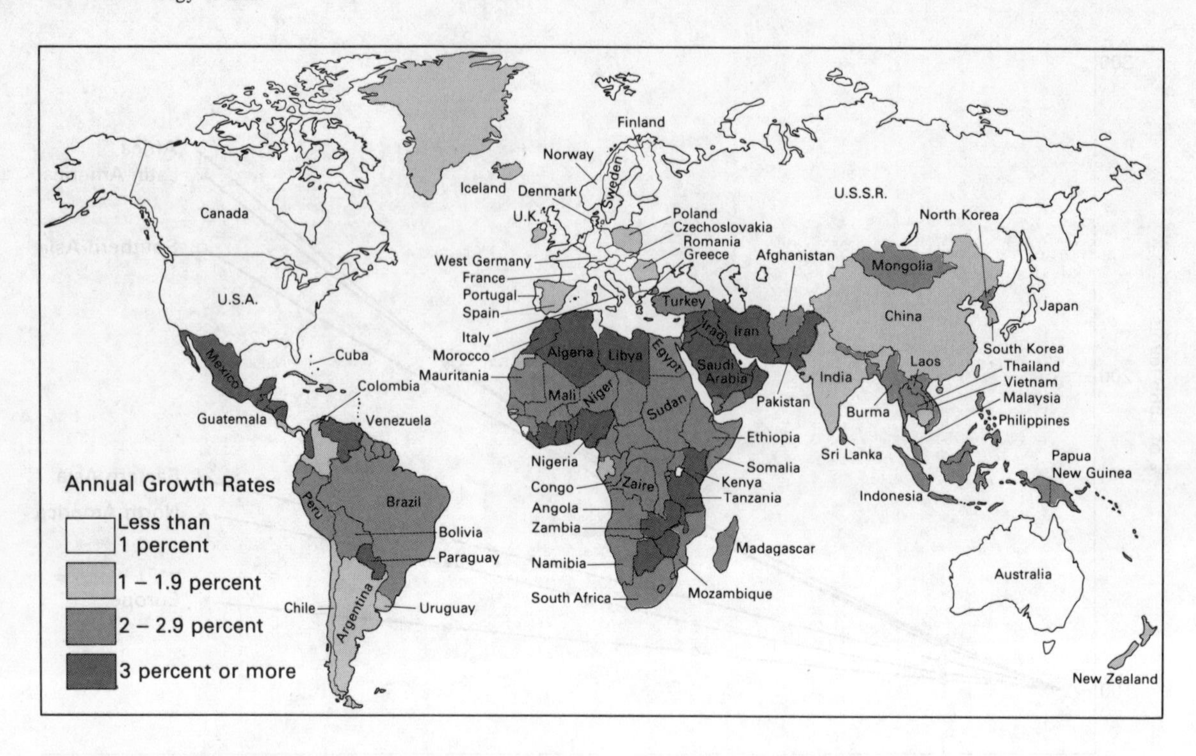

Figure 8.8. The growth rates, or rates of natural increase, for 1980 as obtained by subtracting the country's crude annual death rate from its crude birth rate. (These are called crude rates because they are not adjusted for sex or age.)
(From *Scientific American*, 1980, copyright © 1980 by Scientific American, Inc. All rights reserved.)

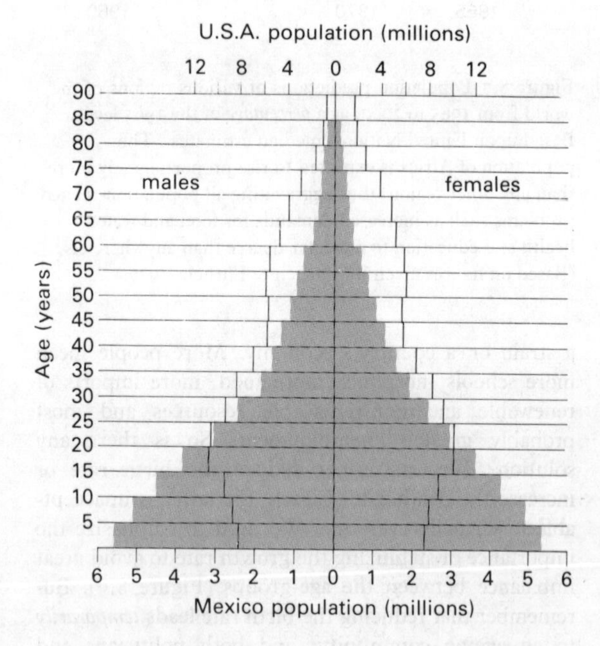

Figure 8.9 The population pyramid of Mexico (stippled) is broad based, a characteristic of many developing countries, showing that the young greatly outnumber older people, whereas the profile for the USA shows a more stable age-structure.
(From *Scientific American*, 1980, copyright © 1980 by Scientific American, Inc. All rights reserved.)

in old age, but better pension schemes should reduce this reason for having large numbers of children. However, while many countries will argue that they are underpopulated, most experts feel that the world's carrying capacity is around the 1980 figure of some 4240 million. They also realize that a sustainable population will not be achieved without intensive education and planning and the willing consent of the people. There is an urgent need in some areas for a move towards what has been termed **zero population growth,** in other words, each couple having two children who attain adulthood.

Impact of man

The changing environment

Man is only one of several million species of organisms living in the world, but he has changed the environment far more than all the other species put together; as Box 8.2 aptly describes it – man tries to live in two worlds.

> **Box 8.2**
> As Ward, B. & Dubos, R., 1972, say in *Only One Earth*, Penguin Books, Middlesex, England: 'Man inhabits two worlds. One is the natural world of plants and animals, of soils and airs and waters which preceded him by billions [thousands of millions] of years and of which he is a part. The other is the world of social institutions and artefacts he builds for himself, using his tools and engines, his science and his dreams to fashion an environment obedient to human purpose and direction.'

Many of the changes man makes are unintentional, like soil erosion, soil exhaustion, desertification and environmental pollution. Others are intentional such as cutting down forests to provide firewood, timber for building and wood pulp for the manufacture of our daily newspapers. We also use more and more land for cultivation, the construction of dams and changing river systems for irrigation and hydroelectric power, but this does not detract from them causing ecological damage. For example, ecologists are greatly concerned at the alarming rate at which forest ecosystems are being destroyed, not just because many animals and plant species will become extinct and their genetic resources lost for ever, but also because such wide-scale destruction causes soil erosion and will probably change the world's climate. The biological effects of environmental changes vary greatly. The depletion of oil and other non-renewable resources have little direct effect on living organisms other than man, but paradoxically it is often the destruction of living resources that cause most damage despite their being potentially renewable. For example, fish are depleted from both rivers and oceans by overfishing, and whale numbers are reduced because they provide us with valuable oils, and in some countries like Japan may also be in demand for food. Excessive hunting of game animals for trophies can similarly lead to serious reductions in their numbers.

Chemical pollution

Most people would regard environmental pollution as a man-made phenomenon, although the natural discharges of sulphur and ash from volcanoes can pollute the atmosphere and prevent plant growth, and some fish contain mercury levels toxic to man which have been concentrated naturally from the oceans. So, it could be argued that natural environmental pollution occurs, but here we are mainly concerned with the much more serious problem of man-made pollution, a definition of which is given in Box 8.3.

> **Box 8.3**
> Pollution can be defined as 'the introduction of man-made substances and energy (e.g. heat, radiation and noise) into the environment that are likely to harm man or other living organisms or the ecosystem, or damage structures and amenites'. This definition covers almost all aspects, from contamination of rivers by sewage, pesticide residues in living organisms, noise created near large factories and airports, industrial atmospheric pollution causing damage to buildings, and radiation dangers.

Man-made pollution is not new; 200 years ago the river Thames in London was an open cesspool, and during the 1800s there was already considerable industrial pollution of the atmosphere and rivers in much of the developed world. In contrast, the widespread environmental contamination by man-made pesticides is more recent, and it is this cause of pollution more than any other that has generated so many emotive claims and counterclaims about pollution.

There can be no argument that synthetic insecticides have saved millions of lives, such as by killing the mosquito vectors of malaria, killing body lice that transmit typhus, and destroying other insects that transmit killing or crippling diseases to man. They are also invaluable in protecting crops from the ravages of pests, and preventing damage during their storage after harvest. If it were not for insecticides, about one-third of the world's food would be lost to pests. The problem is that these insecticides which have brought untold benefits to mankind have also caused considerable environmental damage. Very few insecticides are selective in the species they destroy, so when a crop is sprayed with insecticides to kill its pests, many harmless and even beneficial insects such as bees and the pests' natural enemies are also destroyed. Sometimes the destruction of natural enemies, which previously have probably kept pest numbers below a certain level, results in pest populations increasing to a larger size than prior to spraying.

Biodegradable insecticides, such as malathion and other organophosphates, although killing a wide variety of insects, rapidly break down in the environment into less harmful substances. In contrast the non-biodegradable insecticides such as the organochlorines (e.g. DDT, HCH (=BHC), aldrin and dieldrin) can persist for many months and even years in soil, plant and animal tissues. The quantities of residues generally increase in higher trophic levels of the food chain. Thus DDT spilled on the soil is taken up by earthworms which, if eaten by small mammals, result in higher residues in them, and then, if they are eaten by predatory birds, the birds contain greater quantities of DDT than the earthworms or mammals. The presence of insecticide residues can cause abnormal behaviour, breeding failure or even death. DDT residues have been found in much of man's food such as fish, meat, vegetables, in cows' milk and even in human milk, and this has naturally caused considerable alarm! Residues have also been found in arctic and antarctic waters and in animals living in these remote areas where no insecticides are used. But the effects of this world-wide contamination must not be exaggerated. We may not like the idea that DDT is so widespread in the environment, but the levels are sometimes so minute that it is hard to believe that they are detrimental. There must be a sense of proportion. For example, if DDT spraying enabled a country to eradicate malaria, but as a consequence the people had minute amounts of DDT in their bodies which apparently caused them no harm, would you consider this an acceptable 'price to pay' for getting rid of malaria? We would argue that it was, but admit it is controversial.

Most pollution, whether it is through over-use of nitrogenous fertilizers, pesticides, industrial smoke or oil spillage occurs in the developed and more industrialized world. As yet, environmental pollution in Africa is on a local scale but this does not mean it is absent. In Ghana, some 1100 kg of arsenic oxide are discharged daily in smoke from gold mining camps, whilst in Zambia about 1250 kg of sulphur dioxide gas are emitted daily in some areas of the copper belt. The corrugated roofs of houses near the Jinja copper smelting works in Uganda rusted away much quicker than roofs on houses further away. In Kenya, despite a successful 1974 clean water act, coffee berries are still sometimes washed in water which is discharged into rivers or lakes causing decreased oxygen levels, so that the water has a high Biochemical Oxygen Demand (BOD), which is linked with a form of pollution called eutrophication (page 56). Occasionally the shipwrecks of large oil tankers and other oil spillages have caused marine pollution along the shores of Africa.

Increased carbon dioxide in the air

Burning fossil fuels, firewood and clearing and burning forests releases large quantities of carbon dioxide into the atmosphere, most of which is absorbed by the oceans and converted into carbonates and hydrogen carbonates but some remains in the air (page 52). This has resulted in increases in average carbon dioxide levels in the atmosphere, from about 316 parts per million (ppm) in 1958 to 332 ppm in 1976. The prediction of the World Meterological Organization is that this could increase to 450 ppm by the year 2050. Now, much of the short-wave radiation from the sun that reaches the earth's surface is re-radiated as longer wave infrared radiation, but the carbon dioxide in the atmosphere acts as a filter and traps some of this energy (heat) within the earth's atmosphere. This causes a gradual rise in temperature, and is termed the 'greenhouse effect'.

Some environmentalists believe that increased CO_2 levels will cause mean global temperatures to increase by 2–3 °C. This may not seem much but there could be drastic consequences such as the melting of the antarctic ice cap, raising the sea level as much as 10 m, and catastrophic changes in rainfall. Other environmentalists, however, consider that such predictions of increased world temperature are based on too many unknown factors and that it may not occur. Others argue that increased pollution will cause a haze in the atmosphere which will tend to reduce solar radiation reaching the earth, and thus make the earth cooler; or that the massive destruction of forests (about 11 million ha yr^{-1}) will, through decreased photosynthesis, lead to increases in CO_2 levels. Whatever is the correct prediction, there is no doubt that we should all be concerned that mankind may be capable of altering world climate by his activities, and caution is needed while more research is carried out.

Conclusion

In this chapter we hope to have shown that most of man's dilemmas such as an exponentially increasing population, habitat destruction, environmental pollution, depletion of the supply of fossil fuels and minerals and the general exploitation of the world's living and non-living resources are all ecologically based. We would hope that the general public, as well as those responsible for shaping man's future, will take more notice of ecological constraints. They have to recognize that small families are good families, that we must conserve our resources and recycle our scarce commodities (Figure 8.5) even if in the present 'throw-away-society' of the rich countries this proves inconvenient. It should be remembered that it is only in the last 100 years or so that one-third of the world has become so technically orientated. Admittedly, technology can improve, and even save life, but the benefits gained must be costed against any adverse effects such as damaging the environment and quality of life. We are now approaching the crossroads in the history of the world. We have the chance of taking an alternative route into the future, but if we do this we must be prepared to abandon many of our narrow concepts and self-centred ambitions. Inevitably sacrifices will have to be made by some for the benefit of all if the world is to become, and remain, a fitter place in which to live (Box 8.4).

Box 8.4
'Apparently you cannot convince nations that they should assist others voluntarily. But people should realise that if no solution is found, the future looks rather bleak. If the rich countries will not share their wealth, the poor people of the world will come and take it for themselves' (Tinbergen, J., 1980, *Development Forum*, **6**, 3–10.) – A rather alarmist view that has clear political overtones, but Tinbergen says that the process of taking has started already, and cites the presence in the USA of 7–10 million illegal Mexican workers.

The differences between high and low population growth rates is the choice between quantity and quality. Should we have small, healthy families that are well fed, well housed and well educated, or larger ones that are not?

Suggested reading

Leakey, 1981, has produced a very readable and profusely illustrated book on man's origins, with considerable emphasis on early man in Africa. Climatic changes are discussed briefly in a number of the books, but Gribbin, 1978, has devoted a whole book to the threat of climatic changes. McEwen & Stephenson, 1979, and Holdgate, 1980, write specifically on pollution and pesticide residues, while this subject is also covered more briefly by Murdoch, 1975, together with the energy crisis, diminishing resources, and human population growth.

There are many books dealing with population growth, resources, and destruction of the environment, but one of the problems with these books is that much of their contents are speculative, being based on modelling. Consequently, their predictions on population trends, consumption of resources etc. quickly become outdated. Nevertheless some of the better and more balanced interpretations are by Ehrlich *et al.*, 1977, Murdoch, 1980, Pimental & Pimental, 1979, and Ward, 1979; while Meadows *et al.*, 1972, concentrate mainly on problems of human population growth, and Allaby, 1977, discusses world food resources. The paperback edition of the small book by Allen, 1980, which discusses strategies for conservation is both cheap and readable.

Karmarck, 1980, gives a provocative view that the poor progress in development in the tropics is largely due to its climate, which has hindered agriculture, and to disease which has weakened the people. Owen, 1973, writes on human ecology in tropical Africa, while People Magazine, 1981, has a very good issue describing how the population of Africa is growing faster than any other region of the world. An issue of Scientific American, 1980, is devoted to economic development and has informative articles on human populations, resources and food production etc., and an account of economic development in Tanzania. An issue of the United Nations Educational, Scientific and Cultural Organization's journal Impact of Science on Society, 1982, contains informative articles on the chemistry and physics of the atmosphere, including an account of pollution, and the 'greenhouse effect'. Another useful book on environment and development in Africa is published by the United Nations Environmental Programme, 1981. A small, but useful book on environmental sciences by Anderson, 1981, presents some up-to-date and readable accounts on the biosphere, pollution and human populations.

A very useful reference source on different countries' population density, growth rates, productivity and other statistics is published by the United Nations, the latest edition was issued in 1979, but later ones are to be published. Another informative publication is the Demographic Yearbook, the latest referring to the situation in 1981 (United Nations, 1983).

Essays and problems

1 Make a brief description of a human community (e.g. village, suburb, university campus) with which you are familiar. Classify the components of this environment, providing actual data where available. What are the major environmental problems facing the various members of the community?

2 Which of the climatic changes induced by human activities are most likely to affect his future welfare? What remedies are being considered?

3 Discuss the usage and importance of wood, charcoal, kerosene (paraffin), coal and oil as a source of energy for cooking and lighting in (a) rural communities and (b) urban areas, in your country. Is there any likelihood that any of these resources will soon become scarce?

4 Describe the advantages and disadvantages of large-scale drip irrigation. Describe an irrigation project, preferably in your own country, and assess potential future developments.

5 What types of environmental pollution occur at present in your country? Do you foresee any increase in pollution, and if so of what types? What measures can be taken to reduce environmental damage?

6 Obtain figures from the National Censuses to plot the population increases of your own country. Assuming that present trends continue, what will be the major impacts on the environment of the likely increase between now and the end of this century?

7 Give an objective assessment of the arguments for and against policies aimed at reducing the present rate of population increase in Africa.

8 What traditional methods of environmental conservation (e.g. of forests) have in the past been practised in your community, and what has become of them?

References to suggested reading

Allaby, M. 1977. *World Food Resources: Actual Potential*, Applied Science Publishers: Barking.

Allen, R. 1980. *How to Save the World*, Kogan Page: London.

Anderson, J. M. 1981. *Ecology for Environmental Sciences: Biosphere, Ecosystems and Man*, Edward Arnold: London.

Ehrlich, P. R., Ehrlich, A. H. & Holdren, P. R. 1977. *Ecoscience, Population, Resources, Environment*, Freeman: San Francisco.

Gribbin, J. 1978. *The Climate Threat*, Fontana, Collins: Glasgow.

Holdgate, M. 1980. *A Perspective of Environmental Pollution*, (2nd edn). Cambridge University Press: Cambridge.

Karmarck, A. M. 1980. *The Tropics and Economic Development. A Provocative Inquiry into the Poverty of Nations*,

Leakey, R. E. 1981. *The Making of Mankind*, Michael Joseph: London.

McEwen, F. L. & Stephenson, G. R. 1979. *The Use and Significance of Pesticides in the Environment*, Wiley: New York.

Meadows, D. M., Meadows, D. L., Randers, J. & Behrens, W. W. 1972. *The Limits to Growth*, University Press: New York.

Murdoch, W. W. (ed.) 1975. *Environment, Resources, Pollution & Society*, (2nd edn). Sinauer Association: Stamford.

Murdoch, W. W. 1980. *The Poverty of Nations. The Political Economy of Hunger Population*, John Hopkins: Baltimore.

Owen, D. F. 1973. *Man's Environmental Predicament: An Introduction to Human Ecology in Tropical Africa*, Oxford University Press: Oxford.

People Magazine 1981. *'Black Africa. Population Development and Family Planning in Africa South of the Sahara'*, vol. 8, No. 1.

Pimental, D. & Pimental, M. 1979. *Food Energy and Society*, Resource and Environmental Science Series, Edward Arnold: London.

Scientific American 1980. *Economic Development*, September issue 243, No. 3194.

UNESCO 1982. *Impact of Science on Society*, UNESCO: Paris.

United Nations 1979. *World Statistics in Brief*, (4th edn later ones expected periodically). United Nations: New York.

United Nations 1983. *Demographic Yearbook 1981*, (33rd edn). United Nations: New York.

United Nations Environmental Programme 1981. *Environment and Development in Africa*, Pergamon Press, UNEP studies, vol. 2, Pergamon Press: Oxford.

Ward, B. 1979. *Progress for a Small Planet*, Penguin Books: London.

Index